LANGUAGE IN ACTION
Categories, Lambdas and Dynamic Logic

STUDIES IN LOGIC

AND

THE FOUNDATIONS OF MATHEMATICS

VOLUME 130

NORTH-HOLLAND
AMSTERDAM • NEW YORK • OXFORD • TOKYO

LANGUAGE IN ACTION

Categories, Lambdas and Dynamic Logic

Johan VAN BENTHEM
Faculteit der Wiskunde en Informatica
Universiteit van Amsterdam
Amsterdam, The Netherlands

1991

NORTH-HOLLAND
AMSTERDAM • NEW YORK • OXFORD • TOKYO

ELSEVIER SCIENCE PUBLISHERS B.V.
Sara Burgerhartstraat 25
P.O. Box 211, 1000 AE Amsterdam, The Netherlands

Distributors for the United States and Canada:

ELSEVIER SCIENCE PUBLISHING COMPANY INC.
655 Avenue of the Americas
New York, N.Y. 10010, U.S.A.

ISBN: 0 444 89000 9

Printed in The Netherlands

to the memory of my father

Abraham Karel van Benthem

Table of Contents

PREFACE

This book has arisen out of a series of papers documenting some five years of research into the logical foundations of Categorial Grammar, a grammatical paradigm which has close analogies with Lambda Calculus and Type Theory. The technical theory presented here is a natural outgrowth of an earlier interest in the interface between Logic and Linguistics: in particular, the theory of generalized quantification developed in van Benthem 1986A. Like many of my more linguistic colleagues in The Netherlands, I have gradually found that a categorial framework, provided with a lambda calculus oriented semantics, is a most convenient vehicle for generalizing semantic insights obtained in various corners of natural language into one coherent theory. Moreover, this book is meant to demonstrate to fellow logicians that the resulting applied lambda calculus is not without its intrinsic logical interest either.

Nevertheless, theorizing at this level of abstraction has also led to a general change in my thinking about language. In the final analysis, it seems to me now, the crucial issue is not primarily to 'break the syntactic code' of natural languages, but rather to understand the cognitive functioning of the human mind. And the latter manifests itself not just in the grammatical patterns of our sentences, but also in more global textual arrangement, steered by discourse particles and interpunction, or even in the invisible rules guiding our language games. Fortunately, as it happens, the various formal systems studied here as engines for categorial grammatical analysis also admit of more 'procedural' interpretations, as logics of dynamic interpretation and inference. Thus, motives from Computer Science also make their appearance, resulting in a shift from Categorial Grammar to Dynamic Logic towards the end of the book. For a logician, this is not such a dramatic move as it might seem: for instance, the technical setting for this shift lies in Modal Logic, another earlier research interest of the present author (witness van Benthem 1985). At the moment, I cannot judge the precise significance of this emerging connection between these various logical concerns: perhaps, human beings just cannot help singing the same tune once in a while.

I would like to thank a number of people for their direct help in the actual preparation of this manuscript: in particular, Dirk Roorda, Victor Sanchez and Anne Troelstra have provided many useful comments. As to its incubation period, I have profited for a long time from participation in an active Categorial Grammar community, both in Holland and abroad, including such inventive colleagues as Wojciech Buszkowski, Michael Moortgat and Frans Zwarts. Finally, for a more general intellectual environment, I should also mention the 'Institute of Language, Logic and Information' at the University of Amsterdam, which has provided a very pleasant environment for someone who finds it hard to stay within the usual academic boundaries, and sometimes even more heretically (rumour has it), to share the received value judgments and intellectual dogmas of the logical profession.

I INTRODUCTION

In this first Part, the basic ingredients of this book will be presented, whose interaction, and resulting logical theory, will then unfold subsequently.

1 Two Traditions

The idea that the objects of thought form a hierarchy of categories is an old one in Philosophy. That this hierarchy may be based on function-argument relationships was realized by two eminent mathematicians/philosophers in the last century, namely Gottlob Frege and Edmund Husserl. Their influence may be traced in two subsequent streams of logical investigation. One is that of mathematical ontology, where Bertrand Russell developed his *Theory of Types* describing mathematical universes of individual objects, functions over these, functionals over functions, et sursum. Although set theory in the Zermelo style, rather than type theory, became the eventual 'lingua franca' of Mathematics, the type-theoretic tradition in the foundations of mathematics has persisted to this day, inspiring such important special research programs as the Lambda Calculus in its course. A second stream arose in Philosophy, where Lesniewski and Ajdukiewicz developed a technical theory of categories, which eventually became a paradigm of Linguistics known as *Categorial Grammar*.

In the meantime, both streams have found applications within a single new area, namely that of Computer Science. Type theories play a conspicuous role in the semantics of programming languages, while categorial grammars are prominent in computational processing of natural languages. In fact, a convergence may already be observed in the seminal work of Richard Montague, where mathematically inspired type theories fed into the semantics of natural languages. By now, however, a much richer picture is emerging of possible contacts between the two streams: and the main purpose of this book is to set out the basic ideas of logical syntax and semantics relating Categorial Grammar and Lambda Calculus. In the process, it will also become clear that there remains no natural frontier between the fields of mathematical linguistics and mathematical logic.

One pervasive concern in our treatment may be called *fine-structure*. Notably, Montague's original system employed a very powerful type-theoretic machinery, much of which is not actually involved in providing meanings for linguistic expressions. Thus, it may be compared to a syntactic paradigm like unrestricted rewrite grammars, which generates all recursively enumerable languages, but which has to be 'fine-tuned' at various lower levels (e.g., context-free, or regular) in order to get better estimates of the actual syntactic complexity of linguistic constructions. Likewise, we shall be concerned here with a spectrum of categorial calculi, corresponding to a hierarchy of fragments of the full type-theoretic language, in order to get a more sensitive instrument for semantic analysis. That such an interest in fragments and fine-structure, rather than extension and generalization, of existing logical systems can be crucial in applications is also shown by the example of computation. In contemporary Computer Science, the art has been to chip off suitable pieces from such monoliths as 'Turing-computable functions' or 'predicate-logically definable predicates'.

There is also a second theme in this book. Behind a linguistic paradigm such as Categorial Grammar, there lies a more general perspective. In the final analysis, understanding natural language is just one instance of the more general phenomenon of *information processing*. And indeed, there is an interesting convergence to be recorded nowadays between various logical paradigms directed toward the structural and procedural aspects of information, such as Relevance Logic, Modal Logic or Linear Logic. Indeed, much of the theory that we develop applies to this more general setting too, and hence what started out as a mathematical study of a grammatical theory has gradually turned into a kind of 'dynamic logic', independent from the particular encodings found in linguistic syntax. We shall try to map out this broader perspective as a natural extension of our main topic in the final Part of this book, which may in fact be read independently.

As will be clear from the preceding description, this is a book on the interface of a number of disciplines: Mathematics, Linguistics and Computer Science. What we want to present is a unifying logical perspective between these, developed in some mathematical depth, without making a very detailed examination of the various ingredients. In particular, our purpose is not to explain or motivate Categorial Grammar per se. For that, the reader may consult the anthologies Oehrle, Bach & Wheeler 1988 or Buszkowski, Marciszewski & van Benthem 1988, while van Benthem 1986 also provides some relevant background in the theory of generalized quantification. Also, there is already a number of standard references available for the Lambda Calculus as an enterprise in its own right, which need no copying here: see Barendregt 1981 or Hindley & Seldin 1986, or the earlier Gallin 1975 for more linguistically oriented manifestations. A good introduction to the formal machinery of logical semantics for natural languages is Dowty, Wall & Peters 1981. And finally, some independent basic information on various forms of Modal and

Dynamic Logic may be found in Bull & Segerberg 1984 and Harel 1984, while van Benthem 1985 supplies more extensive mathematical background.

This book has been written for an audience with a certain sophistication in mathematical logic, and non-parochial interests in investigating logical structures across various fields of application. What is needed on the part of the reader is a certain acquaintance with basic techniques from model theory and proof theory, both for classical logic and some of its variations (lambda calculus, intensional logic). The necessary 'iron rations' are surveyed in a special Appendix at the end (Part VII of this book).

The above description of contents for the present volume has emphasized a certain *research programme*. What we would like to put on the map is the perspective of a Categorial Hierarchy with an attendant mozaique of fragments of the Lambda Calculus, viewed as a tool for studying both linguistic structures and dynamic procedures. But there are also more concrete logical contributions to be reported here. The exposition to follow involves a survey of a good deal of the relevant literature supporting the preceding picture, while also presenting a number of technical logical results demonstrating its potential. Explicit credits have been given whenever available: all other results presented would seem to be new.

2 Lambda Calculus and Theory of Types

For a start, here is a short review of some basic notions and results in Lambda Calculus and Theory of Types, that will recur in the Chapters to follow.

2.1 Types, Models and Languages

Types consist of some number of *primitive types* and all those which can be formed out of them by means of a certain group of admissible operations. Common primitive types in logical semantics include

e	individual objects	('entities')
t	truth values	
s	intensional indices	('possible worlds', 'computer states', ...)

Common type-forming operations are

(a, b)	function type	('from a-arguments to b-values')
a•b	product type	('ordered pairs')

In these lectures, the above primitive types will often be used for the sake of concrete illustration: but, most of the theory presented is perfectly general. As for operations, we shall work mainly with function types - although product types are very convenient when it comes to practice.

Next, we introduce semantic structures. A *standard model* is a family of domains D_a, one for each type a, such that, for primitive a, D_a is some arbitrary set (pending further stipulations) and, for complex a,

$D_{(a,b)} = {D_b}^{D_a}$ (function space)

$D_{a \bullet b} = D_a \times D_b$ (Cartesian product)

Now, in order to refer to objects in such structures, we need a suitable *language*. The simplest case is that of a 'lambda calculus', having enough variables (and when desired, also constants) x_a of each type a, as well as the following operations for constructing complex terms:

Application If σ, τ are terms of types (a, b), a respectively, then $\sigma(\tau)$ is a term of type b

Abstraction If σ is a term of type b, x a variable of type a, then $\lambda x \bullet \sigma$ is a term of type (a, b)

We shall be rather carefree in notation for terms henceforth, suppressing brackets whenever convenient.

Remark. Notation.
Notations for types and lambda terms tend to vary across different research communities. For instance, functional types are sometimes written with arrows:

$$(a \to b)\ ,$$

and another well-known notation for application of terms is simple juxtaposition:

$$\sigma\tau\ .$$

Our particular choice in this book reflects the Montague-Gallin tradition.
The latter will tend to generate large clusters of commas and brackets if pursued consistently, so that the reader is urged to use any ad-hoc abbreviations that might appeal to her. For instance, in many contexts, the basic type (e, (e, t)) of 'two-place predicates of individuals' might be shortened to just

$$eet\ .$$

Likewise, an important type like that of 'determiners' (see Chapter 3), whose official notation reads ((e, t), ((e, t), t)) , might be abbreviated to the more convenient

$$(et)\ (et)\ t\ .$$

With product types also present, we may add suitable pairing and projection functions to the language:

Pairing If σ, τ are terms of types a, b respectively,
then $<\sigma , \tau>$ is a term of type $a \bullet b$

Projection If σ is a term of type $a \bullet b$,
then $\pi_L(\sigma)$, $\pi_R(\sigma)$ are terms of types a , b , respectively.

In the language so far, one can form complex descriptions of functions ('procedures'), ascending to a meta-level in order to compare them as to equivalence, etcetera. But, we can also add the basic comparative predicate itself to the object language, being the *identity*:

If σ , τ are terms of the same type,
then $\sigma = \tau$ is a term of the truth value type t .

In the resulting language L_ω , a special primitive type t becomes distinguished.
If enough assumptions are made about the truth value domain, then the usual logical operators can be defined in this 'type theory': notably, the Boolean *connectives* and the

universal and existential *quantifiers* (see Henkin 1963, Gallin 1975 or Lambek & Scott 1986 for full details). For instance, here are some simple equivalences:

$\top$	$\lambda x_t \bullet x_t \ = \ \lambda x_t \bullet x_t$	('true')
$\bot$	$\lambda x_t \bullet x_t \ = \ \lambda x_t \bullet \top$	('false')
$\neg\phi$	$\phi \ = \ \bot$	(negation)
$\phi \wedge \psi$	$\lambda x_{(t, (t, t))} \bullet x(\phi)(\psi) \ = \ \lambda x_{(t, (t, t))} \bullet x(\top)(\top)$	(conjunction)
$\forall x_a \, \phi$	$\lambda x_a \bullet \phi \ = \ \lambda x_a \bullet \top$	(universal quantifier)

Conversely, the new language may also be set up from the start as a higher-order calculus with the usual logical constants

$$\neg\ ,\ \wedge\ ,\ \vee\ ,\ \rightarrow\ ,\ \forall x_a\ ,\ \exists x_a\ ,\ = .$$

(Henkin 1950 even adds such further amenities as a *iota-operator* describing objects via definite descriptions.) The resulting system may also be described as a direct generalization of first-order logic, second-order logic and so on through all finite levels.

Interpretation of these languages in the above standard models is straightforward, using auxiliary assignments to free variables in the standard manner. (An ability to supply such formalities when necessary is a typical illustration of the kind of technical facility presupposed on the part of readers of this book.) Thus, we shall assume that, in any model $\mathbb{M}$, given some assignment of suitable objects u to variables x , a term τ with free variables $x_1, .., x_n$ has an appropriate corresponding semantic value

$$[[\tau]]^{\mathbb{M}} \ [x_1, ..., x_n \,/\, u_1, ..., u_n] \ .$$

2.2 Axiomatic Calculi

Now, in an axiomatic approach to these systems, one sets up some reasonable calculus of inference rules. Notably, we have the following key principle relating application and abstraction:

Lambda Conversion

$$(\lambda x \bullet \tau)\ (\sigma) \ = \ [\sigma/x]\tau\ ,$$

provided that σ is free for x in τ .

In the literature, this is sometimes called 'β-conversion'. Together with the obvious equality of alphabetic variants (known under the name of 'α-conversion'), as well as the usual rules of inference for Identity - in particular, Replacement of Identicals - one obtains the *Typed Lambda Calculus*.

The typed lambda calculus still allows us to think of functions in various ways, ranging from 'intensional' rules of computation to 'extensional' listings of courses of values. The latter option becomes enforced by the following additional principle:

Extensionality

$\lambda x \bullet \tau(x) \;=\; \tau$,

provided that x do not occur in τ .

In the literature, this is sometimes called 'η-conversion'. The resulting system is the *Extensional Typed Lambda Calculus* λ_τ , which shall be the main tool in this Book.

Much is known about the properties of this system. Notably,

1 The 'normalization' process on terms is *confluent*:
if one term can be reduced via the above rules to two others,
then both of these can be further reduced to one common term again.
(This is the familiar 'Church-Rosser property'.)

2 Each term has a unique *normal form* (up to alphabetic variance)
in which no more so-called 'redexes' of the form $(\lambda x \bullet \tau)(\sigma)$ occur.

3 Provable identity of terms is effectively *decidable*.

See Hindley & Seldin 1986 for further details. Later on, we shall often use normal forms for lambda terms, exploiting a number of their special syntactic characteristics.

Nevertheless, the extensional lambda calculus λ_τ can still be strengthened considerably to a so-called *theory of finite types* T_ω by adding the usual rules for the earlier logical constants, including quantifiers over variables of arbitrary types. Thus, unlike the pure lambda calculus, type theories also encode information reflecting the structure of the distinguished truth value domain D_t . In fact, there are several options here: notably, intuitionistic or classical variants (taking D_t to be either a Heyting Algebra, for instance in a topos, or a Boolean algebra). In these lectures, the particular form of type-theoretic axioms and rules will not be important (see Gallin 1975 or Lambek & Scott 1986 for some elegant candidates): as our primary concern will be with semantics and expressive power. Moreover, our central system will be somewhat in between bare lambda calculus and full-fledged type theory, namely, an extensional lambda calculus with a special Boolean type t . Some properties of such a hybrid formalism are developed in more detail in Appendix 2 at the end of this Chapter.

2.3 Basic Model Theory

There are many obvious semantic questions concerning types and domains.

For instance, given that lambda terms describe functions or processes, it is of interest to determine just when the latter have certain important mathematical properties. Here is one simple illustration:

When does a term τ describe a function which is *one-to-one*?

Examples of terms which do are

$\lambda x_e \cdot x_e$, $\lambda x_e \cdot \lambda y_t \cdot \langle x, y \rangle$,

or, provided that D_t contain more than one object,

$\lambda x_e \cdot \lambda y_{(e, t)} \cdot y(x)$.

By contrast, the following terms do not:

$\lambda x_{(e, (e, t))} \cdot \lambda y_e \cdot x(y)(y)$, $\lambda x_{e \bullet t} \cdot \pi_L(x)$.

The question whether this, and similar semantic properties of lambda terms are effectively *decidable* seems open so far.

Another central semantic concern is just which objects in standard models are *definable* by closed terms in the languages introduced here. Obviously, with infinite primitive domains, the exponential growth of function spaces soon outruns the countable supply of type-theoretic terms. But, at least with finite base domains, there is no a priori impossibility in defining even all objects in a standard model. And yet, there are further constraints.

The relevant notion here is one of *invariance*. What makes the denotations of type-theoretic terms special is their being insensitive to certain changes in the underlying model structure. After all, even the full type-theoretic calculus may be viewed as a higher-order logic of identity, and hence it can only detect semantic differences up to identity of basic individuals. This phenomenon can be described technically as follows:

Permutation Invariance

Let $\{\pi_x \mid x \in P\}$ be a family of permutations on the primitive type domains D_x $(x \in P)$. Lift these to permutations π_a on arbitrary type domains D_a by the following recursion

$$\pi_{a \bullet b}(\langle x, y \rangle) = \langle \pi_a(x), \pi_b(y) \rangle$$

$$\pi_{(a, b)}(f) = \pi_b \, f \, \pi_a^{-1}$$

Now, an object $x \in D_a$ is *permutation invariant* if

$\pi_a(x) = x$ for all such permutations π .

By a straightforward induction on the construction of terms τ, it may be shown that, for all such individual permutations π,

$$\pi ([[\tau]] [x_1, ..., x_n / u_1, ..., u_n]) \quad = \quad [[\tau]] [x_1, ..., x_n / \pi(u_1), ..., \pi(u_n)] .$$

In particular, then,

Closed terms τ (having no free variables) always denote permutation invariant inhabitants of their type.

Moreover, on standard models with *finite* base domains, a converse holds too:

All permutation invariant objects are type-theoretically definable (see Chapter 10, or Statman 1982, van Benthem 1989A).

For pure lambda terms, however, this invariance property is still too weak. Even changes of a model that lump individual objects together need not disturb their evaluation:

Relational Invariance

Let $\{R_x \mid x \in P\}$ be a family of binary relations on the basic domains D_x $(x \in P)$. Lift these to relations R_a on arbitrary type domains D_a by this recursion:

$f R_{a \bullet b}\, g$ iff $\pi_L(f) R_a \pi_L(g)$ and $\pi_R(f) R_b \pi_R(g)$

$f R_{(a,\, b)}\, g$ iff for all $x, y \in D_a$, $x R_a y$ only if $f(x) R_b g(y)$

Now, an object $x \in D_a$ is *relation invariant* if

$x R_a x$ for all such relations R.

A second induction on terms shows that (suppressing some indices):

if $u_i R v_i$ for all i $(1 \leq i \leq n)$, then

$[[\tau]] [x_1, ..., x_n / u_1, ..., u_n] \;\; R \;\; [[\tau]] [x_1, ..., x_n / v_1, ..., v_n]$.

In particular, each closed term must have a relation invariant denotation.

No general converse is known, however, not even on models with finite base domains. (Plotkin 1980 has an interesting but complicated partial result in this direction.)

These general relations encompass many special cases of interest. The earlier permutation invariance is the special instance where

$x R_a y$ iff $y = \pi_a(x)$.

Another important case is extension of basic domains D_e^1, D_e^2, with

$x R_e y$ iff $y = x \in D_e^1$

which gets lifted to a general notion of model extension through all domains. Further applications will be found in later Chapters.

In some ways, the basic 'model theory' of the lambda calculus is different from the usual topic which goes under the former name. Notably, this is due to the 'contextual' nature of the lambda operator λ. For instance, a denotation $[[\tau]]^{\mathbb{M}}$ may change in subtle ways when we move from a model $\mathbb{M}$ to some 'submodel' based on a smaller domain of individuals. Thus, comparing denotations of terms across models is a delicate issue.

Even so, it makes sense, e.g., to search for analogues of classical model-theoretic questions such as the following:

Which special types of semantic preservation behaviour
characterize natural syntactic fragments of the full lambda language?

More specifically, on the pattern of the well-known Los-Tarski Theorem, is there a suitable kind of preservation under 'submodels' which determines precisely to the class of so-called *pure combinators* having the following 'universal form':

"prefix of lambdas, followed by a pure application term" ?

We shall return to these questions in Appendix 1 at the end of this Chapter, which takes a look at λ_τ as a more standard logical formalism.

2.4 Completeness

The given semantics generates an independent notion of *validity* for the statements of type theory. Thus, there arises a question of *completeness*:

Do the above axiomatic calculi capture all of semantic validity?

For the pure lambda calculus, a positive answer may be found in Friedman 1975:

An identity $\sigma=\tau$ between two lambda terms σ, τ
is true in all standard models if and only if it is provable in λ_τ.

(Incidentally, this result does not extend to the case of derivability for lambda identities from *premises*.)

For the full type theory T_ω, however, the answer is negative (cf. Henkin 1950). In the latter higher-order language, one can give a categorical definition of the standard natural numbers - and hence all arithmetical truths can be effectively encoded as universally valid statements. But then, by the known non-axiomatizability of arithmetical truth (Tarski's Theorem), there is no adequate axiomatization of the full type theory on standard models either.

Nevertheless, a more modest form of completeness is available after all: provided that one relax the notion of a standard model to that of a *general model*, in which functional domains $D_{(a, b)}$ need no longer be full set-theoretic function spaces. This was done in Henkin 1950, where a general method is presented for establishing completeness of higher-order theories with respect to properly generalized classes of models. (See also the survey Doets & van Benthem 1983.) In the setting of the lambda calculus, the best-known examples are so-called *term models*, whose objects in arbitrary types are equivalence classes of terms in λ_τ identified modulo provable equality, with an obvious choice of application and abstraction.

Perhaps the most austere re-interpretation of this move is as follows. Type theory becomes a *many-sorted logic* dealing with a family of 'types' on which we have an 'application' predicate

App (f, x, y) (" f(x) = y ") .

The graph of an object f is then the set of all pairs $<x, y>$ such that App (f, x, y) holds. On top of this, one can make models more 'standard' by requiring such behaviour as *extensionality*:

objects with the same graphs must be equal.

Also, additional *comprehension* postulates will have to be introduced if it is to be guaranteed that each lambda term receive a denotation represented by an object in the model.

These stark many-sorted models seem rather far removed from our original motivation, even though they have their technical uses. But still, it is not so unreasonable to consider models for our functional language which have the basic machinery allowing function application as well as abstraction, while varying in their ontological commitment as to how many functions are actually supposed to exist. Such multiplicity is found, in particular, in the category-theoretic approach to lambda calculus, where the proper models of λ_τ are *Cartesian-closed categories* (see Lambek & Scott 1986). The proper models of higher type theories will then be *topoi*, or similar categories having a distinguished truth value object with enough logical structure, but otherwise quite varying supplies of functional 'arrows'.

2.5 Variants

As has already become clear by now, there are various alternatives in setting up lambda calculus and type theory, having to do with choices of types, axioms and models. Moreover, the whole format of presentation may be shifted. In particular, for several purposes, it turns out convenient to use hierarchies of *relations* rather than functions, as was already proposed in Orey 1959. (See Doets & van Benthem 1983 or Muskens 1989 for details and motivation.)

More importantly, 'lambda calculus' can also be (and usually is) performed in a *type-free* setting, with one sort of variables ranging over all objects in the domain, viewed indiscriminately as functions or arguments. Barendregt 1981 has the full theory of this approach, which will only be tangentially relevant in these lectures (but see the account of polymorphic 'variable types' in Chapter 13). Scott 1980 is an interesting discussion of the proper way of construing the conceptual priorities between the typed and untyped variants. The perhaps unconventional reversal in this book stems from the following consideration. Lambda Calculus may be viewed in two ways. In one sense, it is an all-purpose logical tool which provides a 'general combinatorics' that can be used to extend and streamline various logical systems, across many different areas of application. In most cases of interest, types would seem to be an essential feature here. In a narrower sense, however, lambda calculus is a specific applied logical theory in the foundations of mathematics, embodying a particular view of the nature of functions and their effective computability in which type-freedom seems quite natural. The two perspectives do not necessarily suggest the same presentation, or the same central concerns. And although the latter is historically prior, it is the former which is at issue here.

2.6 Appendix 1: **What is Lambda Calculus?**

Even though the lambda calculus is a somewhat unorthodox logical system, it can be analyzed by ordinary model-theoretic tools.

To see this, first consider the embedding from the language of λ_τ into that of T_ω suggested earlier. To each identity $\sigma = \tau$ between typed lambda terms, there corresponds an equivalent formula of L_ω via the following procedure:

It suffices to translate identities of the special form $\sigma = x$ into formulas '$\sigma = x$', since $\sigma = \tau$ is equivalent to $\exists x\, (\sigma = x \wedge \tau = x)$.
Now, for the latter purpose, the following recursion on σ may be used:

$x = y$: $x = y$

$x = \sigma(\tau)$: $\exists y\, \exists z\, ('y = \sigma' \wedge 'z = \tau' \wedge x = y(z))$

$x = \lambda u_a \bullet \sigma_b$: $\forall u_a\, \exists z_b\, ('z_b = \sigma' \wedge x(u_a) = z_b)$.

Thus, one can ask for characteristic semantic features of the *lambda calculus fragment* of our full type theory. That this question makes sense may be seen by an analogy. Within the full language of predicate logic, one can locate the 'algebraic fragment' consisting of all universally quantified algebraic identities between terms, viz. as being those statements (modulo logical equivalence) which are preserved under the formation of homomorphic images, direct products and subalgebras (cf. Chang & Keisler 1973).

Something similar also holds here. The above transcriptions of lambda term identities are constructed in the following restricted 'positive syntax':

"identities between application terms, $\wedge$, $\exists$ and $\forall$ " .

And this means again, by standard model theory, that

> Lambda identities are preserved under *homomorphic images* and also under *direct products* of general models for full type theory.

Both of these semantic operations have a straightforward definition on general models, similar to the first-order case. There is no direct analogue for preservation under 'submodels', however, to establish a closer fit with those special formulas in positive syntax that occur as translations of lambda identities. (These are the ones that admit of removing their existential quantifiers via successive lambda abstractions.) The reason is that, although with ordinary algebra, submodels are closed under all relevant algebraic constructions, imposing a similar restriction here to 'lambda-closed submodels' seems to beg the question at issue. Thus, a more precise model-theoretic characterization of the lambda calculus formalism remains open.

Nevertheless, one can ask similar questions for more tractable fragments of the full language of λ_τ .

For instance, the earlier-mentioned class of *pure combinators* naturally involves the very austere fragment of

"all lambda-free pure application terms" .

The latter may be characterized using the ordinary model-theoretic notion of 'submodel', with respect to closure under the application function:

Proposition. A lambda term τ with free variables $x_1, ..., x_n$ is equivalent to a pure application term if and only if it has the following invariance property on general models $\mathbb{N}$ in which it possesses a denotation for certain parameters $u_1, ..., u_n$:

whenever $\mathbb{M}$ is a *submodel* of $\mathbb{N}$ containing $u_1, ..., u_n$,
$[[\tau]]^{\mathbb{M}} [x_1, ..., x_n / u_1, .., u_n]$ exists and is equal to
$[[\tau]]^{\mathbb{N}} [x_1, ..., x_n / u_1, .., u_n]$.

Proof. That all application terms have this invariance property is clear. Conversely, suppose that τ were equivalent to no application term σ. Then, for each such term, there exists a model and an assignment falsifying the identity $\tau = \sigma$. But then, consider the direct product $\mathbb{N}$ of all these models, with the obvious 'product assignment' A . By the above general observation on preservation under products, $[[\tau]]^{\mathbb{N}}$ exists with respect to the assignment A . Now, let $\mathbb{M}$ be the smallest submodel of $\mathbb{N}$ containing all objects assigned by A to the free variables in τ. By its definition, $[[\tau]]^{\mathbb{M}} [A]$ must be equal to a denotation $[[\sigma]]^{\mathbb{M}} [A]$ for some pure application term σ, whereas it cannot have this form in the larger model $\mathbb{N}$. This is a contradiction with the above invariance property.
■

But quite different kinds of model-theoretic analysis are possible too. Sometimes, one may have to change the usual semantics in order to bring out special characteristics of fragments.

For instance, later on in this book, the central fragment of λ_τ to be considered will be that where each lambda operator binds *one* unique *occurrence* of a free variable. The semantic implications of this restriction may be demonstrated as follows (after Schellinx 1990, which has a related analysis in type-free lambda calculus). Let us interpret our variables as running over *sets*, and lift application to sets that are intuitively located at the same level:

$$X(Y) = Z \quad \text{iff} \quad Z = \{ z \mid \exists y \in Y: \langle y, z\rangle \in X \}.$$

Then, pure combinators may be viewed as natural operations on sets, such as

$\lambda X \bullet \lambda Y \bullet X$	'left projection'
$\lambda X \bullet \lambda Y \bullet X(X(Y))$	'squaring'
$\lambda X \bullet \lambda Y \bullet Y(X)$	'dualization'.

Now, the single bind lambda terms among these have the following characteristic mathematical property:

Proposition. Single-bind pure combinators F define operations on sets that are *continuous*, in the sense of commuting with arbitrary unions of each of their arguments.

Proof. The reason for this behaviour lies in the syntactic form of the relevant definitions. The condition for membership of $F(X_1, ..., X_n)$ can be written using only existential quantifiers stating the existence of certain pairs, and in general longer sequences, in the argument sets. And such statements will induce the above continuous behaviour just in case they involve only one existential quantifier for each set argument X_i . As an illustration, compare the transcriptions for the above three examples:

- $\exists x \in X: z = x$
  Y is not mentioned here, whence continuity fails in this argument: consider the empty union.
- $\exists x_1 \in X\, \exists x_2 \in X\, \exists y \in Y: \pi_R(x_1) = z \wedge x_2 = \langle y, \pi_L(x_1)\rangle$.
  Here, two objects are required to exist in X :
  something which invalidates continuity.
- $\exists y \in Y\, \exists x \in X: y = \langle x, z\rangle$
  Only here, continuity is guaranteed.

●

Of course, a cogent proof of the converse would require further argument.

In any case, many other set operations that can be described in this format are continuous too, without being lambda-definable at all.

2.7 Appendix 2: **Boolean Lambda Calculus**

The extensional typed lambda calculus that will be used in many of our applications has a distinguished truth value type t in addition to the individual type e , and also various built-in Boolean constants with a standard interpretation, such as

0_t	(falsity)	1_t	(truth)
$\neg_{(t,\, t)}$	(negation)	$\wedge_{(t,\, (t,\, t))}$	(conjunction) .

Thus, it may be considered as a natural extension from 'Boolean Algebra' to 'Boolean Calculus' (see Chapter 11).

But then, what is the appropriate calculus of inference here, that will produce all valid identities? In particular, is there a generalization of Friedman's Completeness Theorem? This question is not entirely trivial, since it will not do merely to add all ordinary Boolean axioms as 'premises': as was observed before, the Friedman result does not hold under addition of arbitrary premises. So, the outcome must depend on special features of the Boolean case.

Our analysis starts with a new calculus λ_τ^* of 'sequents'

$\Delta \vdash \alpha = \beta$ where Δ is any finite set of lambda identities.

Its principles are obvious generalizations of those for λ_τ. For instance, 'Replacement of Identicals' returns in the axioms

$$\alpha = \beta \vdash \gamma(\alpha) = \gamma(\beta) \qquad \alpha = \beta \vdash \alpha(\gamma) = \beta(\gamma)$$

as well as the inference rule

if $\Delta \vdash \alpha = \beta$ where x does not occur free in Δ,

then $\Delta \vdash \lambda x \bullet \alpha = \lambda x \bullet \beta$.

Moreover, Extensionality has the following form:

if $\Delta \vdash \alpha(x) = \beta(x)$ where x does not occur in Δ, α, β,

then $\Delta \vdash \alpha = \beta$.

Finally, the usual structural rules hold, such as Reflexivity, Monotonicity and Cut.

As to the intended corresponding semantic notion, the relevant sense of 'validity' for sequents is here that

in each model, each assignment verifying all identities in Δ also verifies $\alpha = \beta$.

Now, by a straightforward adaptation of the usual completeness proof for lambda calculus, this time with respect to term models divided out by provable identity under the premises Δ, we have

Proposition I. A sequent $\Delta \vdash \alpha = \beta$ is derivable in λ_τ^* if and only if it is *valid* in all extensional general models for the lambda calculus.

Whether this equivalence extends to validity on all *standard* models depends on the specific form of the premise set Δ. To understand this, one must examine Friedman's completeness proof in more detail. Its crucial construction starts from some general term model $\mathbb{M}$ where some non-derivable identity $\alpha = \beta$ is falsified under a certain assignment, and then constructs a 'partial homomorphism' from the full standard model $\mathbb{M}^+$ over the base domains of $\mathbb{M}$ onto $\mathbb{M}$ itself, which preserves evaluation of lambda terms. As a consequence, the failed identity $\alpha = \beta$ also fails in the standard model $\mathbb{M}^+$. But, what is not ensured is that *true* identities in $\mathbb{M}$ will be transferred to $\mathbb{M}^+$ in this process.

Nevertheless, applying this kind of analysis to the above language with Boolean constants, and taking Δ to be the finite complete set BA of all Boolean axioms, one gets the following

Proposition II. A sequent $BA \vdash \alpha = \beta$ is provable in $\lambda_\tau{}^*$ if and only if the identity $\alpha = \beta$ is valid in all standard models having some Boolean algebra for their ground domain D_t.

Proof. The reason involves direct inspection of the above construction. The standard interpretations of the above Boolean constants in $\mathbb{M}^+$ will be mapped onto their counterparts in the general model $\mathbb{M}$ by Friedman's partial homomorphism. (This argument is in fact more general, and will work for any set of algebraic conditions Δ on non-nested 'first-order types' over ground domains.)

Still, our standard models were more special than this. If validity is to be obtained on all standard models having $D_t = \{0,1\}$, then additional principles are required. For instance, the following identity is valid in the latter case, but not with arbitrary Boolean algebras for D_t:

$$x_{(t,\,t)}(x_{(t,\,t)}(x_{(t,\,t)}(y_t))) \;=\; x_{(t,\,t)}(y_t)\,.$$

Thus, Boolean Calculus is different from Boolean Algebra, where restriction to the two-element algebra does not yield any gain in valid identities.

Now, we define a formal system $\lambda_\tau{}^*B$ as a calculus of sequents having the above principles of $\lambda_\tau{}^*$ as well as all Boolean axioms, together with the following rule of

Bivalence

if $\Delta,\ \sigma_t = 1 \vdash \alpha = \beta$ *and* $\Delta,\ \sigma_t = 0 \vdash \alpha = \beta$,
then $\Delta \vdash \alpha = \beta$.

Moreover, for technical reasons, we add *identity predicates* $=_{(e,\,(e,\,t))}$ for each ground domain e: requiring also principles of

Individual Identity

$(\alpha_e =_{(e,\,(e,\,t))} \beta_e) = 1 \vdash \alpha_e = \beta_e$,
$\alpha_e = \beta_e \vdash (\alpha_e =_{(e,\,(e,\,t))} \beta_e) = 1$.

This system is obviously sound for its intended semantic interpretation, involving the above notion of validity for sequents. Moreover, we can prove the stronger

Proposition III. An identity $\alpha = \beta$ is provable in $\lambda_\tau{}^*B$ if and only if it is valid in all standard models having the ground domain $D_t = \{0,1\}$.

Proof. Enumerate all terms of the truth value type t as well as all identities in our language, and interleave the following two steps in the construction of a 'maximally consistent' set of identities. Here, *consistency* of a set of identities means that no finite subset heads a derivable sequent whose conclusion is $0 = 1$. Likewise, 'derivability' from a possibly infinite set of premises is understood as derivability from one of its finite subsets.

Suppose that some identity $\sigma = \tau$ is not derivable. We find an infinite set of identities Δ^* such that

1 $\Delta^* \vdash \sigma = \tau$ is not derivable ('$\Delta^* \nvdash \sigma = \tau$')

2 if $\Delta^* \nvdash \alpha = \beta$, then $\Delta^* \nvdash \alpha(x) = \beta(x)$ for some variable x

3 for each term σ_t, either $\sigma_t = 1 \in \Delta^*$ or $\sigma_t = 0 \in \Delta^*$.

Construction:

Start with $\Delta = \emptyset$, and choose successive finite consistent sets Δ^n through the following double stages:

- Select the next underivable identity $\alpha = \beta$ on the list (starting with the original $\sigma = \tau$ in the first step) and add a suitable identity of the form

  $(\alpha(X) =_e \beta(X)) = 0$.

More precisely, X is to be a tuple of *new* variables not occurring in Δ^n, α, β, whose length is such that after their addition in successive argument positions, $\alpha(X), \beta(X)$ have become of ground type, say e.

Analysis: Suppose that Δ^n now becomes inconsistent, i.e.

$$\Delta^n, (\alpha(X) =_e \beta(X)) = 0 \vdash 0 = 1 .$$

Now use the following rule, which is easily derivable in $\lambda_\tau{}^*B$, using Bivalence:

$$\textit{if } \Delta, \gamma = 0 \vdash 0 = 1 , \textit{ then } \Delta \vdash \gamma = 1 .$$

So, $\Delta^n \vdash (\alpha(X) =_e \beta(X)) = 1$, whence $\Delta^n \vdash (\alpha(X) =_e \beta(X))$, and then $\Delta^n \vdash \alpha{=}\beta$ by successive applications of Extensionality. This is the required contradiction.

- Select a new truth value term σ_t on the list and fix its truth value: by the Bivalence rule, either $\Delta^n + \sigma_t = 1$ qualifies, or $\Delta^n + \sigma_t = 0$ does.

Now, the term model constructed in the usual fashion, by identifying terms modulo provable equivalence from $\Delta^* = \bigcup_n \Delta^n$, will have its D_t isomorphic to the Boolean algebra $\{0, 1\}$, while extensionality is still guaranteed by our construction. (The usual trick to the latter effect in the completeness theorem for term models is no longer available, due to the infinite size of Δ^*.)

Then, the Friedman construction will work as before, and one gets a counter-example for $\sigma = \tau$ on a standard model with a two-element Boolean algebra. ■

So, $\lambda_\tau{}^*B$ is an appropriate deductive system for an applied lambda calculus with Boolean parameters. This system has some further useful features. For instance, other common notions of Boolean inference can be formulated within it too, such as *inclusion*:

$$\alpha \subseteq \beta \qquad : \qquad \alpha = \alpha \wedge \beta \,.$$

Also, if desired, sequents may be compressed to simpler forms (cf. Friedman 1975), in particular, to shapes '$\alpha = \beta \vdash$' with an empty conclusion.

There are also other natural meta-logical questions here. For instance, does $\lambda_\tau{}^*B$ have a Church-Rosser property? The answer is affirmative, by the general results surveyed in Klop 1987.

The crucial feature of Boolean parameters which explains many of the above outcomes is their first-orderness in the hierarchy of types. And in fact, a similar analysis may be given for ordinary first-order *predicate logic* with a lambda calculus superstructure. In this case, the Booleans are supplemented with one more constant, namely the binary existential quantifier

some

of the generalized quantifier type ((e, t), ((e, t), t)) (cf. van Benthem 1986A).

Without going into details of the proof, we state our final

Proposition IV. The valid identities on standard models, with *some* interpreted as standard set intersection, are axiomatized by $\lambda_\tau{}^*B$ together with the following quantifier principles:

- $\mathit{some}\ \alpha\beta = \mathit{some}\ \beta\alpha$ — Symmetry
  $\mathit{some}\ \alpha\beta = \mathit{some}\ \alpha(\beta\wedge\alpha)$ — Conservativity
  $\mathit{some}\ \alpha\beta \subseteq \mathit{some}\ \alpha(\beta\vee\gamma)$ — Monotonicity
  $\mathit{some}\ 00 = 0$ — Non-Triviality
- $\mathit{some}\ \alpha\beta = 0 \vdash \alpha(x)\wedge\beta(x) = 0$ — for all variables x
  if $\Delta \vdash \alpha(u)\wedge\beta(u) = 0$ — with u not occurring in Δ, α, β,
  then $\Delta \vdash \mathit{some}\ \alpha\beta = 0$.

3 Categorial Grammar

3.1 The Basic Framework

According to the Polish categorialists in the thirties, such as Ajdukiewicz, categories expressed in natural language exhibit a function / argument structure, that can be symbolized in the pattern (A, B) of expressions needing an A-expression to yield one in category B . The principal rules of combination are then

$$(A, B) \ + \ A \ = \ B$$

$$A \ + \ (A, B) \ = \ B \ ,$$

stating that two adjacent expressions of suitable function and argument type combine to an expression of the value type.

As it stands, however, this formalism fails to capture a prominent syntactic peculiarity of most natural languages, namely that their functors often have a preferred direction from which they want to consume arguments. (For instance, in Milan, on the evening of June 24th, 1990, Klintzmann "scored (a goal)" , not "(a goal) scored".) Accordingly, in the fifties, Lambek proposed and Bar-Hillel adopted a *directed* variant of the calculus, employing the following two slashes:

A\B	left-looking functor
B/A	right-looking functor

with corresponding application rules

$$A \ + \ A\backslash B \ = \ B$$

$$B/A \ + \ A \ = \ B \ .$$

Thus, the formalism became suitable, at least, for describing the syntax of formal languages as well as part of natural languages.

Example. Polish Propositional Formulas.

Description of a set of expressions starts with an initial assignment of types to primitive symbols. In the case of Polish prefix notation, we take

proposition letters	t
unary connectives	(t/t)
binary connectives	((t/t)/t)

The recognizable strings will be all and only those sequences of symbols for which a matching sequence of types exists which can be reduced to the distinguished sentence type t via repeated function application. For instance, here is one parse tree:

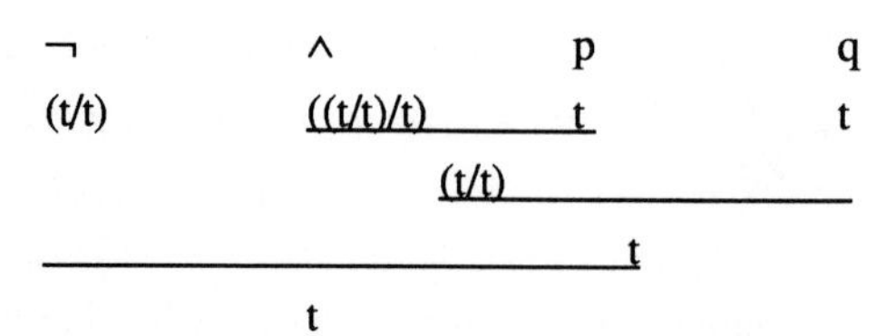

As soon as one changes to infix notation, however, both slashes will be needed in recognition, witness the following string:

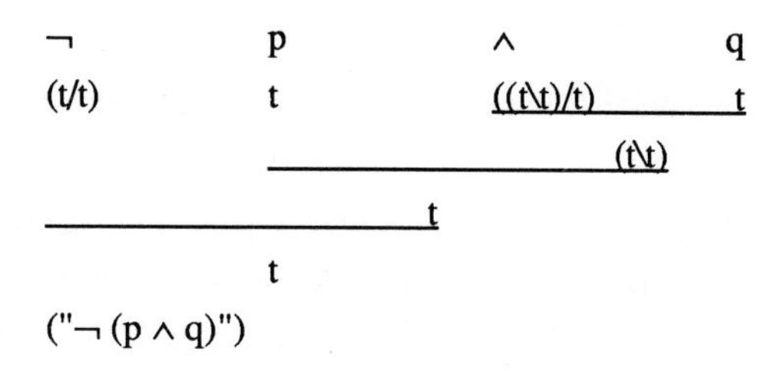

("$\neg$ (p $\wedge$ q)")

But, there is another construal for this string too, corresponding to a different *reading* in which the conjunction takes wide scope over the negation:

¬ p ∧ q

(t/t) t ((t\t)/t) t

t (t\t)

t

("$\neg$p $\wedge$ q")

Such ambiguities are quite typical of *natural* languages: and indeed, categorial grammars turned out to work in the latter realm as well. In logical formalisms, of course, such ambiguities would be removed by obligatory bracketing.

Example. Complex Noun Phrases.

This time, the initial assignment has syntactically motivated category symbols

"woman", "child", "doll"	(common nouns)	N
"every", "no", "a"	(determiners)	(NP/N)
"with", "without"	(prepositions)	((N\N)/NP)

A typical string recognized in the noun phrase type NP will then look like this:

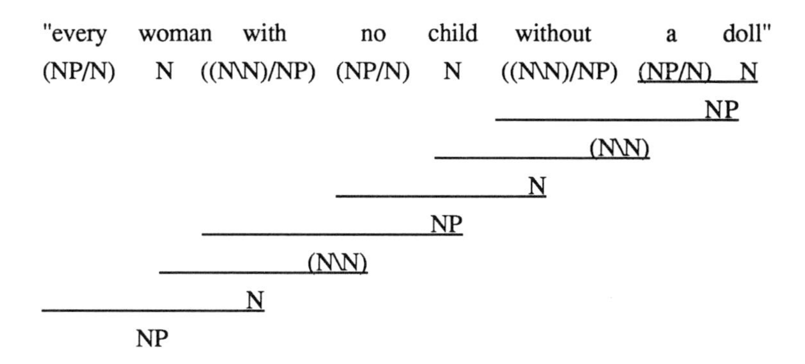

This parse is reminiscent of more traditional grammatical analyses. And indeed, the same class of noun phrases would have been produced by the following *context-free* rewrite grammar: (A definition of this technical notion, and similar ones across the text, may be found in Part VII.)

NP	$\Rightarrow$	Det N	
Det	$\Rightarrow$	"every", "no", "a"	
N	$\Rightarrow$	"woman", "child", "doll"	
N	$\Rightarrow$	N PP	(Common Noun + Prepositional Phrase)
PP	$\Rightarrow$	Prep NP	
Prep	$\Rightarrow$	"with", "without"	

Moreover, the special 'rightward' constituent structure in the above parse tree is an indication of the fact that this particular context-free grammar can even be replaced by an equivalent purely iterative *regular* one.

As a matter of fact, Bar-Hillel, Gaifman & Shamir 1960 proved the following well-known equivalence result:

Categorial grammars and context-free grammars
recognize exactly the same family of languages.

Admittedly, to obtain this equivalence, one has to allow the possibility of finite *multiple* initial type assignment to one single lexical item: but, that seems reasonable in any case, in view of the widespread grammatical phenomenon of 'homonymy'.

Given its origins, Categorial Grammar is a formalism accommodating both syntactic *categories* of expressions and more semantic *types* of objects. The former will be used in the syntactic component of a linguistic application, making use of directed slashes, while the latter will occur rather in its semantic component, probably in the undirected format. Even so, the suggestion of the framework is that the two perspectives are systematically related. By way of illustration, here is a list of grammatical categories with corresponding semantic types, that are frequently encountered in the literature:

proper name	e	"Mary"
intransitive verb	(e, t)	"laugh"
(i.e., a one-place predicate of individuals)		
transitive verb	(e, (e, t))	"beat"
(a two-place relation between individuals)		
complex noun phrase	((e, t), t)	"every doll"
(i.e., a second-order property of predicates)		
adverb	((e, t), (e, t))	"slowly"
(a one-place operation on predicates)		

Now, syntactic categories need not correspond one-to-one with semantic types: and this slack enables us to capture certain semantic analogies between different grammatical kinds of expression. For instance, common nouns are similar to verbs in various ways, and hence we stipulate no new type for them:

common noun	(e, t)	"girl"
adjective	((e, t), (e, t))	"blonde"

Other, more complex types can be derived from this list. For instance, determiners ("every", "some", "five") will be found to require the type of binary relations between unary predicates

((e, t), ((e, t), t)) .

And for instance, prepositions, which take noun phrases to prepositional phrases (as in the string "with" "a book") would be assigned the type

(((e, t), t), ((e, t), (e, t))) ,

where the resulting prepositional phrases behave semantically like adverbial expressions.

Although Categorial Grammar is by no means committed to the particular correspondence displayed here, the latter does serve to give an impression of what is after all one of the main attractions of the enterprise: namely, a certain systematic semantic way of thinking about grammatical categories.

Nevertheless, the prediction of the above assignment is not that all semantically permissible combinations will actually be encountered. It is the proper business of the syntax of particular languages to constrain the combinatorics of expressions in such a way as to block, e.g., "every child doll" or "blonde weep" . Compare also the discussion at the end of this chapter.

3.2 Extensions: Type Change and Flexibility

In the sixties, context-free recognizing power was widely considered too poor for serious linguistic purposes. (A modern reappraisal of this complexity debate takes place in Gazdar & Pullum 1987, W. Savitch et al., eds., 1987.) Therefore, various proposals were made for strengthening the basic framework of Categorial Grammar. Thus, Curry 1961, Lewis 1972 and Montague 1974, each in their own way, proposed a combination of categorial analysis with a much free-er syntax. Instead of working only with adjoining strings for functors and arguments, one could now allow any combination of the form

"syntactic construction from 'components' X, Y to expression Z",

"categorial combination of associated types a, (a, b) to b".

As has been shown by various authors, however, this richer paradigm will allow recognition of all recursively enumerable languages (cf. Buszkowski 1988) - so that we must have over-shot the mark.

Hence, linguists and logicians have explored another, more cautious (and principled) route of extending the basic categorial framework, leading to richer so-called 'flexible' versions. The underlying idea may be described in several ways.

First, let us observe that we need additional *modes of categorial combination* to do justice to the observed facts of natural language. Here is an illustration, involving a well-known problem for traditional categorial grammar, treated in a semantic mode.

Example. Transitive Verbs with Complex Direct Objects.

Certain sentence patterns combine smoothly on the earlier assignment of types to traditional grammatical categories, witness such cases as the following

1 "Mary cries"

e (e, t)

t

2 "Every baby cries"

((e, t), ((e, t), t)) (e, t) (e, t)

((e, t), t)

t

Likewise, simple transitive sentences present no difficulty:

3 "Wanda flatters Wendy"

e (e, (e, t)) e

(e, t)

t

But, in combination with complex noun phrases, the transitive verb has no possibility of attachment on the standard categorial view:

4	"No girl	feeds	every pelican"
	((e, t), t)	(e, (e, t))	((e, t), t)

This difficulty has led several authors, including Lewis and Montague, to 'upgrade' the initial type assignment to transitive verbs

from (e, (e, t)) to (((e, t), t), (e, t)) :

a move with many unpleasant combinatorial repercussions elsewhere in the system. ◆

By contrast, what the seminal paper Geach 1972 suggested as a remedy is simply this. Evidently, the verb phrase "feeds every pelican" should form a semantic unit, of type (e, t) . Thus, we want to allow a combination mode of 'parametrized function application': or , in terms of an immediate mathematical analogy, in addition to function application, we now also want to allow function *composition*:

$$(a, b) \quad + \quad (b, c) \quad \Rightarrow \quad (a, c) .$$

For instance, in the above example of transitive sentences, the relevant transition would be

$$(e, (e, t)) \quad + \quad ((e, t), t) \quad \Rightarrow \quad (e, t) .$$

There is also another way of describing what goes on here (cf. Cresswell 1973). In the above so-called 'Geach Rule', the functional type has undergone what may be called a *type change*, so as to accommodate an argument with type (a, b) :

from (b, c) *to* the parametrized form ((a, b), (a, c)) .

For instance, in transitive sentences, the object noun phrase ((e, t), t) would then change its type to ((e, (e, t)), (e, t)) .

And in fact, this points at a second source of linguistic evidence for more flexible categorial calculi, namely the phenomenon of *polymorphism* in natural language. That is, expressions can modify their categories or types as required by the needs of syntactic combination or semantic interpretation.

A striking example of such polymorphism is found with the *Boolean* particles "not", "and", "or" which may occur in a potential infinity of syntactic contexts (cf. Keenan & Faltz 1985), witness such cases as

"not walk"	predicate negation	((e, t), (e, t))
"not every girl"	noun phrase negation	(((e, t), t), ((e, t), t)))
"not with a knife"	prepositional phrase negation,	etcetera.

Such diversity cannot be handled by means of the earlier finite multiple initial assignment to Booleans: and indeed, there does not seem to be any homonymy involved here at all. Nevertheless, there is one simple rule which generates all these cases from one initial type (t, t) for sentential negation: being the type changing formulation of, again, the Geach Rule, suitably iterated if necessary.

Thus, Geach 1972 (and more implicitly, Cresswell 1973) may be viewed as proposing a more flexible calculus of categorial grammar, supplemented with a system of rules licensing additional categorial combinations or type changes. Around 1980, this movement gathered momentum, as various linguists arrived at similar ideas independently. For instance, coming from the richer Montagovian setting, Bach 1984 proposed a more restricted categorial calculus than Montague's, allowing only some minimum over and above the standard basis. Moreover, Partee & Rooth 1984 proposed simplifying Montague's complex system of semantic types by having simple assignments only, while letting some calculus of type change produce more complex ones if and when actually needed. Another noticeable contribution is the paper Ades & Steedman 1982 as well as Steedman's many subsequent publications on the syntactic application of flexible categorial grammars with (a few) combinatorial extras. Finally, we mention a group of Dutch linguists with related ideas, of whose work Zwarts 1986, with its emphasis on the much free-er constituent structures made possible in the new categorial grammar, is a good example. See also Moortgat 1988 for an up-to-date treatment with a strong computational slant.

This survey is by no means complete. There has been a veritable explosion of similar papers around 1980, written by linguists, philosophers and logicians - who all, for varying reasons, took up the theme of a more flexible categorial grammar. For instance, Buszkowski 1982 was a logical study of such grammars, inspired mainly by their mathematical interest. On the other hand, van Benthem 1984 developed analogous systems rather to obtain a notion of semantic interpretability independent from syntax, so as to explain the empirical phenomenon of meaningfulness of syntactically ill-formed material. Finally, for a more philosophical angle, see the monograph Levin 1982.

By now, most of these separate currents have merged. A good impression of the resulting area of logico-linguistic research is conveyed by the two anthologies Oehrle, Bach & Wheeler 1988 and Buszkowski, Marciszewski & van Benthem 1988. Later on in this Book (Part VI), even more recent confluences will be encountered, with the research programs of Relevant Logic and Linear Logic (see also Dosen & Schröder-Heister, eds., to appear).

What we have seen so far is a traditional categorial grammar, whose basic rule was one of Function Application

$$a \quad (a, b) \quad \Rightarrow \quad b\,,$$

supplemented with a new rule of Function Composition

$$(a, b) \quad (b, c) \quad \Rightarrow \quad (a, c)\,,$$

or equivalently, a rule of type change for parametrizing functional types. Before passing on to further proposed combinatorial mechanisms, let us consider some further examples illustrating the range of innovation due to the latter Geach Rule alone.

The first of these is concerned with linguistic *syntax*.

Example. Noun Phrases with Relative Clauses.

Complex noun phrases with prepositional complements, such as "no person with a hat", are usually construed as follows:

Det N Prep Det N

((e, t), ((e, t), t)) (e, t) (((e, t), t), ((e, t), (e, t)) ((e, t), ((e, t), t)) (e, t)

((e, t), t)

((e, t), (e, t))

(e, t)

((e, t), t)

But, in some languages (cf. Hoeksema 1984 on Iraqi Arabic), there is serious syntactic evidence for another analysis, requiring the following constituent structure:

$((\text{Det}\ \ \text{N})_{NP}\ (\text{Prep}\ \ \text{Det}\ \ \text{N})_{PP}\ )_{NP}$

And, using Function Composition, this is indeed possible:

((e, t), ((e, t), t)) (e, t) (((e, t), t), ((e, t), (e, t))) ((e, t), ((e, t), t)) (e, t)

((**e**, **t**), t) ((e, t), t)

((e, t), (**e**, **t**)) FC

((e, t), t)

What we cannot judge at this stage, however, is whether this move preserved the correct *meaning* for this noun phrase (understood in an intuitive sense). And in fact, it does not - as will be explained in Chapter 5 - although a slightly more complex derivation exists producing indeed the desired effect (cf. van Benthem 1988C). ◆

Our next illustration concerns compositional *morphology*.

Example. Argument Inheritance with Nominalization.
Categorial typing may be used to describe the behaviour of various word-forming operations. For instance, a nominalizing suffix "-ing" takes an intransitive verb ("cry") to an entity ("crying"): that is, it lives in type

((e, t), e).

But, nominalizations can also inherit arguments, for instance, when the verb is transitive: "leav-ing the country" . What this requires is a semantic transition described in Moortgat 1984:

from ((e, t), e) to ((e, (e, t)), (e, t)) ,

which is again provided by the Geach Rule.

Actually, there is no sharp boundary between categorial syntax and categorial morphology, since much of what is traditionally regarded as syntactic information, to be formulated in grammatical rules, becomes encoded in the lexically assigned categories.

As for other modules of Linguistics, there is even a categorial *phonology* (cf. Wheeler 1981). But, perhaps the most important illustrations come from categorial *semantics* - or rather, the syntax/semantics interface.

Example. Montagovian Metamorphoses.
One of the famous features of Montague 1974 is its baroque semantic formula for the verb "be" , known to many suffering students, which had the virtue of serving both as individual identity ("Mary is the Queen of Sheba") and as the copula of predication ("Julius is a fool"). Again, the Geach Rule provides a simple explanation of what is going on here. Individual identity represents the basic type (e, (e, t)) for the transitive verb "be" , which can also combine, as we have seen, with complex noun phrase objects:

from (e, (e, t)) ("be") and ((e, t), t) ("a fool") to (e, t) ("be a fool") .

And, as we shall see later in Chapter 5, the natural meaning computed for this transition yields precisely Montague's denotational recipe for free.

Of course, these few examples can hardly convey the reality of application of the newer categorial techniques: for which the reader is referred again to the earlier-mentioned anthologies. Moreover, the above methods also are also beginning to play a role in the mechanical *parsing* of natural language: for which we must refer to Moortgat 1988, Koenig 1989 or Klein & van Benthem, eds., 1988.

3.3 Categorial Grammar To-Day

Modern systems of Categorial Grammar usually provide facilities for both function application and function composition.

Next, let us turn to some further proposed rules of category or type change.

The first of these arises again in Montague's work, where proper names were assimilated syntactically to complex noun phrases, in order to account for coordination phenomena as in "Mary and every child" . Semantically, this means that proper names, which intuitively denote individual objects in type e , now denote 'bundles' of properties:

from e to ((e, t), t) .

The general pattern behind this transition has become known as the *Montague Rule* for 'raising' denotations:

from a *to* ((a, b), b) .

This is a new rule of type change, independent from that of Geach.

Further examples may be found in Partee & Rooth 1984, such as the following rule of *Argument Lowering*:

from (((a, b), b), c) *to* (a, c) .

If only by symmetry, one might then also expect a matching rule of 'argument raising'

from (a, c) to (((a, b), b), c) .

But in fact, no such rule has been proposed in the literature: except with some special cases of the type c (cf. Groenendijk & Stokhof 1988B). A justification for this intuitive caution will be given in Chapter 4 below.

Finally, let us consider another well-known linguistic study where type change turns out to be of the essence.

Example. Boolean Homomorphisms.

Although intransitive verbs have been taken to occur in type (e, t) so far, there may be syntactic arguments for promoting them to some higher type. Notably, one might wish to guarantee their operator status vis-a-vis subject noun phrases, even when the latter have type ((e, t), t) . For this purpose, the well-known study Keenan & Faltz 1985 (in effect) classifies intransitive verbs in the following type:

(((e, t), t), t) .

Moreover, the authors observe with delight that the corresponding denotations will become very special inhabitants of this higher type domain, namely Boolean *homomorphisms* respecting the Boolean structure of their arguments (cf. Chapter 11 below). For the moment, we just observe that this move has the following combinatorial background. The Montague Rule sanctions the transition

from (e, t) to (((e, t), t), t) .

(Note that this could also be regarded as a valid special case of Argument Raising.)
And actually, the converse is also derivable: by Argument Lowering. Thus, the two given domains for intransitive verbs are indeed intimately related. Later on, using the methods of Chapter 5, we shall also be able to explain how the 'homomorphism' property emerged in the process of lifting.

The picture so far may be summarized as follows. Nowadays, there exists a more powerful, flexible paradigm of Categorial Grammar, allowing various further rules of combination over and above the traditional one of function application. Nevertheless, the obvious question arising here must surely be this:

What is the *system* behind this flexible apparatus:
or is there just a collection of ad-hoc rules?

And also, one would like to see a more principled semantic motivation for the flexible paradigm, allowing us to study its systematic properties. These questions are addressed in the next two chapters, which are devoted, respectively, to the logical system behind Categorial Grammar (Chapter 4) and its semantic meaning (Chapter 5). After that, we shall develop the logical theory of this paradigm in much more detail - including its various possible refinements.

In all this, there will be a curious historical inversion - in that the categorial calculus which turns out to occupy a central position is in fact an old proposal, dating back to Lambek 1958, which had lain dormant for about two decades before any prince bothered to come along.

To conclude the present chapter, it may be appropriate to state a few disclaimers. As is usual in the application of an analytic tool to empirical reality, the actual business of linguistic description is more complex than the preceding account may have suggested. In particular, the study of different components of natural language may involve a number of different categorial calculi, with suitably chosen interfaces. This is true in particular, as we have observed, of the border line between syntax and semantics: where a directed categorial calculus may be needed for the former, appropriately linked to some non-directional calculus of types for the latter. This second calculus may in fact naturally produce a 'logical space' of meanings properly including those immediately encoded in syntactic constituent structures: so that we can, and must, formulate empirical *semantic constraints* on the expressive power of natural languages. In addition, we have already seen how an interface like a category-to-type mapping itself could store useful linguistic information, such as the semantic analogy between adverbs and adjectives. But also, different calculi might interact at another interface, namely that between phonology and syntax. Ejerhed & Church 1983 has the interesting suggestion that phonetic information from incoming discourse already allows us to perform some finite-state preprocessing, resulting in an initial chunking of parts of sentences into some rough constituent structure, which can then be used by a categorial calculus as a guide through its possible derivations.

Moreover, such more concrete applications also suggest new kinds of logical question about the calculi employed. For instance, current applied computational research on categorial *parsing* raises many interesting issues. One concerns the phenomenon of so-called 'incrementality': Can we always construct constituent structures for expressions progressing from left to right? Chapter 4 contains some further discussion of this issue from a logical point of view. Another is the emergence of 'variable types' suggested by current unification techniques in parsing. This would add another level of polymorphism to our system, calling for something like a 'second-order lambda calculus' (see Chapter 13 for elaboration). And finally, e.g., Moortgat 1988 also finds occasion to introduce new category-forming operations into Categorial Grammar, for the purpose of handling expressions having gaps inside. This possibility of extending of the paradigm serves as a pointer towards the more general logics of categories and types presented in Chapter 14 and Part VI in general.

Even so, why all this attention for what is, after all, a rather minor linguistic enterprise? Here is the historical verdict of Gazdar & Pullum 1987 (p. 404):

> "Categorial grammars, which were developed by Bar-Hillel and others
> in the 1950's, have always had a somewhat marginal status in linguistics.
> There has always been someone ready to champion them, but never
> enough people actually using them to turn them into a paradigm."

Perhaps the main reason for our persistence lies in the following observation. If there are any two guiding ideas which have become widely established by now as common ground among workers in the field of natural language processing, these would seem to be the following:

- a commitment to some form of *parsing as* logical *deduction*,
- the necessity of providing a *systematic semantics* for any grammatical framework underlying the chosen parser.

But, precisely these two features have always been at the heart of Categorial Grammar: as will be demonstrated more extensively in the following Chapters. There, we shall develop the proof theory of possible 'categorial engines' of derivation, as well as a corresponding type-theoretic semantics and its model theory, linking the enterprise with central questions of logical semantics.

Thus, we have come to the main topic of this Book, being the logical system behind Categorial Grammar, and its wider ramifications: a subject, as we shall see, which also presents a good deal of intrinsic interest.

II A LOGICAL PERSPECTIVE

The desired systematic perspective behind Categorial Grammar may in fact be found in the realm of Logic: as will be explained now, both from a proof-theoretic and from a semantic point of view.

4 The Hierarchy of Implicational Logics

The process of categorial combination described so far involved a number of rules, such as the following:

from	a (a, b)	to	b	(Ajdukiewicz)
from	(a, b) (b, c)	to	(a, c)	(Geach)
from	a	to	((a, b), b)	(Montague)
from	(((a, b), b), c)	to	(a, c)	(Partee & Rooth)

A principled motivation behind this mechanism lies in the following crucial analogy, observed by several authors in the fifties (in particular, Curry & Feys):

4.1 Function Types and Implications

Functional types behave very much like logical implications:

(a, b) $\quad\quad\quad\quad\quad\quad$ $a \rightarrow b$

This may be seen by comparing the Ajdukiewicz combination with the familiar inference rule of *Modus Ponens*:

$$\frac{a \qquad (a,\, b)}{b} \qquad\qquad \frac{a \qquad a \rightarrow b}{b}$$

And this analogy extends to more recent rules of flexible combination too.
For instance, Geach's Rule is the well-known Hypothetical Syllogism - and the other examples in the above list express valid principles of implicational logic as well.

What we shall mean here by 'implicational logic' is, to a first approximation, the *constructive* (or 'intuitionistic') propositional calculus of pure implications.

This system can be axiomatized in the usual Hilbert style using Modus Ponens over the following two axiom schemas

$$(A \to (B \to A))$$

$$((A \to (B \to C)) \to ((A \to B) \to (A \to C))) \,.$$

More perspicuous for later purposes, however, is its presentation as a tree-based *natural deduction* calculus, whose derivation trees are formed according to the rules

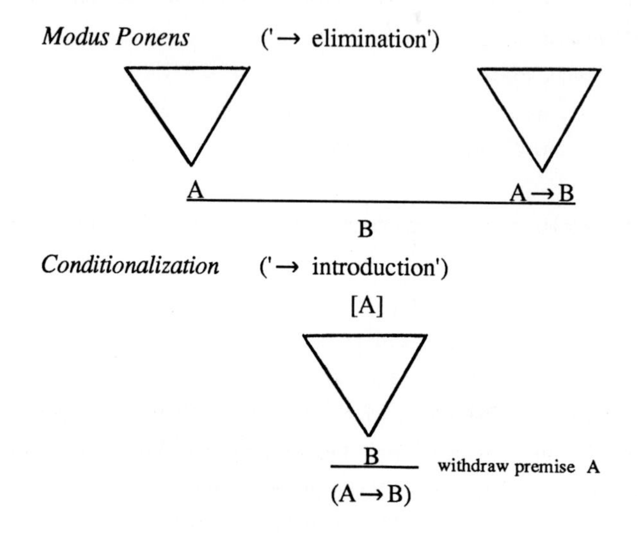

In a sequent format, these rules look like this:

- $X \Rightarrow A \quad \text{if } A \in X$
- $$\frac{X \Rightarrow A \qquad Y \Rightarrow (A \to B)}{X, Y \Rightarrow B}$$
- $$\frac{X, A \Rightarrow B}{X \Rightarrow (A \to B)}$$

We shall adopt this double arrow notation for derivable sequents henceforth, whenever convenient.

Example. Derivation of Proposed Categorial Combinations.

Geach

$$\cfrac{\cfrac{\cfrac{a^1 \qquad (a, b)}{b}\,\text{MP} \qquad (b, c)}{c}\,\text{MP}}{(a, c)}\,\text{COND}, -1$$

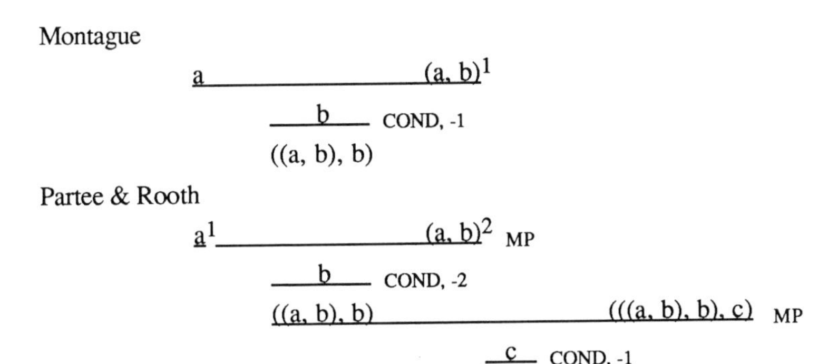

Other cases can be derived similarly - as was observed already in Chapter 3 in connection with a proposal by Keenan & Faltz 1985:

$$(e, t) \quad \Leftrightarrow \quad (((e, t), t), t) \, .$$

But, could not there also be plausible categorial rules which fail to correspond to laws of implication? At least, practical evidence does not point that way: and Chapter 5 provides further semantic evidence for this observation.

Admittedly, there are some potential counter-examples, such as the claimed rule for taking proper names to unary predicates which should operate in languages like German or Spanish. The phrase

"der Heinrich"

has a determiner "der" operating on a proper name "Heinrich" , so that the latter might have undergone a shift as follows:

from e to (e, t) .

Evidently, the latter transition is not logically valid. But then, it is alo debatable as a general principle of categorial combination.

Example. Argument Raising.

Some authors have considered a 'dual' to the above rule of argument lowering, being, in principle, the transition

from (a, c) to $(((a, b), b), c)$.

As it stands, however, this is not a logically valid inference. A classical counter-example is given by the propositional valuation making b true and a, c false. Nevertheless, what can be derived in the above calculus is the following special case:

$(a, (c, d)) \quad \Rightarrow \quad (((a, d), d), (c, d))$.

And the latter is in fact the form in which argument raising is found employed in the literature. (Compare Groenendijk & Stokhof 1988B on type changes connected with the behaviour of questions and answers.) ●

Another piece of evidence for the logical analogy is that it generalizes smoothly to other type-forming operations. In particular, *product types* turn out to correspond to logical *conjunctions*

$a \bullet b \qquad a \wedge b$.

This becomes especially clear in the following two sequent rules of inference:

$$\frac{X, a, b, Y \Rightarrow c}{X, a \bullet b, Y \Rightarrow c} \qquad \frac{X \Rightarrow a \qquad Y \Rightarrow b}{X, Y \Rightarrow a \bullet b}$$

Example. Flattening Iterated Functions.

The following equivalence allows us to curry and uncurry sequences of arguments for functions:

$(a, (b, c)) \Leftrightarrow (a \bullet b, c)$.

Here are derivations in the appropriate natural deduction format:

$$\cfrac{\cfrac{\cfrac{a^1 \qquad (a, (b, c))}{(b, c)}\,\text{MP} \qquad b^2}{c}\,\text{MP}}{(a \bullet b, c)}\,\text{COND}, - \{1,2\}$$

There is a simultaneous withdrawal of two premises in this derivation.

$$\cfrac{\cfrac{\cfrac{\cfrac{a^1 \qquad b^2}{a \bullet b} \qquad (a \bullet b, c)}{c}\,\text{MP}}{(b, c)}\,\text{COND}, - 2}{(a, (b, c))}\,\text{COND}, - 1$$

Compare this with the standard logical equivalence between

$(a \rightarrow (b \rightarrow c)) \qquad$ and $\qquad ((a \wedge b) \rightarrow c)$.

In Chapter 5, an alternative natural deduction rule for products will be considered too, which may be somewhat more perspicuous in practice.

●

4.2 Alternative Implicational Logics

Does all this mean that the logic of categorial combination is the standard constructive implicational logic? Or, are there still further options?

For a start, even within standard logic itself, there is no single candidate for a calculus of implication. Notably, the above system, however elegant, does not yet contain all truth functional tautologies concerning material implication which belong to *classical logic*. In particular, it lacks 'Peirce's Law'

$(((a \rightarrow b) \rightarrow a) \rightarrow a)$.

Is the latter a plausible principle of categorial combination? To judge this, we can restrict attention to instances with primitive types only, as all others are derivable from these (van Benthem 1987A). Thus, what would be the sense of a transition

from $((e, t), e)$ to e ?

This would turn a 'choice function' assigning individual objects to predicates (compare the nominalizer "ing" from Chapter 3) into an object itself. Although this is not totally incomprehensible (one might select the value uniformly at some distinguished predicate: say, 'true'), it certainly does not qualify as a general linguistic principle.

But more importantly, not even all constructively valid implicational laws make sense as principles of categorial combination. For instance, the logically innocuous transition which contracts redundant premises

$a, a \Rightarrow a$

would then express that, say, any two sentences juxtaposed again form a sentence. Or, the rule of vacuous conditionalization ('weakening')

$a \Rightarrow (b \rightarrow a)$

would let sentences (type t) also function as intransitive verbs (type (e, t)). Finally, using Contraction again, one can also derive a transition like

$(a \rightarrow (a \rightarrow b)) \Rightarrow (a \rightarrow b)$:

whence, say, transitive verbs would also automatically become intransitive ones.

Fortunately, there are other options. A natural calculus of implication which comes much closer to the mark may be found already in the pioneering paper Lambek 1958, a precursor to the recent work on flexible categorial grammar, which is gradually receiving proper recognition. We shall work extensively with this *Lambek Calculus* in these lectures - though the presentation and further theory will deviate progressively from the original.

4.3 Lambek Calculus

Our basic idea is this. Although categorial derivation is like logical derivation, it is a kind of inference which pays much closer attention to the actual *presentation* of premises. In particular, what needs to be recorded precisely is which *occurrences* of premises were used where. Thus, we get a calculus L of occurrences of premises, which can be defined as follows in the earlier natural deduction format:

> A sequent $X \Rightarrow b$ is derivable if there exists a proof tree for b in which exactly the occurrences of the premises mentioned in the sequence X remain in force, such that each Conditionalization has withdrawn *exactly one* occurrence of its antecedent.

Later on, we shall introduce some further notation and remove some vagueness in this formulation.

All categorial derivations presented so far satisfy the above restrictions, and hence they are justified by the Lambek Calculus. For an example of a derivation beyond its reach, we have to look at the earlier-mentioned phenomenon of Boolean polymorphism (cf. Chapter 3):

Example. Raising Sentence Conjunction to Predicate Conjunction.

The following sequent, which reflects the transition from propositional "and" to predicate conjunction, is not Lambek derivable:

$$(t, (t, t)) \quad \Rightarrow \quad ((e, t), ((e, t), (e, t))) \ .$$

Its natural deduction tree must use multiple withdrawal in Conditionalization:

$$\dfrac{\dfrac{\dfrac{\dfrac{\dfrac{\dfrac{\dfrac{e^1 \quad (e,t)^2}{t}\,\text{MP} \quad (t,(t,t))}{(t,t)}\,\text{MP} \quad \dfrac{e^1 \quad (e,t)^3}{t}\,\text{MP}}{t}\,\text{MP}}{t}\,\text{COND}, -1\ (!)}{(e,t)}\,\text{COND}, -2}{((e,t),(e,t))}\,\text{COND}, -3}{((e,t),((e,t),(e,t)))}$$

Incidentally, there are also other versions of this derivation, whose semantic differences will become clear in Chapter 5. ◆

As this calculus is more sensitive to the actual structure of proofs, certain differences between derivations of the same sequent become even more important than in the full constructive conditional logic. (See also Lambek & Scott 1986 on the importance of proof structure in a category-theoretic setting.) The following two illustrations of such proof-theoretic phenomena will be used later on.

Example. Deriving an Axiom.
The following sequent is axiomatically derivable, in a one node tree:

(e, (e, t)) ⇒ (e, (e, t)) .

But, it can also be derived as follows, by 'permuting arguments':

e^1 (e, (e, t)) MP

e^2 (e, t) MP

t COND, -1

(e, t) COND, -2

(e, (e, t))

This difference will turn out to have semantic repercussions later. ●

Example. Incremental Derivation.
The sequent e e (e, t) ((e, t), (e, (e, t))) ⇒ t has one derivation as follows:

(e, t) ((e, t), (e, (e, t))) MP

e (e, (e, t)) MP

e (e, t) MP

t

But now, suppose that we are actually processing the premises, drawing preliminary conclusions 'dynamically' as successive premises come in. Then, we would probably 'bite' too early: combining the (e, t) with an e as soon as it came in - after which one cannot proceed any further. The Lambek Calculus, however, allows us to find at least one non-deadlocking rearrangement for the original proof, using the following three derivable sequents:

e + e ⇒ ((e, t), (((e, t), (e, (e, t))), t))

((e, t), (((e, t), (e, (e, t))), t)) + (e, t) ⇒ (((e, t), (e, (e, t))), t)

(((e, t), (e, (e, t))), t) + ((e, t), (e, (e, t))) ⇒ t

Admittedly, this is an analysis with the benefit of hindsight. But, one can also devise algorithms for making optimal *predictions* as to preliminary inferences from premises seen so far. (See Moortgat 1988 on this issue as it occurs in categorial parsing, which inspired the present passage.)

Two further issues in this process of 'incremental inference' are worth mentioning as pointers toward later Chapters.

First, in some cases, it makes sense to 'bite' at once, leaving possible backtracking to a later stage. In other cases, a natural policy might be to postpone inference. For instance, with the above initial sequence e e , we might just *store* these two pieces of data. For this purpose, an 'inference' is available after all, this being precisely the point of our *product* types: $e \bullet e$.

Next, if our policy is to insist on inferences at every step, a certain room for manoevering could also be preserved by employing *variable* types, to be specified only later (cf. Chapter 3, Chapter 13). For instance, the sequence e e might then give rise to the following indeterminate or 'open' inference, with x still to be determined:

$$e \;\; e \;\; \Rightarrow \;\; ((e, (e, x)), x) \, .$$

The next premise (e, t) does not fit the antecedent slot of the latter type yet, whence it has to be encoded as follows in its variable x :

$$((e, (e, x)), x) \quad (e, t) \quad \Rightarrow \quad ((e, (e, ((e, t), y))), y) \, .$$

Finally, however, the premise ((e, t), (e, (e, t))) will fit the antecedent slot (after some permutation), forcing the substitution $y = t$ for the conclusion.

This is basically the kind of parsing strategy found in recent so-called 'categorial unification grammars', to be discussed in Chapter 13 .

4.4 Directed and Non-Directed Systems

Let us now take a closer look at the actual formulation of the Lambek Calculus. The system introduced so far was based on our earlier notion of undirected functional types

(a, b) :

which implies a certain disregard of *order* of premises. This will come out even more clearly if we use the format of a Gentzen style *sequent calculus* for it, with a, b standing for expressions, and X, Y for finite sequences of these:

Axiom $\quad a \Rightarrow a$

Rules 1 $\quad \dfrac{X \Rightarrow a}{\pi[X] \Rightarrow a}$ $\qquad$ for any permutation π of X

$$2 \quad \frac{X, a \Rightarrow b}{X \Rightarrow (a, b)} \qquad \text{(right introduction)}$$

$$3 \quad \frac{X \Rightarrow a \qquad Y, b \Rightarrow c}{Y, X, (a, b) \Rightarrow c} \qquad \text{(left introduction)}$$

$$4 \quad \frac{X \Rightarrow a \qquad Y, a \Rightarrow b}{Y, X \Rightarrow b} \qquad \text{(Cut Rule)}$$

It is easy to see that this calculus proves exactly the same sequents as those derivable in the earlier natural deduction format. We shall call it

L*P ,

for reasons to become clear presently.

In the original Lambek calculus, there is one further restriction on derivations, namely, that premise sequences must always be *non-empty* throughout a derivation. Thus, there will be no 'categorically provable' laws such as (a, a) , or inferences depending on these: such as

$((a, a), b) \Rightarrow b$.

Linguistically, this excludes 'phantom categories' derived out of empty expressions.

With this restriction in force, we obtain a calculus to be called

LP.

This system consists essentially of two introduction rules for implication, as well as two so-called *structural rules*, being Permutation and Cut . Actually, in Chapter 7, it will be shown that the Cut Rule is redundant in both LP and L*P : sequents derivable with it can also be derived without it. Thus, the sequent calculus formulation for these systems becomes very perspicuous, as it involves no length-decreasing rules.

On the other hand, the Permutation Rule is essential here: and it expresses an abstraction from the order in which our premises occur. In Lambek's original calculus, the latter aspect of presentation is crucial, however: and hence, it is a system without any structural rules, based on directed types (cf. Chapter 3)

a\b b/a .

Natural deduction rules then assume the following shape:

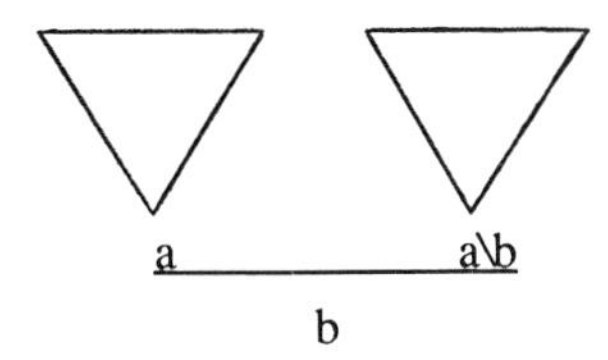

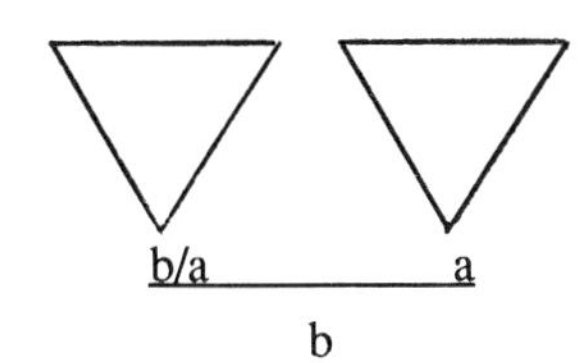

In the latter two introduction rules, the picture is meant to convey that a is the left-most (respectively, right-most) active premise occurrence.

Thus, we obtain the *directed Lambek Calculus* L (together with an obvious variant L* allowing empty premise sequents).

Accordingly, there will now be a number of directed variants for earlier non-directed type transitions: not all of which need be derivable in L .

Example. Montague Raising.

The following directed variant of the Montague Rule $a \Rightarrow ((a, b), b)$ is not derivable:

$$a \Rightarrow ((a\backslash b)\backslash b) .$$

This may be seen by an ad-hoc analysis of possible derivation trees, or by an application of the combinatorial methods presented in Chapter 6.

With mixed slashes, however, we get a derivable case

$$a \Rightarrow ((b/a)\backslash b) :$$

$$\dfrac{\dfrac{(b/a)^1 \qquad a}{b}\ \text{MP}}{((b/a)\backslash b)}\ \text{COND, -1}$$

♦

Example. Geach Rule.

Here is a derivable variant, requiring parallel slashes

$$(a\backslash b) \quad (b\backslash c) \quad \Rightarrow \quad (a\backslash c) :$$

$$\dfrac{\dfrac{\dfrac{a^1 \qquad (a\backslash b)}{b}\ \text{MP} \qquad (b\backslash c)}{c}\ \text{MP}}{(a\backslash c)}\ \text{COND, -1}$$

Non-derivable would be such heterogeneous cases as

$$(a\backslash b) \quad (c/b) \quad \Rightarrow \quad (a\backslash c) .$$

♦

Example. Argument Switching.

The following derivable principle allows us to shift temporal, though not spatial, order of consumption of arguments:

$$((a\backslash b)/c) \;\Rightarrow\; (a\backslash(b/c))$$

$$\dfrac{a^2 \quad \dfrac{((a\backslash b)/c) \qquad c^1}{(a\backslash b)}\,\text{MP}}{\dfrac{\dfrac{b}{(b/c)}\,\text{COND, -1}}{(a\backslash(b/c))}\,\text{COND, -2}}$$

Whether actual reasoning respects every detail of the ordering of premises is doubtful. Likewise, some linguistic phenomena suggest that a certain amount of permutation tolerance is common in natural languages. (For instance, Romance languages like Latin or Spanish have relatively free word order.) This makes the relation between directed and non-directed Lambek calculi of special interest. Moreover, another possible connection would be that directed systems express local syntax-driven reasoning, while non-directed systems might reflect more global reasoning after a while, when details of presentation have been forgotten. (See also Chapters 14, 16 on the general issue of 'short-term' versus 'long-term' reasoning.)

Nevertheless, there are some technical surprises here. For instance, adding even slight, apparently harmless, ways of loosening up directionality may already collapse L into LP .

Example. Harmonic Lifting.

Suppose that the earlier harmonic form of Montague Raising is addded to L as an axiom:

$$a \quad\Rightarrow\quad ((a\backslash b)\backslash b)\,.$$

Then, full Permutation becomes derivable.

Here is a special case illustrating the principle:

if $a\ b \Rightarrow c$ is derivable, then so is $b\ a \Rightarrow c$.

The required manipulation is illustrated in the following tree

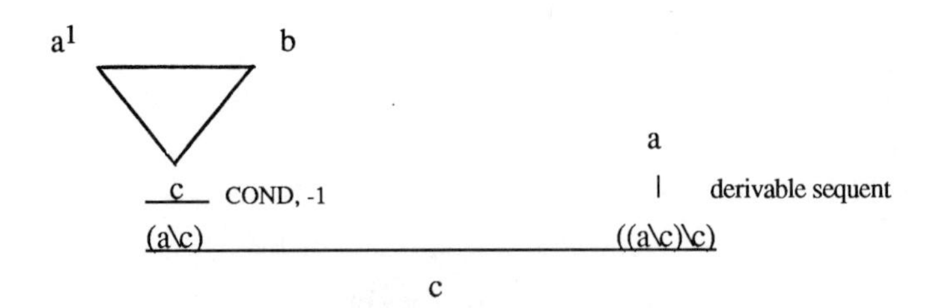

◆

Indeed, the question remains whether any interesting calculi lie in between directed L and non-directed LP .

Remark. Another interesting question is if every LP-derivable sequent has at least one directed version derivable in L . The answer is negative.
For instance, in LP we can derive

$(e, (s, t)) \Rightarrow (s, (e, t))$.

But, no way of replacing commas by slashes will make this into a derivable sequent of the system L . ◆

Even so, the directed calculus is not without logical interest. For instance, natural language itself has two variants of the conditional construction, one forward looking:

" if A , (then) B" ,

another backward-looking:

" B if A " .

And the two differ, at least, in their anaphoric behaviour:

" If the queen comes, Marvin will avoid her. "

? " Marvin will avoid her if the queen comes. "

In the second case, no direct coreference can be enforced between the "her" of the conclusion and "the queen" .

And also theoretically, the directed calculus brings some interesting questions. For instance, suppose that some derivable sequent involves only leftward slashes \ . Must it then have a derivation in L involving such slashes only, or can the interplay with rightward slashes be unavoidable sometimes? The answer for L is that some pure \ derivation must exist (see Chapter 7, or Buszkowski 1982: the full calculus L has the 'subformula property'). But, for other directed calculi the outcome may be different.

4.5 The Categorial Hierarchy

The difference between the Lambek Calculus and standard implicational logics may be described as follows. Both have the same logical rules of inference, governing introduction and elimination of logical operators. But, they differ in their structural rules, which do the 'book-keeping' of premises and conclusions. The latter are often regarded as lacking logical interest. But here, they have become crucial - and debatable.

Here are some major options as to structural rules for sequents $X \Rightarrow b$, where X is any sequence of types:

P $\dfrac{X \Rightarrow b}{\pi[X] \Rightarrow b}$ for each permutation π of X *Permutation*

C $\dfrac{X, a, a \Rightarrow b}{X, a \Rightarrow b}$ *Contraction*

Actually, in the absence of Permutation, Contraction needs a more general formulation, allowing us to suppress any occurrence of some premise, provided that at least one remains. And a similar observation applies to the other rules to follow.

E $\dfrac{X, a \Rightarrow b}{X, a, a \Rightarrow b}$ *Expansion*

Or, more generally,

M $\dfrac{X \Rightarrow b}{X, a \Rightarrow b}$ *Monotonicity*

The general picture now becomes one of a hierarchy of implicational calculi, which can be defined by adding one or more structural rules to the Lambek Calculus L . For instance, L itself may be regarded as a logic of *ordered sequences* of premises, whereas LP is a logic of *bags* of premises (counting only occurrences). A richer logic like LPCE (not yet the full constructive implicational logic!) then becomes a logic of the familiar *sets* of premises. But, one could also descend below L , by making certain *groupings* in a sequence of premises different from others (cf. Dosen 1988/1989). And, there can be many other principles of creating implicational calculi. For instance, Prijatelj 1989 proposes the use of Contraction or Expansion only for certain restricted classes of types: which gives another degree of freedom, as structural rules then no longer serve as an all-or-nothing device.

Thus, the landscape arises of a *Categorial Hierarchy* of engines for driving categorial combination. At the lower end, there is the original Ajdukiewicz system A , allowing only Modus Ponens for a logical rule - and its extensions go all the way up to the full constructive conditional logic I (for 'intuitionistic'), and beyond to the classical conditional logic C :

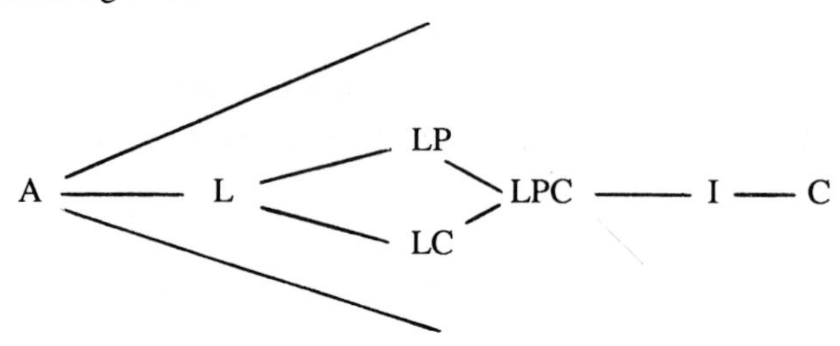

Note that this is not a linear order, but a branching pattern of logical systems.

Many of these calculi have potential uses in linguistic description - and no doubt, other scientific applications will be found in due course. (See Chapters 14, 15 for some examples in the realm of programming and computation.) So, there is a lot of 'underground' life, below what is usually considered the rock bottom of standard logic. The following Chapters will be devoted to exploring the proof theory and model theory of this new territory, which can be developed by standard logical techniques, even though the outcomes are often different from the usual.

Example. Reversibility of Rules.

One of the convenient features of the structural rules of classical logic is that they allow for formulations of the logical rules in terms of *equivalences*. For instance, with the multiple (disjunctive) conclusion format for the classical calculus C , the left introduction rule for implication can be taken to be

$$\frac{X \Rightarrow a, Y \qquad\qquad X, b \Rightarrow Y}{X, a \to b \Rightarrow Y}$$

Here, the conclusion also implies both premises. Therefore, when searching for derivations, it does not matter which conditional on the left-hand side is tackled first.

With the constructive calculus, however, this does matter: and even more so with the Lambek system in its earlier Gentzen sequent formulation. For instance, decomposition at (e, t) will not lead to a succesful derivation of the valid sequent

s (e, t) ((e, t), (s, t)) $\Rightarrow$ t ,

since no non-empty subset of { s, ((e, t), (s, t)) } L-implies e . So, we have to be careful with peculiarities of formulation. On the other hand, matters also become easier in going down the Hierarchy. In particular, searching for L-derivations leads to a much faster decrease in the size of potential proofs: as the parameter types X in a conclusion are not reproduced in both premises, but distributed over them.

Actually, the above discussion has not produced a genuine exact *definition* of the Categorial Hierarchy. For that, one would at least have to make a choice: whether or not to employ directed types. And also, the general notion of a 'logic' itself should be defined more rigorously. One current practice is to identify logics with their sets of provable sequents. But then, there may still be various proof formats generating the same logic: and such different formats may actually give us different ways of carving up the categorial landscape into natural zones. The above Gentzen style is precisely one very natural tool for this purpose, bringing to light interesting options.

Another caveat concerns the word 'hierarchy' itself. When arranged in the above landscape, there is a natural sense of 'ascending strength' among these various categorial engines. But, this need not exclude the possibility that apparently 'weaker' logics could have the power of simulating 'stronger' ones via some suitable *translation* (as is the case, in the standard realm, with intuitionistic versus classical logic).

Remark. Predecessors.

Although the constructive conditional calculus forms the usual lower bound in standard logic, it is not true that no weaker alternatives have ever been considered. After all, some of the principles rejected above, such as the inference

from b to (a, b) ,

correspond to the much-debated 'paradoxes of material implication'. And at least in the area of *Relevant Logic* (cf. Dunn 1985), weaker implicational calculi have been studied, located below I in the above landscape. This fits in with the previous discussion. In effect, the very nature of 'relevance' requires that one pay suitable attention to the syntactic structure of premises used in an argument. Thus, the Categorial Hierarchy provides a meeting ground for logics which are more sensitive to premise structure than the usual ones, and - coming from the other direction - categorial engines which are less sensitive to rigid syntactic structure than the original paradigm. This theme will be elaborated in Chapter 14.

Another source of weaker implicational systems lies in *Modal Logic*, with its modal entailments $\square(\phi \rightarrow \psi)$. For instance, S4-entailment has a conditional logic strictly in between LP and I . The same system can also be motivated as the 'basic non-monotonic logic' underlying C and I as identified in van Benthem 1986B. This modal perspective too will be pursued in this Book, witness Chapter 15 below.

Finally, the idea of having a landscape of possible calculi for a field of inference, rather than one single distinguished one, may seem strange to certain logicians, used to clear-cut winners such as "Frege's Predicate Calculus" or "Heyting's Arithmetic". But in fact, even in some areas of Logic itself, the situation is similar: witness the 'lattice of modal logics' that lies at the heart of intensional logic (cf. Bull & Segerberg 1984). The core of a reasoning practice usually does not lie encoded in any particular set of axiomatic principles written on Stone Tablets, but rather in certain 'styles of inference'. This even holds for the above 'standard systems', whose often-asserted 'unicity' as canons of classical or constructive reasoning owes as much to text-book propaganda as to genuine philosophical or mathematical necessity.

Thus, the grammarian's perspective, envisaging a family of possible grammatical engines for linguistic description seems the more sensitive strategy for the description of reasoning too, when all is said and done. For instance, in the Categorial Hierarchy, the Lambek Calculus is one reasonably stable all-purpose system. But, various subsystems, or extensions, may be needed in order to measure the exact complexity of specific phenomena such as quantifier scoping (cf. the type change calculus H employed in Hendriks 1989), or well-behaved backtracking when parsing (cf. the "monotone" subsystem M in Moortgat 1988). But this is no different in Logic. We have a family of formal systems related to a certain ratiocinative activity - different members of which may be relevant to different aspects of the latter - and hence our task is to arrive at a comparative understanding of their proof-theoretic strength, computational properties and semantic peculiarities.

5 Proofs, Terms and Meanings

The systems introduced so far were purely syntactic or proof-theoretic, although they had a certain semantic motivation. Now, we must turn to the question of their more intrinsic semantic *meaning*.

5.1 Semantics for Sequents

One way of approaching the issue is the standard route in Logic. The calculus L generates a set of derivable sequents: and it is our task to find some matching notion of semantic validity and consequence. Now, this is indeed possible, if we take our cue from the well-known existing modelling for the constructive conditional logic I in *possible worlds* semantics. There, structures are of the form

$\mathbb{M} = (W, \subseteq, V)$,

with W a set of 'information states', ordered by a relation of 'inclusion' $\subseteq$, which may or may not verify atomic propositions: the record being kept by the 'valuation' V sending proposition letters to their truth ranges in W . Then, the truth definition for

$\mathbb{M} \models \phi[w]$ (ϕ *is true in* $\mathbb{M}$ *at* w)

has the following clauses:

$\mathbb{M} \models p[w]$ iff $w \in V(p)$

$\mathbb{M} \models \alpha \to \beta[w]$ iff *for all* $v \in W$ *such that* $w \subseteq v$, *if* $\mathbb{M} \models \alpha[v]$, *then* $\mathbb{M} \models \beta[v]$.

And, if desired, one can add a clause for conjunction:

$\mathbb{M} \models \alpha \wedge \beta[w]$ iff $\mathbb{M} \models \alpha[w]$ *and* $\mathbb{M} \models \beta[w]$.

Moreover, attention is restricted to *hereditary* propositions in the atomic case, whose truth ranges are closed under extension along $\subseteq$. This property is then inducted upward to all formulas.

Remark. Variations.

Dropping the Heredity condition, one gets a perhaps more basic, 'non-monotonic' conditional logic below I (cf. Chapter 4 , and van Benthem 1986B). With an additional closure condition of 'cofinality' on the other hand, the full classical logic C is arrived at, including Peirce's Law. Cofinality says that, if a formula α fails at some state, then there must be some successor state 'refuting' it (in the sense of lacking α-successors altogether). ■

Now, an attractive generalization of this setting to calculi like L has been proposed by several authors (cf. Dosen 1988/9, Buszkowski 1986, van Benthem 1988C, Ono & Komori 1985), in the line of earlier work on the semantics of relevant logics by Alasdair Urquhart (see Dunn 1985). Models will now be algebraic structures

(W, +, V)

where + is an associative operation of *addition* of information pieces.

The crucial truth conditions then become

$\mathbb{M} \models (a, b)\ [w]$ iff *for all* v *such that* $\mathbb{M} \models a\ [v]$, $\mathbb{M} \models b\ [w+v]$

$\mathbb{M} \models a \bullet b\ [w]$ iff $w = u+v$ *for some* u, v *with* $\mathbb{M} \models a\ [u]$, $\mathbb{M} \models b\ [v]$.

Valid consequence can then be defined as follows:

$a_1, ..., a_n \models b$ iff *for all models* $\mathbb{M}$ with $\mathbb{M} \models a_1\ [w_1], ..., \mathbb{M} \models a_n\ [w_n]$, it holds that $\mathbb{M} \models b\ [w_1+...+w_n]$.

Here is one simple result, illustrating the desired connection with our calculi.

Theorem. For the product-free fragment,

$a_1, ..., a_n \models b$ if and only if $a_1, ..., a_n \Rightarrow b$ is derivable in L .

Starting from here, various stronger systems than L can be modelled by imposing various further restrictions on the addition operator + .

Other, more algebraically oriented semantics for categorial calculi exist too: see Buszkowski 1982, 1988. These various forms of semantics will return in Part VI below. In particular, in Chapter 14, the algebraic viewpoint becomes a wider perspective on Categorial Grammar as a logic of language families. The information-based theme is developed further in Chapter 15. And finally, Chapter 16 will develop certain analogies with Relational Algebra, shedding additional light upon L and its relatives.

Nevertheless, for our central purposes in this Book, the more important semantic question lies somewhere else.

5.2 Semantics for Derivations

One of the virtues of a semantics in the classical vein lies in the fact that it liberates us from 'details of syntax'. But in our present perspective, it is exactly the latter which matter. More precisely, one of the crucial issues is this. It is all right for the earlier proposed categorial combinations or type changes to congregate in elegant calculi like Lambek's: but what is their *meaning*, in terms of denotational effects? This meaning must lie one level below what has been considered so far, since we have seen already how one valid transition can have quite different derivations, with intuitively different effects.

Example. Scope Distinctions.
The propositional formula $\neg p \wedge q$ is recognized via the transition

$$(t, t) \quad t \quad (t, (t, t)) \quad t \quad \Rightarrow \quad t .$$

But different derivations of this sequent intuitively produce two different readings, being

$$\neg(p \wedge q) \qquad \text{and} \qquad (\neg p \wedge q) .$$

●

Here is where we can use another basic insight from logical Proof Theory and Lambda Calculus:

> There exists an effective one-to-one correspondence between natural deduction *derivations* in the constructive conditional logic and typed *lambda terms* .

The broad idea is that Modus Ponens corresponds to *function application*, in line with the original motivation of Categorial Grammar, while instances of Conditionalization are encoded by *lambda abstraction* . Details may be found in Hindley & Seldin 1986 (the idea goes back to Curry, Howard, De Bruyn and others in the sixties): but, a few concrete examples will readily demonstrate what is going on. We shall present a number of the earlier categorial derivations, now with corresponding lambda terms computed. The latter may be viewed as the corresponding 'meaning recipes':

> A derivation P for a transition $a_1, ..., a_n \Rightarrow b$ encodes a uniform instruction for obtaining a denotation in type b from a sequence of denotations in types $a_1, ..., a_n$, which is given by a lambda term
>
> $$\tau_b{}^P\{u_{a1}, ..., u_{an}\} .$$

Here, the 'u_{ai}' are unique free variables, indicating 'parameters' of the procedure.

In special cases, this was already the original intuition, of course. For instance, one motivates a Montague raising

$$e \Rightarrow ((e, t), t)$$

by displaying the canonical recipe for 'bundles of properties' associated with individual objects:

$$\lambda P_{(e, t)} \bullet P(u_e) .$$

And likewise, a Geach transition turning sentence negation into predicate negation

$$(t, t) \Rightarrow ((e, t), (e, t))$$

really amounts to an instruction for 'pointwise computation' of a set complement:

$$\lambda P_{(e, t)} \bullet \lambda y_e \bullet NOT_{(t, t)}(P(y)) .$$

These ideas were already at work in the 'lambda-categorial languages' of Cresswell 1973.

Example. Derivations and Instructions.

The first case goes back to Chapter 3, showing how standard categorial derivations correspond to pure application terms:

- The Propositional Formula $\neg(p \wedge q)$

$$\wedge_{(t, (t, t))} \qquad q_t$$
$$p_t \qquad \wedge(q)$$
$$\neg_{(t, t)} \qquad \wedge(q)(p)$$
$$\neg\wedge(q)(p)$$

Next, lambda abstraction comes in with the new rules for flexible categorial grammars (compare Chapter 4 for the relevant derivations):

- Montague Raising

$$u_a \qquad x_{(a, b)}$$
$$x(u)$$
$$\lambda x \bullet x(u)$$

i.e., the usual definition of a *functional dual.*

- Geach Composition

$$x_a \qquad u_{(a, b)}$$
$$u(x) \qquad v_{(b, c)}$$
$$v(u(x))$$
$$\lambda x \bullet v(u(x))$$

i.e., the usual definition of *function composition.*

- Argument Lowering

$$\dfrac{\dfrac{\dfrac{x_a \qquad y_{(a,b)}}{y(x)}}{\lambda y\bullet y(x)} \qquad u_{(((a,b),b),c)}}{\dfrac{u(\lambda y\bullet y(x))}{\lambda x\bullet u(\lambda y\bullet y(x))}}$$

For the type change of parametrization $(b, c) \Rightarrow ((a, b), (a, c))$, proposed in Chapter 4 as an alternative to Geach Composition, one more Conditionalization step in the above derivation will yield an instruction

$\lambda u\bullet \lambda x\bullet v(u(x))$.

This explains the above shift for negation from a truth function in type (t, t) to set-theoretic complement in type $((e, t), (e, t))$. With the other possible Conditionalization step, however, one obtains the transition $(a, b) \Rightarrow ((b, c), (a, c))$. Amongst others, the latter licenses a similar shift for one-place operations on individuals:

$(e, e) \Rightarrow ((e, t), (e, t))$,

whose meaning becomes

$\lambda v_{(e,t)}\bullet \lambda x_e\bullet v(u_{(e,e)}(x))$.

Example. Ambiguity from Different Derivations.

A first example concerns operator scoping. In addition to the one meaning computed above, the propositional formula $\neg(p\wedge q)$ also has a second reading with 'small scope' for the negation:

$$\dfrac{\dfrac{\neg_{(t,t)} \qquad p_t}{\neg(p)} \qquad \dfrac{\wedge_{(t,(t,t))} \qquad q_t}{\wedge(q)}}{\wedge(q)(\neg(p))}$$

Next, here is a case of permutation of arguments.

The one-step proof of the sequent $(e, (e, t)) \Rightarrow (e, (e, t))$ mentioned in Chapter 4 amounts to the trivial recipe

$u_{(e,(e,t))}$.

The longer proof given as well has the following effect, however:

$$\dfrac{y_e \qquad \dfrac{x_e \qquad u_{(e,(e,t))}}{u(x)}}{\dfrac{\dfrac{u(x)(y)}{\lambda x \bullet u(x)(y)}}{\lambda y \bullet \lambda x \bullet u(x)(y)}}$$

i.e., $u_{(e, (e, t))}$ has been transformed into its *converse* relation. ■

Finally, we make good on a promise from Chapter 3, by 'defusing' Montague's well-known formula for (extensional) identity.

Example. Montagovian Being Derived.
Starting from the identity relation $=$ between individuals, in type $(e, (e, t))$, we can derive a relation acting on complex noun phrases, that is, in type $(((e, t), t), (e, t))$, as follows:

$$\dfrac{\dfrac{\dfrac{\dfrac{e^1 \qquad (e,(e,t))}{(e,t)}\,\text{MP} \qquad ((e,t),t)^2}{t}\,\text{MP}}{(e,t)}\,\text{COND, -1}}{(((e,t),t),(e,t))}\,\text{COND, -2}$$

Its meaning is this:

$$\dfrac{\dfrac{\dfrac{x_e \qquad u_{(e,(e,t))}}{u(x)} \qquad y_{((e,t),t)}}{y(u(x))}}{\dfrac{\lambda x_e \bullet y(u(x))}{\lambda y_{((e,t),t)} \bullet \lambda x_e \bullet y(u(x))}}\ .$$

In Montague's particular case, this results in the formula

$$\lambda y_{((e, t), t)} \bullet \lambda x_e \bullet y(=_{(e, (e, t))}(x)) ,$$

or, more colloquially:

" be Y " holds of those individuals x for which $\{x\} \in Y$.

And this was indeed Montague's semantic proposal. ■

The lambda correspondence developed so far does not depend on idiosyncracies of implication. For, it lends itself to generalization involving other logical constants.

Example. Extension to Products.

Derivations involving product types can be handled by an appropriate extension of the type-theoretic language, introduced already in Chapter 2. Written in the earlier sequent format, the relevant recipes look like this

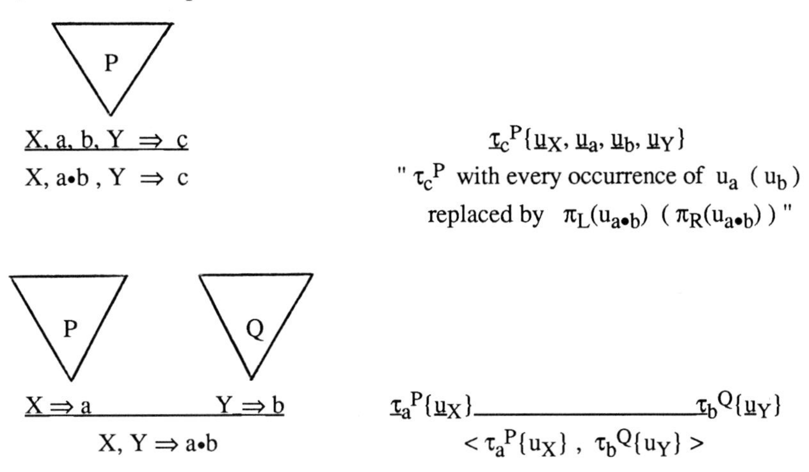

For the earlier derivations of currying and uncurrying presented in Chapter 4, this will produce the following two (correct) recipes:

$\lambda x_{a \bullet b} \bullet\, u_{(a,\,(b,\,c))}(\pi_L(x))(\pi_R(x))$

$\lambda x_a \bullet\, \lambda y_b \bullet\, u_{(a \bullet b,\,c)}(\langle x, y\rangle)$.

With natural deduction trees, another formulation of the correspondence may actually be easier to use in practice:

Introduction Rule

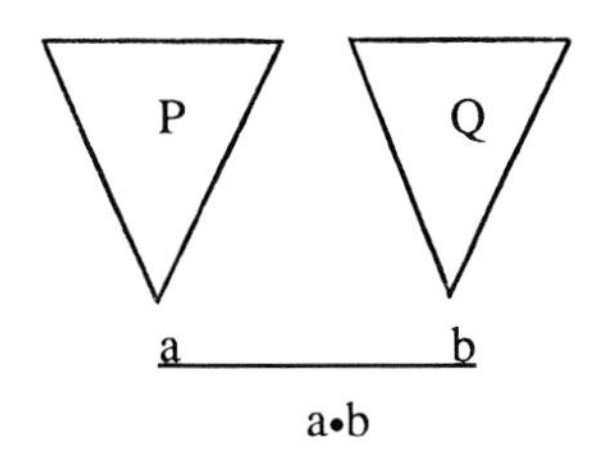

$\underline{\tau_a{}^P \qquad\qquad \tau_b{}^Q}$

$\langle \tau_a{}^P, \tau_b{}^Q \rangle$

Elimination Rule

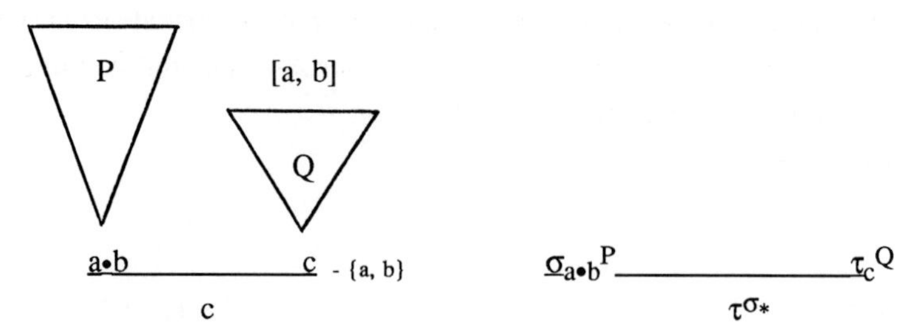

where $\tau^{\sigma *}$ is τ with the relevant occurrence of u_a replaced by $\pi_L(\sigma)$ and that of u_b by $\pi_R(\sigma)$.

Once in the possession of this correspondence, we can use it for many more general purposes. For instance, in Chapter 10 we shall see how it enables us to lift a good semantic account found for expressions occurring in one linguistic category to many other contexts where it occurs.

Example. Reflexives.

The basic use of the reflexivizer "self" is to turn binary relations into unary properties: that is, it acts as a 'relation reducer' in type

$((e, (e, t)), (e, t))$.

The paradigm for this phenomenon is a syntactic context with a transitive verb, such as "know oneself" . But, reflexives can also appear in quite different environments, such as prepositional phrases:

" [Mary kept The Bible] *to herself* " .

As we shall see in Chapter 10, there is a straightforward categorial derivation here, which produces a computed meaning as desired:

$\lambda P_{(e, t)} \bullet \lambda x_e \bullet TO_{(e, ((e, t), (e, t)))}(x)(P)(x)$,

using no more than the base meaning of reflexivization.

5.3 Different Proofs and Different Readings

As we have seen, the basis for computation of meanings is formed by derivations, not by flat expressions or type sequences. In a slogan, then:

"The unit of meaning is the derivation, not the expression" .

In particular, different derivations for one sequent will correspond with different lambda terms. But, contrary to what is assumed sometimes by working syntacticians, not all graphical differences in derivational structure lead to genuine semantic differences: as the corresponding lambda terms may be *logically equivalent*.

Thus, the map from proofs to readings is not one-to-one.

Example. Coordinating Conjunction.
A phrase like "radio and television fan" is ambiguous between a reading with Boolean conjunction ('both radio fan and television fan') and one with coordinating non-Boolean conjunction: 'fan of the combination'. Now, it has been proposed in the linguistic literature to account for this in terms of the difference between the following two derivations [employing coarse syntactic categories for convenience]:

Boolean

```
(N, N)     (X, (X, X))       (N, N)
____________________________________
          (N, N)                    N
          ___________________________
                      N
```

Non-Boolean

```
                                      N             (N, N)^1
                                      ______________________
(N, N)    (X, (X, X))    (N, N)                N
________________________________          __________ -1
         (N, N)                           ((N, N), N)
         ____________________________________________
                          N
```

But, as is easily computed, the corresponding lambda terms are logically equivalent, by performing one *lambda conversion* (cf. Chapter 2). In technical proof-theoretic terms (compare Prawitz 1965), the first proof arises out of the second by one *normalization* step: and normalization on proofs is a transformation which preserves meaning. ◆

Remark. Equating different derivations with different readings is in fact admissible in the original categorial grammar. For, in the pure application language, two terms are logically equivalent if and only if they are identical. ◆

Thus, from a semantic point of view, we are only interested in equivalence classes of lambda terms up to provable identity in the Lambda Calculus.

In this connection, there are other types of irrelevant derivational structure too, corresponding to another inference rule in this calculus.

Example. η-Conversion.

The following two different Lambek proofs

$$(e, t) \qquad\qquad \dfrac{\dfrac{e^1 \quad (e, t)}{t}\ \text{MP}}{(e, t)}\ \text{COND}, -1$$

have these associated terms:

$$u_{(e, t)} \qquad\qquad \dfrac{\dfrac{x_e \quad u_{(e, t)}}{u(x)}}{\lambda x \bullet u(x)}$$

which are equivalent by η-conversion. ●

Remark. Semantics for Sequent Calculus.

It is easy to formulate our lambda semantics also directly for derivations in the earlier *sequent calculus* format for categorial calculi (one way is shown in Moortgat 1988). But, derivations in sequent calculus may have more semantic redundancy than those in our natural deduction format. Here is an illustration of two different sequent derivations corresponding even to the same pure application term:

$$\dfrac{e \Rightarrow e \qquad \dfrac{t \Rightarrow t \qquad s \Rightarrow s}{t\ \ (t, s) \Rightarrow s}}{e\ \ (e, t)\ \ (t, s) \Rightarrow s}$$

$$\dfrac{\dfrac{e \Rightarrow e \qquad t \Rightarrow t}{e\ \ (e, t) \Rightarrow t} \qquad s \Rightarrow s}{e\ \ (e, t)\ \ (t, s) \Rightarrow s}$$

●

The general issue then becomes *how many* readings a given derivable transition can have. This issue will be investigated more systematically in Chapter 9. For the moment, it willl be clear that there is a whole range of numerical possibilities here, from unambiguous cases to highly ambiguous ones.

Example. Basic Rules of Type Change.

By closer inspection of their possible lambda *normal forms*, it may be shown that both the Montague and the Geach transition are unambiguous. Modulo provable equivalence in the lambda calculus, they have just one meaning: being the ones computed earlier. Thus, when discussing flexible categorial grammars, one can often concentrate on the patterns of type change, without worrying about the possibility that these might still stand for semantically different procedures. ♠

By contrast, many categorial patterns in natural language show various ambiguities. A case in point are transitive verbs with complex subject and object

NP_1 TV NP_2 .

In the directed Lambek Calculus, with suitable meanings computed for it (the relevant mechanism is an obvious variant of the semantics presented so far) , the corresponding sequent

$$(t/(e\backslash t)) \quad (e\backslash(t/e)) \quad ((t/e)\backslash t) \quad \Rightarrow \quad t$$

has two different readings, representing the two scope orderings for the quantified noun phrases. In the non-directed calculus, however, the matching sequent

$$((e, t), t) \quad (e, (e, t)) \quad ((e, t), t) \quad \Rightarrow \quad t$$

is even fourfold ambiguous, as we can also decide to convert the binary predicate. (Moreover, if the internal structure of the NPs, being " Det N " , is taken into account, another doubling takes place, as we can also convert the semantic determiner relations.)

Could this ambiguity even be *infinite*? For an answer to such more general questions we refer again to Chapter 9.

5.4 A Semantic Hierarchy of Fragments

Up till now, we have employed general features of the Lambda Calculus and constructive implicational logic. Against this background, weaker calculi in the Categorial Hierarchy then turn out to correspond to certain *fragments* of the full lambda/application language. Notably, various restrictions on the rule of Conditionalization will be reflected in various constraints on lambda binding.

The case of the Lambek Calculus L itself will illustrate the phenomenon.

Let Λ be that fragment of the lambda calculus whose terms satisfy the following constraint:

Each occurrence of a λ binds *exactly one* free variable occurrence.

An easy induction then establishes the following result (see van Benthem 1983, 1986A for a proof).

Theorem. L*P-derivations of sequents $a_1, ..., a_n \Rightarrow b$ in the natural deduction calculus correspond effectively to Λ-terms τ of type b having exactly n distinct occurrences of free variables: $u_{a1}, ..., u_{an}$. This correspondence is one-to-one and onto.

For the stricter class of LP-derivations, an additional restriction appears on the corresponding terms, namely that

Each subterm should have at least one free variable.

Remark. We can also extend this lambda correspondence to the case of L with products, using lambda terms with projection and pairing. But then, the resulting class of terms has a less perspicuous description. In particular, lambdas may now bind more than one occurrence of a variable, provided that these occurrences are 'projected' differently. A precise description, in terms of 'balanced trees' for occurrences of variables in the relevant lambda terms, may be found in Roorda 1989B.

The above correspondence has been stated here for permutation-closed calculi. As was observed before, an extension to the case of directed non-permuting systems is straightforward. Of course, in the latter case, there is no longer uniqueness: since, e.g., both a\b and b/a will translate into the same functional type.

For an illustration of the Theorem, it may be checked that all lambda terms computed so far fall indeed within Λ .

Another, more encompassing example would be the calculus L*PC allowing also the structural rule of Contraction. The latter corresponds precisely to all lambda terms having *non-vacuous* lambda bindings: i.e., Church's so-called 'λI' fragment.

Example. Polymorphic Conjunction.
Unlike Boolean negation, conjunction needs multiple binding for an adequate semantic account of its polymorphism. This shows in the lambda term matching the derivation of ((e, t), ((e, t), (e, t))) from (t, (t, t)) given in Chapter 4, which fell indeed within LPC (though not within LP) :

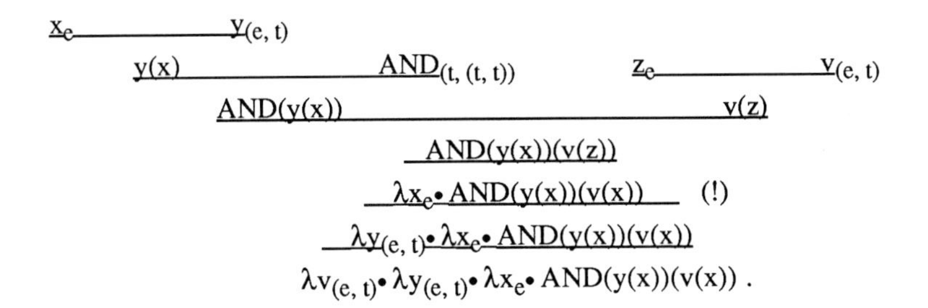

Incidentally, there is a number of alternatives here. For instance, withdrawing the two premise occurrences (e, t) in a different order would result in an interchange of the abstractions $\lambda y_{(e,t)}$ and $\lambda v_{(e,t)}$. But also, at least in the constructive conditional logic I, both ocurrences of (e, t) might have been withdrawn at the same time, making the last Conditionalization vacuous, resulting in a 'reading'

$$\lambda v_{(e,t)} \bullet \lambda y_{(e,t)} \bullet \lambda x_e \bullet \text{AND}(y(x))(y(x)) .$$

These fragments are restricted formats for defining denotations of linguistic expressions. Thus, they provide an opportunity for what may be called "semantic fine-tuning" . That is, we can measure the complexity of some proposed denotation in terms of its position in a *Semantic Hierarchy*, running from pure application terms to the full language of the Lambda Calculus. Thus, categorial semantics admits of the same kind of layering which has been so characteristic for linguistic syntax, in the usual hierarchy of rewrite grammars.

Example. Reflexivization Revisited.
Many 'invisible', i.e., non-lexically marked, type changes are captured by the Lambek Calculus, which employs only one binding per lambda. By contrast, the earlier reflexivizer "self" had a double bind lambda definition

$$\lambda R_{(e,(e,t))} \bullet \lambda x_e \bullet R(x)(x) .$$

And the latter formula is easily shown to be undefinable in the fragment Λ : it represents an essentially more complex kind of semantic operation. Linguistically, one interesting issue is then to which extent such a procedure can be generated freely (so that the above calculus LPC would be allowed in certain cases as a categorial engine), and when explicit lexical instructions for identification of argument roles are necessary. This is a central issue in the syntax of pronominalization: witness Szabolcsi 1989.

This development may be seen in the following light. The initial Montagovian approach in the seventies put almost no constraints on the abstract denotations which could be assigned to expressions of natural language. In particular, even though it emphasized Fregean *Compositionality* reflecting syntactic construction, this constraint remained virtually empty, since there were no restrictions at the semantic end (cf. van Benthem 1984). But now, one constraint implicit in the present set-up is this:

> Semantic denotations should be defined from atomic items
> in the lambda/application language.

Within this vehicle of interpretation then, semantic fine-tuning can take place, searching for the lowest levels of complexity for linguistic expressions of processes.

In practice, one readily acquires a feeling for this 'positioning'.

Example. Boolean versus Non-Boolean Coordination.
In many languages, there is a non-Boolean coordinator "and" which joins individuals into groups, as in

"Peter and Paul met" ,

which does not mean that "Peter met and Paul met" . Unlike the Boolean conjunction, which was of type (t, (t, t)) , this conjunction lives in type

(e, (e, e)) ,

with a denotation indicating (to a first approximation) the formation of a set:

$\lambda x_e \bullet \lambda y_e \bullet \{x, y\}$.

Here, sets are again taken to be (complex) individual objects.

Now, in Hoeksema 1988, Krifka 1989, the question is raised how the polymorphism of the latter conjunction is related to that of its Boolean counterpart. This issue can be analyzed with our present apparatus, with the following outcome:

- Non-Boolean conjunction does not support significant *predicate* conjunction. This follows from a logical observation:

  The sequent (e, t) (e, (e, e)) (e, t) $\Rightarrow$ (e, t)

  is underivable in the calculus LP , and even in LPC .

The argument for this is by proof-theoretic analysis, showing by induction that

$(e, t)^i \; t^j$ + 'pure e-types' LPC-derive t only if i=1 & j=0 or j=1 & i=0 ,

$(e, t)^i \; t^j$ + 'pure e-types' LPC-derive e only if i=j=0 .

(Here, ' a^i ' stands for i consecutive occurrences of the type a .)

- Non-Boolean conjunction does support *noun phrase* conjunction, and that even without being forced upward from the basic calculus LP.

Here is a derivation of the relevant sequent

$$((e, t), t) \quad (e, (e, e)) \quad ((e, t), t) \quad \Rightarrow \quad ((e, t), t):$$

$$\dfrac{\dfrac{\dfrac{\dfrac{\dfrac{\dfrac{\dfrac{\dfrac{e^1 \quad (e, (e, e))}{(e, e)}\text{MP} \quad e^2}{e}\text{MP} \quad (e, t)^3}{t}\text{MP}}{(e, t)}\text{COND, -1} \quad ((e, t), t)}{t}\text{MP}}{(e, t)}\text{COND, -2} \quad ((e, t), t)}{t}\text{MP}}{((e, t), t)}\text{COND, -3}$$

The corresponding Λ-term may be compiled as

$$\lambda x_{(e,t)} \bullet w_{((e,t),t)}(\lambda y_e \bullet v_{((e,t),t)}(\lambda z_e \bullet x(u_{(e,(e,e))}(z)(y)))) .$$

More concretely, a noun phrase like "every sigh and every tear" would work out to the following, intuitively correct, reading:

$$\lambda X \bullet \forall y\, (Sy \rightarrow \forall z\, (Tz \rightarrow X(\{y, z\}))) .$$

◆

5.5 Further Transformations on Types

Finally, although attention up till now has focused on type changes expressible in the pure Lambda Calculus, there may be cases where the richer type-theoretical language with *identity* is needed.

For denotations of lexicalized expressions, this is clear anyway: for instance, as was observed in Chapter 2, many logical constants find their definition here. But, is identity ever needed to express 'emergent' genuine type changes? One putative example in Chapter 3 was that of German or Spanish constructions of the form " Determiner Proper Name ", where an underlying shift seems to take place from individuals, in type e, to unary predicates, in type (e, t). And an obvious meaning for the latter would be, say, 'Quine's Rule' (reflecting Quine's well-known predicative reduction of proper names):

$$\lambda x_e \bullet \lambda y_e \bullet y = x .$$

Thus, there may be a case for studying, at least, fragments of the full Type Theory too.

Another possible extension arises as follows.
With many authors, there is a tendency to allow 'type changes' not only of a functional character, expressible in Lambda Calculus, but also of a Boolean character: in particular, via *conjunction* (cf. Groenendijk & Stokhof 1988B). Thus, we may also add an axiom

$$t, t \Rightarrow t$$

to our systems, providing it with the standard meaning of Boolean conjunction. This is in fact one instance of an option mentioned earlier in Chapter 4, namely to allow certain restricted versions of more powerful structural rules (such as, in this case, Contraction). In Chapter 12 we shall meet this possibility again, this time for basic intensional types.

Example. From Predicates to Adjectives.
One way of analyzing so-called 'intersective adjectives' like "blonde" or "Dutch" is by assuming that these are really individual predicates A , having undergone a type raising

$$(e, t) \Rightarrow ((e, t), (e, t)) .$$

This may be shown using the above 'free conjunction', provided that we employ the appropriate categorial calculus. For,

$$(e, t) , (t, (t, t)) \Rightarrow ((e, t), (e, t))$$

is derivable in LPC , with an associated meaning for the adjective of

$$\lambda P_{(e, t)} \bullet \lambda x_e \bullet \wedge_{(t, (t, t))}(A_{(e, t)}(x))(P(x)) .$$

◆

Digression. Witnesses for Transitions.
Without proof, we mention which type transitions can be 'witnessed' via type-theoretic terms with identity, in the $\{e, t, s\}$ - based framework. These are all sequents $a_1, ..., a_n \Rightarrow b$ for which the type $(a_1, (a_2, ... (a_n, b) ...))$ contains permutation-invariant items (in the sense of Chapter 2). Such types will be characterized effectively in Chapter 10.
◆

The preceding discussion raises a more general issue, as to structural connections between various type domains. There is a lot of 'coherence' in function hierarchies, which leads to natural transformations from one type domain to another. For instance, Partee 1986 stresses the linguistic similarities between proper names, predicates and complex noun phrases, relating them in the following diagram:

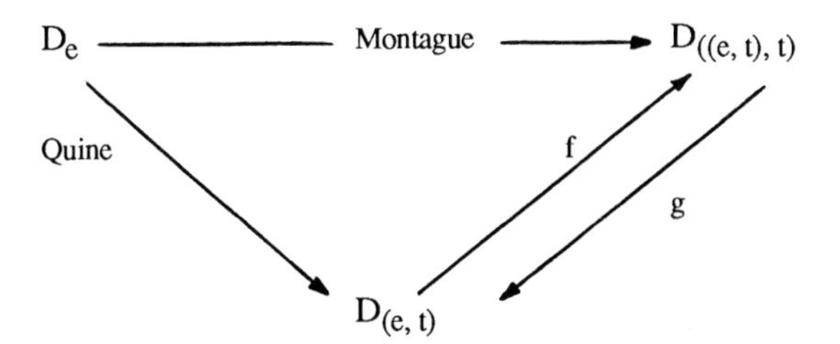

The question here is which mappings f , g make the diagram commute. Note that f has the type ((e, t), ((e, t), t)) of determiners: and indeed, we can take the determiner "the" to play this role. Conversely, it is shown in van Benthem (1986, chapter 3) that the only Boolean homomorphism which qualifies for g is the Montagovian "be" of this Chapter.

Another way of viewing the general situation is this. We have studied 'purely combinatorial' type changes, expressible in pure lambda terms. But, there can also be type changes having more mathematical content: say, of a set-theoretic nature. For instance, Quine's Rule takes individuals to their set-theoretic singletons. And yet other forms may occur too.

Now, what Quining does, in a sense, is to embed the individual universe one level up in the type hierarchy, among the sets of individuals. Diagrams such as the above then represent the question how maps from the old domain can be related systematically, and 'conservatively', to maps from the lifted domain. (If commuting diagrams are to exist always, then the lifting function must be necessarily one-to-one.)

To continue the example, raising individuals to singletons is often a preliminary to the treatment of *plural* expressions in natural language, which naturally call for the introduction of predicates on sets or groups. This process provides further illustrations of type changes not covered by our earlier calculi.

Example. Collectivizing Determiners.
A distributive determiner as used in

" All [boys whistled] "

has an obvious meaning in type ((e, t), ((e, t), t)) . But in a plural context such as

" All [boys gathered] " ,

it seems to function rather in type ((e, t), (((e, t), t), t)) : since "gather" is a predicate of groups, rather than individuals.

Is there any systematic sense in which the latter plural reading of the determiner D_{new} is derived from the original individual one D_{old} ? Van Benthem 1989F presents two set-theoretic proposals to this effect, one being

$$D^1_{new} \quad = \quad \lambda A_{(e,\,t)} \bullet\ \lambda B_{((e,\,t),\,t)} \bullet\ D_{old}(A)((\bigcup B) | A)\ ,$$

and the other

$$D^2_{new} \quad = \quad \lambda A_{(e,\,t)} \bullet\ \lambda B_{((e,\,t),\,t)} \bullet\ D_{old}(A)(\bigcup (B \cap pow(A)))\ .$$

On the first account, 'all boys gathered' if the set of boys is contained in the set of all people taking part in gathering groups (restricted to boys). On the second account, they must be contained in the union of all gathering groups consisting entirely of boys. The former would be true, and the latter false, in case each boy only got together with a girl.

Again, both these clauses may be formulated in a simple fragment of the full Type Theory with identity. ◆

As a final illustration, we mention the existence of various 'type lowerings' which can occur for semantic objects satisfying certain special *denotational constraints*.

For instance, if a function on some domain preserves enough Boolean structure, it may sometimes be represented by means of some object living in a lower type. Thus, e.g., Boolean homomorphisms in types $((a, t), (b, t))$ may be 'condensed' to arbitrary functions in the lower type (b, a) . This kind of lowering phenomenon will be investigated in Chapter 11. Likewise, 'absolute' adjectives correspond to unary predicates, thereby displaying a 'non-standard' type transition

from $((e, t), (e, t))$ to (e, t) .

Nevertheless, alternative categorial analyses are sometimes possible too. For instance, adjective lowering may also be analyzed by converting an earlier example (as has been observed by Victor Sanchez).

Example. From Absolute Adjectives to Predicates.

Let us assume that adjectival lowering is performed lexically, namely, via a predicative verb "is" : i.e., by the copula of traditional logic. Then, the latter would possess type

$$(((e, t), (e, t)), (e, t))\ ,$$

and the only issue is to provide a suitable denotation for this process. One pertinent proposal may be found in Chapter 11 (see also van Benthem 1986, chapter 3), relying on the 'continuity' and 'introspection' of absolute adjectives. Here is its description in type-theoretic terms:

$\lambda A_{((e,\, t),\, (e,\, t))} \bullet \lambda x_e \bullet A(=_{(e,\, (e,\, t))}(x))(x)$.

Note that this amounts to a semantic derivation of predicative "be" from the earlier Montagovian identity reading. ●

For the moment, it will suffice to have indicated the open-endedness of the polymorphism inherent in linguistic denotations.

III PROOF THEORY

In this part, we shall take a closer look at the proof theory of categorial calculi, considering both more logical and more linguistical topics.

6 Exploring Categorial Deduction

First, here are some further illustrations of categorial deduction, in order to increase familiarity with the earlier systems.

Example. 'Flattening Rules'.
The following derivable sequent, 'flattening' a third-order antecedent to a first-order one, is taken from Koenig 1989. In an undirected version:

$$(((x, y), (z, y)), u) \;\Rightarrow\; ((z, x), u) :$$

```
z__________________(z, x)
          x____________________________________(x, y)
                        y
                   ____(z, y)____
                   ((x, y), (z, y))____________________________(((x, y), (z, y)), u)
                                                u
                                            ((z, x), u)
```

One correct directed version, which may be derived similarly, reads as follows:

$$(((z\backslash y)/(x\backslash y))\backslash u) \;\Rightarrow\; ((z\backslash x)\backslash u) .$$

Note how it would already be difficult to judge its validity with the naked eye. ∎

Next, here are some simple useful properties of our calculi.

Fact. The pure / fragment lies faithfully embedded in the full \, / calculus L.

This follows from the Cut Elimination theorem to be stated in the next Chapter.

Conversely, however, there is no faithful embedding from the two-directional Lambek calculus into the one-directional one. In particular, replacing all slashes \ by / will not do:

$((a\backslash b)/c) \Rightarrow_L (a\backslash(b/c))$, but not
$((a/b)/c) \Rightarrow_L (a/(b/c))$, as may be shown by some simple proof analysis.

Of course, outcomes will vary here for different members of the Categorial Hierarchy. For instance, the Ajdukiewicz calculus is much poorer than Lambek's, due to the great power of Conditionalization.

Fact. There is no faithful embedding of the Ajdukiewicz calculus into Lambek's, or vice versa.

Proof. Suppose that a homomorphic translation σ embedded A into L. Then, since L derives $\sigma(a) \Rightarrow ((\sigma(a)/\sigma(a))\backslash\sigma(a))$, it would follow that A derives the sequent $a \Rightarrow ((a/a)\backslash a)$: quod non.

Conversely, suppose that τ embedded L into A. Again, the L-derivable sequent $a \Rightarrow ((a/a)\backslash a)$ would have no τ-translated counterpart in A. ●

As for stronger calculi than L, one central candidate in these lectures has been the undirected variant LP, where both slashes have been collapsed into one operator (a, b). This calculus can derive some surprising types out of a given type sequence, witness the provable sequents

$(e, t)\ (t, e) \Rightarrow (e, e)$, but also $(e, t)\ (t, e) \Rightarrow (t, t)$.

In fact, is there any reasonable restriction on this variety? For instance, might it be possible that some sequence of types, construed in one way, would represent a sentence (type t), but in another, an individual object (type e)?

Here is one useful *invariant* of proofs in occurrence-based systems of logic.

Definition. Primitive Type Counts.

For each basic type x, one defines the x-count $\#_x(a)$ of any type a by the following recursion

$\#_x(x) = 1$,
$\#_x(y) = 0$, for all basic types y distinct from x,
$\#_x((a, b)) = \#_x(b) - \#_x(a)$.

For sequences of types, counts are obtained by simply adding those for their individual members. Moreover, when necessary, products may be treated as well:

$$\#_x(a \bullet b) \;=\; \#_x(a) + \#_x(b)\,.$$

●

Now, the following invariance holds:

Proposition. For each derivable sequent $X \Rightarrow a$ in LP, all primitive type counts must be equal on both sides of the arrow.

Proof. By a straightforward induction on the length of derivations.
For instance, if $Y \Rightarrow a$ and $Z \Rightarrow (a, b)$ are both derivable, then the computation for their conclusion $Y\,Z \Rightarrow b$ runs as follows:

$$\#_x(YZ) \;=\; \#_x(Y) + \#_x(Z) \;=\; \#_x(a) + \#_x(a, b) \text{ (by the inductive hypothesis)}$$
$$=\; \#_x(a) \;+\; (\#_x(b) - \#_x(a)) \;=\; \#_x(b)\,.$$

●

Thus, we can see at least that such strange outcomes as both $X \Rightarrow e$ and $X \Rightarrow t$ for the same sequent X are impossible here.

Remark. Further Invariants.
Other invariants of Lambek derivation exist too. For instance, following a suggestion by Theo Janssen, the 'operator count' $\$(a)$ of a type a may be defined as follows:

$$\$(x) \;=\; 0 \qquad \text{for primitive types } x$$
$$\$(\,(a, b)\,) \;=\; \$(b) + 1 - \$(a)$$

Operator counts of sequences of types are then computed via the stipulation that

the empty sequence has operator count 0

$$\$(X, Y) \;=\; \$(X) + \$(Y) - 1\,.$$

Now, at least for derivations in the Lambek Calculus LP allowing no empty premise sequences, operator counts on both sides of derivable sequents must always be the same. This is a slight improvement over what we had so far. E.g., the operator count rejects the sequent $\Rightarrow (t, t)$, that would pass the above count test for primitive types. On the other hand it may be shown that, when comparing two non-empty sequences of types, having equal counts for all primitive types implies equality of functional counts. ●

The primitive type count invariant does not hold, however, for stronger systems having Contraction. And indeed, for instance, the following sequents are derivable in LPC :

- e (e, t) (t, e) ⇒ e

(this is even derivable in LP) ,

- e (e, t) (t, e) ⇒ t

(using the LP-derivability of the sequent e (e, t) (e, t) (t, e) ⇒ t) .

In fact, such observations can be used to better understand the position of L in the Categorial Hierarchy. For instance, comparing it with the most central standard system, we have the following situation (cf. the discussion of possible 'translation' in Chapter 4).

Fact. The intuitionistic conditional logic I cannot faithfully embed L .

Proof. The following types are all non-equivalent in L :

(e, t) (e, (e, t)) (e, (e, (e, t))) ...

because of their different e-counts. By contrast, I satisfies 'Diego's Theorem' (Diego 1966) stating that, on any finite number of atomic formulas, only finitely many non-provably equivalent formulas exist. Thus, I lacks the sensitivity needed to embed L .

Even for occurrence-based systems, count invariance cannot match derivability: if only, because the former notion is *symmetric*. For instance, all counts are equal on both sides in the sequents

e ⇒ ((e, t), t) and ((e, t), t) ⇒ e ,

and yet, only the former is derivable in LP . (As to the latter, a truth table counter-example shows that it cannot even be derived in the full Classical conditional logic.) Stronger count invariants, doing better in this respect, will be found in Chapter 14.

Nevertheless, already the simple property discovered here has many applications, which will be found in subsequent Chapters.

Remark. Geometric Invariants for Directed Systems.

The notion of 'count' can be extended in a more 'geometric' direction to fit the directed calculus L , as has been observed in Roorda 1989B.

To see this, first define the following operations on marked functional types:

$(a\backslash b)^{+} \quad = \quad b^{+}\ a^{-}$

$(a\backslash b)^{-} \quad = \quad a^{+}\ b^{-}$

$(b/a)^{+} \quad = \quad a^{-}\ b^{+}$

$(b/a)^{-} \quad = \quad b^{-}\ a^{+}$

Next, the 'atomic marking' of any type sequence with + or - markers throughout is defined as the result of successsive application of these four rules, down to the atomic level. For instance, here are three atomic markings:

1 $\quad (a\backslash b)^{-}\ (b\backslash c)^{-}\ (a\backslash c)^{+} \quad = \quad a^{+}\ b^{-}\ b^{+}\ c^{-}\ c^{+}\ a^{-}$

2 $\quad (a\backslash b)^{-}\ (b\backslash c)^{-}\ (c/a)^{+} \quad = \quad a^{+}\ b^{-}\ b^{+}\ c^{-}\ a^{-}\ c^{+}$

3 $\quad a^{-}\ (b/(a\backslash b))^{+} \quad = \quad a^{-}\ a^{+}\ b^{-}\ b^{+}$

Now, call such a marked sequence of atoms 'balanced' if there is some way of linking all atoms in pairs a^{+}, a^{-} in such a way that the linking pattern, written down on one side of the sequence, has no crossing lines. For instance, for the above three cases, the patterns are as follows:

1
```
|____|____|____|____|
     |____|
```

2 no balanced connection exists

3
```
|____|    |____|
```

This observation motivates the following result. For a more 'semantic' motivation, one can think of negative and positive monotonic occurrences of types, in the sense of Chapter 11 below.

Proposition. If a sequent $X \Rightarrow a$ is derivable in the calculus L, then the atomic marking of the marked sequence $X^{-}\ a^{+}$ is balanced.

Proof. This follows by a straightforward induction on derivations, using the auxiliary observation that

Cyclic permutations of balanced sequences are themselves balanced.

Finally, the new geometric invariant itself raises several further questions.

Notably, any given sequent produces one atomic marking, but the latter may still have various different geometric balanced patterns. Might these correspond precisely to the different *readings* of the sequent (in the sense of Chapters 5, 9) ? Unfortunately, the answer is negative, witness the following case:

The sequent $(t/t)\ t\ (t\backslash t) \Rightarrow t$ is two-fold semantically ambiguous.
But, there are more than two different balancings for the atomic marking of its associated sequence $(t/t)^-\ t^-\ (t\backslash t)^-\ t^+$, being $t^-\ t^+\ t^-\ t^+\ t^-\ t^+$:
namely, at least

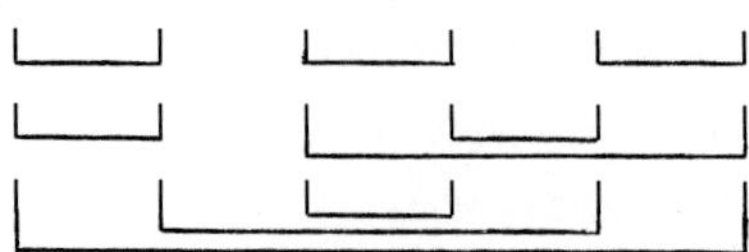

🍎

Derivability is by no means the only notion concerning types whose behaviour is important in applications. Here is another natural linguistic concept, which gives rise to some interesting questions.

When describing the behaviour of coordinating particles, such as Boolean conjunction, it is important to locate so-called *conjoinable* types a, b satisfying the condition that

there exists some c such that $a \Rightarrow c$, $b \Rightarrow c$ are derivable.

For classical logical calculi, such a question lacks interest, since a, b always imply their common *disjunction*. But here, the matter is not trivial.

Example. Conjoinable Types.

The types e and ((e, t), t) are conjoinable in LP : both reduce to ((e, t), t) . Also conjoinable are (e, ((t, t), t)) and (((e, t), t), t) : both reduce to (((e, t), t), ((t, t), t)) . But, e and t are not conjoinable, as their counts do not agree. 🍎

At present, the following obvious question arising in computational practice still seems open:

Is conjoinability a *decidable* notion in the Lambek Calculus?

Fortunately, in an important special case, the answer is positive. Let us assume that some coordinating particle # has the syntactic function of combining two adjacent expressions having the same type into a new compound one of that type. Now, let X be a sequence of the form Y # Z where Y, Z consist of ordinary types.

Claim. The question whether $X \Rightarrow a$ is L-derivable is decidable for all types a .

Proof. Any such derivation must have the form

$$Y_1, Y_2, \#, Z_2, Z_1 \Rightarrow a,$$

with Y_2, Z_2 both deriving some type b such that

$$Y_1, b, Z_1 \Rightarrow a.$$

But then, by the Cut Rule:

$$Y_1, Y_2, Z_1 \Rightarrow a: \quad \text{i.e.,} \quad Y_2 \Rightarrow (Y_1, (Z_1, a)),$$

$$Y_1, Z_2, Z_1 \Rightarrow a: \quad \text{i.e.,} \quad Z_2 \Rightarrow (Y_1, (Z_1, a)).$$

Therefore, we might have used $(Y_1, (Z_1, a))$ instead of b as the type of the coordination. In other words, the search space for coordinating types surrounding the particle # need only be *finite*: being bounded by the sequent $X \Rightarrow a$ itself. And, each separate trial can be decided via the existing decision method for the calculus L (see Chapter 7).

Finally, this argument may be extended to include cases with multiple occurrences of coordinating particles. ●

Remark. The conjoinability problem is also reminiscent of the situation in Linear Logic (cf. Chapter 14), some of whose connectives encode precisely information about equal premise sequents to be identified. ●

Next, we return to the matter of equivalent formats for our categorial calculi. As we have seen before, the calculus L or LP may be set up using *natural deduction* trees (our favourite format from a practical point of view), but also as a *calculus of sequents* having a Cut Rule - but finally also, as an *axiomatic* system. We shall consider the connections between these three formats.

Here is a demonstration, for the *-variants allowing empty premise sequents. The relevant axiomatic calculus has the following principles:

Axioms	$A \rightarrow A$	Identity
	$((A \rightarrow (B \rightarrow C)) \rightarrow (B \rightarrow (A \rightarrow C)))$	Permutation
	$((A \rightarrow B) \rightarrow ((C \rightarrow A) \rightarrow (C \rightarrow B)))$	Composition
Rule	from A and $A \rightarrow B$ to B	Modus Ponens

Here, 'derivability' of a conclusion from a set of premises is defined in the usual Hilbert-style, but for the following provisos:

- each premise should be used at least once ,
- no assumption is used in more than one application of Modus Ponens .

Example. Axiomatic Derivations.

First, here is a correct derivation:

1	e	assumption
2	$(e \rightarrow t)$	assumption
3	$(t \rightarrow s)$	assumption
4	t	Modus Ponens (2, 1)
5	s	Modus Ponens (3, 4)

There is a natural corresponding tree shape

$$\cfrac{\cfrac{1 \quad 2}{4} \quad 3}{5}$$

Next, here is an incorrect derivation:

1	e	assumption
2	$(e \rightarrow (e \rightarrow t)$	assumption
3	$(e \rightarrow t)$	Modus Ponens (2, 1)
4	t	Modus Ponens (3, 1)

whose corresponding structure is rather a graph

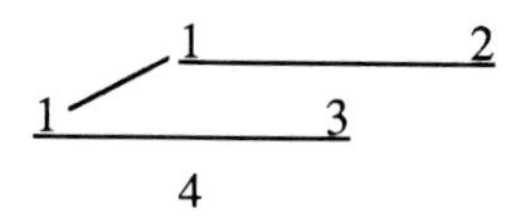

One important property of this axiomatic calculus is that it has successfully absorbed the effect of Conditionalization into its axiom schemata: i.e., it satisfies the following 'Deduction Theorem':

Theorem. A type b is derivable from a premise sequence X, a if and only if $(a \rightarrow b)$ is derivable from X alone.

Proof. From right to left, this is obvious by one (admissible) application of Modus Ponens.

Conversely, the argument is by induction on the length of axiomatic derivations. In case the derivation consists of a single formula b , the latter can be only an assumption: and so $a = b$, and the Identity axiom provides the desired conclusion. The only other case of interest is that of a final application of Modus Ponens. Now, the relevant occurrence of a must have occurred over just one of its premises, and hence we are in either of the following cases:

1 X_1, a derive c — X_2 derives $(c \to b)$

2 X_1 derives c — X_2, a derive $(c \to b)$

Ad 1. By the inductive hypothesis, X_1 derives $(a \to c)$, and hence $X = X_1X_2$ derives $(a \to b)$ using Modus Ponens and the axiom of Composition.

Ad 2. By the inductive hypothesis, X_2 derives $(a \to (c \to b))$, and hence $X = X_1X_2$ derives $(a \to b)$ using Modus Ponens and the axiom of Permutation. ◆

Remark. Note how this axiomatization differs from the usual one for the Intuitionistic conditional logic, whose axioms are

$(A \to (B \to A))$

$((A \to (B \to C)) \to ((A \to B) \to (A \to C)))$,

with Identity becoming a derivable theorem. ◆

Now, the following connection holds.

Theorem. For any sequent $X \Rightarrow a$, the following three notions coincide:

1. derivability in natural deduction trees for L*P
2. derivability in the sequent calculus for L*P (allowing Cut)
3. axiomatic derivability in the above system .

Proof. The argument requires a number of obvious inductions. Here, we only draw attention to two non-trivial steps.

In going from 2 to 3 , the Deduction Theorem is involved in making the inductive step for applications of Conditionalization.

In going from 1 to 2 , the Cut Rule is essentially involved in proving the admissibility of Modus Ponens:

from the sequent derivability of $X \Rightarrow a$ and $Y \Rightarrow (a \to b)$

to that of $X, Y \Rightarrow b$.

This matter will be taken up again in the next Chapter. ◆

These equivalence results do not mean that all proof-theoretic approaches to categorial deduction merely amount to the same. For, each format presented so far has its own conceptual and computational peculiarities, and relative advantages. Indeed, recently, various new proof formats have been proposed, for greater computational ease in searching for derivations, such as 'proof nets' (Girard 1987, Roorda 1989B) or 'quantum graphs' (Roorda 1990). Moreover, the search for 'unambiguous' search spaces for categorial derivations (cf. Chapter 9) has also inspired a number of new 'normal forms' for proofs (cf. König 1988, Hepple 1990, Moortgat 1990). One can draw an analogy with mathematical linguistics here (cf. Chapter 8): focusing exclusively on derivable sequents in Logic is like focusing on flat strings of words in linguistic description. This is the area of 'weak recognizing power', which is of interest, but which leaves out many important questions of 'strong recognizing power', concerning grammatical structure of expressions, that determine the functioning of language. Likewise, strong recognizing power in Logic is concerned with the structure of arguments producing valid sequents, as a reflection of human competence in reasoning.

Another interesting aspect of these various proof formats is that they can be put to new uses, beyond their original motivation. For instance, in Chapters 10 and 11, we shall use 'annotated categorial derivations' as a tool for computing semantic properties and logical inferences. Moortgat 1988 views Gentzen sequent derivations as a locus for processing phonological information too. And Sanchez 1990 uses natural deduction trees as a representation level of 'logical form' where natural constraints may be formulated on scope assignment for operators like quantifiers, and also on possible anaphoric relationships, like binding across various constituents. More concretely, Sanchez restricts the use of hypothetical assumptions in categorial derivation to those which are 'immediately related' (in some suitable sense) to active premises for the conclusion derived : something which rules out many a priori possible proof shapes. Also, his trees have a special indexing mechanism for assumptions which serves as a filter governing possible anaphoric identifications between parts of an expression.

Finally, one can use these proof formats for introducing other useful logical constants too. For instance, in the description of quantifier scoping, Moortgat 1989B proposes an 'exponential' operator

$a{\uparrow}b$

with the following two rules of inference:

- $\dfrac{X \Rightarrow a}{X \Rightarrow a{\uparrow}b}$ $\qquad\qquad$ $\dfrac{\tau_a[u_X]}{\lambda u_{(a,\, b)} \bullet\, u(\tau_a[u_X])}$

- $\dfrac{X, a \Rightarrow b \qquad Y, b, Z \Rightarrow c}{Y, X, a{\uparrow}b, Z \Rightarrow c}$ $\qquad\qquad$ $\dfrac{\sigma_b[u_X, u_a] \qquad \tau_c[u_Y, u_b, u_Z]}{\tau_c[u_Y, u_{a{\uparrow}b}(\lambda u_a \bullet\, \sigma_b), u_Z]}$

This new connective 'encodes' a version of Montague's lifting rule allowing a certain amount of permutation, without making any change in the structural rules of L itself. Further possible categorial connectives will be found in Chapter 14.

Appendix Combinators

There is in fact another, more semantic way of viewing axiomatic calculi of implicational inference. Derivations in the above system correspond effectively to terms from *Combinatory Logic*, which may be regarded as a variable-free alternative notation for the Lambda Calculus employed in Chapter 5.

Here are the relevant combinators for L*P:

I	$\lambda x \bullet\, x$
P	$\lambda xyz \bullet\, x(z)(y)$
C	$\lambda xyz \bullet\, x(y(z))$.

For instance, the heuristics for C derives from the third axiom:

$$(A \rightarrow B)\ [=x] \rightarrow (\ (C \rightarrow A)\ [=y] \rightarrow (C\ [=z] \rightarrow B)\ )\ .$$

Here is an illustration of a combinator analysis for an earlier principle of type change (cf. Chapters 4 and 5).

Example. Argument Lowering.

Axiomatic derivation:

1	$(a \rightarrow b) \rightarrow (a \rightarrow b)$	axiom 1
2	$((a \rightarrow b) \rightarrow (a \rightarrow b)) \rightarrow (a \rightarrow ((a \rightarrow b) \rightarrow b))$	axiom 2
3	$(a \rightarrow ((a \rightarrow b) \rightarrow b)$	Modus Ponens (2, 1)
4	$(((a \rightarrow b) \rightarrow b) \rightarrow c)$	assumption

5	$(((a\to b)\to b)\to c)\to((a\to((a\to b)\to b))\to(a\to c))$	
		axiom 3
6	$((a\to((a\to b)\to b))\to(a\to c))$	Modus Ponens (5, 4)
7	$(a\to c)$	Modus Ponens (6, 3)

Combinator analysis:

1 I
2 P
3 P(I)
4 u
5 C
6 C(u)
7 C(u)(P(I))

As a check, we may compute that this has the same effect as the lambda term found earlier on in Chapter 5:

$$C(u)(P(I)) =$$
$$[\lambda xyz\bullet x(y(z))\ (u)]\,[\lambda xyz\bullet x(z)(y)\ (\lambda x\bullet x)] =$$
$$[\lambda yz\bullet u(y(z))]\,[\lambda yz\bullet (\lambda x\bullet x)(z)\,(y)] =$$
$$[\lambda yz\bullet u(y(z))]\,[\lambda yz\bullet z(y)] =$$
$$\lambda z\bullet u(\lambda y'z'\bullet z'(y')\ (z)) =$$
$$\lambda z\bullet u(\lambda z'\bullet z'(z)) .$$

The above perspective has been advocated for linguistic purposes in Steedman 1988, Szabolcsi 1989.

Note that, although the combinatorial approach is largely equivalent to the more standard one in terms of Lambda Calculus, there may in fact be cases where the former is preferable. For instance, in technical investigations, the two are sides of the same coin, with combinators often giving some advantages when setting up the model theory (cf. Barendregt 1981, Hindley & Seldin 1986).

There are also some other 'intensional' differences. For instance, not all natural categorial calculi correspond to natural, finite sets of combinators. Zielonka 1981 has proved that the original Lambek calculus L itself, with its restriction to non-empty sets of premises throughout, is precisely one case in point.

7 Cut Elimination and Decidability

As is usual in Logic, one does not just formulate systems of inference, but one also wants to make a systematic study of their general properties and behaviour. Historically, the first significant result concerning a categorial calculus as presented here was the *decidability* of the basic Lambek calculus L . We shall take this as a point of departure for further investigation of proof-theoretic properties of categorial calculi, deferring questions having to do with their model theory until Part IV. The methods to be used are partly standard logical ones, partly also some new ones designed especially for this type of calculus.

Using a well-known method of Gentzen's, Lambek 1958 proved the following result.

Theorem. Derivability in the calculus L is decidable.

7.1 Cut Elimination

The strategy of the proof is as follows. As we have noted in Chapter 6, there are several equivalent formats for derivation in L . Here, we concentrate on its manifestation as a calculus of sequents. Now, all of the rules involved in this calculus are introduction rules, on the left-hand side and right-hand side of sequents, for the relevant connectives, except for the *Cut Rule*

$$\frac{X \Rightarrow a \qquad\qquad Y, a, Z \Rightarrow b}{Y, X, Z \Rightarrow b}$$

which rather reflects the ordinary deductive practice of injecting auxiliary or already derived information as a 'lemma' into an ongoing argument. Now, we could simplify the calculus considerably, at least from a theoretical point of view, if, after all, Cut turned out to be a derived rule, whose use can be dispensed with in principle. And in fact, we shall demonstrate how this may be done, not for L itself, but for its undirected variant LP . In this case, the Cut pattern may be simplified to

$$\frac{X \Rightarrow a \qquad\qquad Y, a \Rightarrow b}{Y, X \Rightarrow b}\,.$$

The main idea is this. Consider any sequent derivation in which occurrences of Cut occur. There must be 'uppermost' occurrences of Cut then, none of whose two premises still depend on Cut in their upward proof trees. Thus, in order to 'improve' the situation, it suffices to establish the following assertion, which allows us to remove an 'upper Cut' :

Claim. Any sequent derivation whose only occurrence of Cut is its final step may be effectively transformed into one for the same conclusion in which Cut does not appear at all.

Proof. The essence of the proof consists in performing a number of interchanges which move the application of Cut upward. These moves decrease 'complexity' of the relevant derivations, when this property is measured in the following way:

$$(\,\mathrm{rank}(a), \mathrm{trace}(a)\,),$$

where the *rank* of a is its logical complexity (i.e., the number of type-forming operators occurring in it) , and the *trace* of a is the sum of the lengths of the two unique paths in the trees over both premises of the Cut rule which connect the occurrences of a with their position of first introduction into the tree. Here, the ordering of these two natural numbers is the lexicographic one, i.e.,

$$(m, n) < (k, l) \qquad \text{iff} \qquad m<k \quad \text{or} \quad (\,m=k \;\&\; n<l\,)\,.$$

Now, for the interchanges. Here are some characteristic cases.

- $X \Rightarrow a$ was derived by Permutation.

$$\frac{\dfrac{\pi[X] \Rightarrow a}{X \Rightarrow a} \qquad\qquad Y, a \quad b}{Y, X \Rightarrow b}\ \text{Cut}$$

Now, one can perform the Cut first, and the Permutation afterwards:

$$\dfrac{\dfrac{\pi[X] \Rightarrow a \qquad\qquad Y, a \Rightarrow b}{Y, \pi[X] \Rightarrow b}\ \text{Cut}}{Y, X \Rightarrow b}$$

Note how trace(a) has decreased, so that the inductive hypothesis applies to the upper part of the proof.

- $X \Rightarrow a$ was derived by left-introduction.

Say, $X = X_1 , X_2 , (c, d) , X_3$ with

$$\frac{\dfrac{X_2 \Rightarrow c \qquad\qquad X_1, d, X_3 \Rightarrow a}{X_1, X_2, (c, d), X_3 \Rightarrow a} \qquad\qquad Y, a \Rightarrow b}{Y, X \Rightarrow b}\ \text{Cut}$$

The transformed proof becomes

$$\dfrac{X_2 \Rightarrow c \qquad \dfrac{X_1, d, X_3 \Rightarrow a \qquad Y, a \Rightarrow b}{Y, X_1, d, X_3 \Rightarrow b}\ \mathrm{Cut}}{Y, X \Rightarrow b}$$

Next , here are some cases on the right-hand side.

- $Y, a \Rightarrow b$ was derived by left introduction.

  Say, $Y = Y_1 , Y_2 , (c, d) , Y_3$ with an introduction

  $$\dfrac{Y_2 \Rightarrow c \qquad Y_1, d, Y_3, a \Rightarrow b}{Y, a \Rightarrow b}$$

  The transformed proof becomes

  $$\dfrac{Y_2 \Rightarrow c \qquad \dfrac{X \Rightarrow a \qquad Y_1, d, Y_3, a \Rightarrow b}{Y_1, d, Y_3, X \Rightarrow b}\ \mathrm{Cut}}{Y, X \Rightarrow b}$$

- $Y, a \Rightarrow b$ was derived by right-introduction.

  Then, say, $b = (b_1, b_2)$ such that there was an introduction

  $$\dfrac{Y, a, b_1 \Rightarrow b_2}{Y, a \Rightarrow b}$$

  The transformed proof becomes

  $$\dfrac{\dfrac{X \Rightarrow a \qquad Y, a\ b_1 \Rightarrow b_2}{Y, X, b_1 \Rightarrow b_2}\ \mathrm{Cut}}{Y, X \Rightarrow b}$$

Finally, the most interesting case is this.

- Both occurrences of the cut formula were introduced right before the Cut step.

  Then, the pattern must have looked as follows:

  $$\dfrac{\dfrac{X, a_1 \Rightarrow a_2}{X \Rightarrow a} \qquad \dfrac{Y_2 \Rightarrow a_1 \qquad Y_1, a_2 \Rightarrow b}{Y, a \Rightarrow b}}{Y, X \Rightarrow b}\ \mathrm{Cut}$$

  This time, *two* new cuts appear, but both of lower rank:

  $$\dfrac{\dfrac{Y_2 \Rightarrow a_1 \qquad \dfrac{X, a_1 \Rightarrow a_2 \qquad Y_2, a_2 \Rightarrow b}{Y_1, X, a_1 \Rightarrow b}\ \mathrm{Cut}}{Y_1, X, Y_2 \Rightarrow b}\ \mathrm{Cut}}{Y, X \Rightarrow b}\ \mathrm{Perm}$$

In all these cases, one-cut derivations emerge of lower complexity, which may be disposed of by the inductive hypothesis. (The latter is to be used twice in the last step displayed above.)

Finally, it remains to be mentioned that there is one degenerate case, when the Cut has been pushed to the brink of the abyss, so to speak. If either of the two Cut premises is a sequent axiom, of the form $x \Rightarrow x$, then it may be dropped, since the conclusion will then be identical to the remaining premise. ◆

This method of proof can be adapted to a whole family of categorial calculi, also involving other connectives and structural rules: witness Dosen 1988/9, Ono & Komori 1985.

Remark. 'Don't Eliminate Cut' ?
The above procedure is basically the standard method of cut elimination, found for instance, in Schwichtenberg 1977. Still, there are some differences. Notably, a recently advertized practical disadvantage of the usual procedure is that it *increases* size of derivations exponentially. (See Boolos 1984 for a telling example of a predicate-logical inference having a simple perspicuous proof involving Cut, whose shortest cut-free derivation requires more physical bits than are available in the Cosmos.) The cause is the potential duplication of proof subtrees in the final reduction step mentioned in the above argument.

But, in our *occurrence* logic, no such duplication takes place. In each step of the procedure for LP , the same amount of proof nodes is merely rearranged in new patterns. Hence, to each derivation involving Cut, there corresponds a cut-free one of essentially the same size. Moortgat 1988 applies this insight to categorial parsing, using inference engines with and without Cut in tandem, without any dramatic consequences for efficiency. ◆

Finally, one further extension of the above analysis remains to be stated.
The Lambek Calculus also enjoys the proof-theoretic property of so-called

'Strong Normalization':

Each process of successive removal of Cut inferences is bound to terminate.

A semantic counterpart of this observation is this:

Each sequence of lambda conversions on lambda terms
(corresponding to derivations) is bound to end in finitely many steps.

For the full typed lambda calculus, strong normalization involves a somewhat delicate argument. But for the Lambek fragment Λ with single binding, the result is immediate, since each successive lambda conversion reduces complexity in a straightforward sense.

7.2 Applications

Cut elimination has a number of useful consequences.

One is the earlier-mentioned *Decidability*. Since it now suffices to search for derivations none of whose inference rules are shortening, and there is only a finite number of possible premise patterns for each sequent, the complete search space for any given sequent may be taken to be *finite*. (Of course, there might be irrelevant loops involving endless Permutation: but these can be easily removed without loss of generality.) Thus, Lambek's result may be considered as

Corollary 1. Derivability in the Lambek Calculus is decidable.

In practice, this abstract decidability still needs to be improved, using various additional techniques. For instance, the categorial parser of Moortgat 1988 uses the count invariants of Chapter 6 for the purpose of early rejection of hopeless attempts in the search for categorial derivations. (See also Heylen & van der Wouden 1988 on the use of these invariants in optimizing lexical retrieval.)

Example. Detecting Impending Failure.
The following sequent is derivable in LP :

$$e \quad (e, t) \quad (e, s) \quad (t, ((e, s), s) \quad \Rightarrow \quad s \, .$$

In the initial stage, there are 24 possible premise patterns for derivation, representing the various possible comma introductions with partitions for the inactive remainder. Out of these, only the following 4 have both premises satisfying all count restrictions:

1	$e \Rightarrow e$	$t \quad (e, s) \quad (t, ((e, s), s)) \Rightarrow s$
2	$e \Rightarrow e$	$(e, t) \quad s \quad (t, ((e, s), s)) \Rightarrow s$
3	$e \quad (e, s) \quad (t, ((e, s), s)) \Rightarrow e$	$s \Rightarrow s$
4	$e \quad (e, t) \Rightarrow t$	$(e, s) \quad ((e, s), s) \Rightarrow s \, .$

Next, for instance, there are stil 8 possible patterns for deriving the right-hand sequent of case 1, of which count leaves only 2 possibilities, namely

1.1 $t \Rightarrow t$ (e, s) $((e, s), s) \Rightarrow s$

1.2 t $(t, ((e, s), s) \Rightarrow e$ $s \Rightarrow s$.

Here, count fails to reject only one failure, namely the attempt

$((e, s), s) \Rightarrow e$

in the derivation of 1.2. ♠

Another consequence is the so-called *Subformula Property*:

Corollary 2. If a sequent is derivable, then it has a derivation in which all types that appear are subtypes of those already occurring in the sequent.

In particular, once we make an obvious extension of the above argument to the directed case with two slashes, an earlier observation in Chapter 6 receives an explanation:

No valid sequent for pure leftward functionality \ can depend essentially on principles for the rightward slash / .

And once we have made another straightforward extension to the calculus having product types, we also see that:

No valid principle for functional types (a, b) can depend essentially on the use of product types.

The latter observation allows us to switch back and forth between stacked functional types (a, (b, c)) and flattened forms $(a{\bullet}b, c)$ in derivations, without losing any generality.

A more semantic effect of the subformula property is the following form of 'Predicativity':

Corollary 3. Any term $\tau_b [\, u_{a1}, ..., u_{an} \,]$ as computed in Chapter 5 defines a denotation whose semantic computation involves a survey of only those levels in the hierarchy of type domains which are already 'contained', in a sense, in the domains for the parameters and the target type.

7.3 Stronger Calculi

The above methods and results remain applicable also to implicational logics lying higher up in the Categorial Hierarchy. In particular, of course, they hold for the standard intuitionistic and classical systems.

Nevertheless, such results are not always obvious generalizations. Indeed, there is one rather natural system for which decidability requires much more effort, namely the undirected Lambek Calculus LPC allowing *Contraction*:

$$\frac{X,\, a,\, a \Rightarrow b}{X,\, a \Rightarrow b}\ .$$

Cut elimination may be performed here much as in the above argument, adding a straightforward case for interchanging Cuts with Contractions, and suitably adapting the earlier complexity measure.

But, unlike before, this does not automatically guarantee decidability. For, the Contraction Rule is shortening, and in searching backward through the proof space, one might have to try ever larger repetitions of the same types on the left-hand side. (With stronger classical systems above LPC , this is no longer an issue, because they postulate an *equivalence* between the conclusion and the premise of the Contraction rule.) Therefore, a much more elaborate argument is needed in this case (cf. Dunn 1985). It may be of interest to note its main steps:

- First, remove the Contraction rule from the system, while modifying the operational rules by allowing a little contraction. More precisely, at most one occurrence of the introduced type may be contracted out of each adjoining premise sequent in the conclusion, and also at most one contraction is allowed on occurrence pairs in the latter sequents, which have become juxtaposed in the conclusion.
- Next, one proves 'Curry's Lemma', to the effect that this change does not affect provable sequents: they remain provable with at most the same number of steps.
- Surveying the proof space for any sequent, we see that it is finitely branching, as far as potential premises are concerned. Now, we must enquire into its depth.
- To this end, call two sequents 'cognate' if they can be contracted (using unlimited Contraction) to the same sequent. By the subformula property for LPC , there are only finitely many cognation classes relevant to the derivation of any given sequent. Moreover, in the new calculus, no premise sequent occurring on a branch in a derivation which occurs can be an 'extension' (in the obvious sense) of another on that branch. In other words, although there may be more than one member of the same cognation class on the branch, all of these must be incomparable in the obvious numerical ordering.
- Now, finally, one can apply 'Kripke's Lemma' (in fact, a form of *Kruskal's Theorem* that will also be used in Chapter 8) to show that, therefore, the number of members of each cognation class occurring on any branch must be finite. More precisely,

Kripke's Lemma says that, in each cognation class, the above 'extension' ordering of sequents is a 'quasi-well-ordering': that is, each sequence $S_1, S_2, \ldots$ of cognate sequents must have at least one member S_j extending S_i (where $i<j$).

• So, both the finite width and the finite depth condition of Koenig's Lemma are satisfied, and the complete search space for possible derivations of any given sequent must be finite. Hence, we have decidability for LPC .

Remark. This argument has been applied to systems related to Linear Logic in Roorda 1989A.

Evidently, the above argument is quite abstract and non-constructive. Therefore, deciding validity of concrete sequents in LPC is usually a matter of additional ad-hoc techniques. Here is an illustration of this more down to earth phenomenon, which also shows the kind of question that may be illuminated using the present categorial analysis.

Example. Boolean and Non-Boolean Conjunction Revisited.

In Chapter 5, it was observed how a proper treatment of the polymorphism of Boolean conjunction presupposed derivation in LPC rather than LP . Somewhat more systematically, the underlying question is this.

For which types a is the following sequent L-derivable:

$a \quad (t, (t, t)) \quad a \Rightarrow a$?

At least a necessary condition is provided again by a simple numerical count argument:

$\#_t(a)$ must be equal to 1 ,

all other counts must have value 0 .

This still leaves a number of succesful candidates, such as

$a = t$ or $a = ((t, t), t)$.

But it excludes, for instance, the case of predicate conjunction, where $a = (e, t)$.

Therefore, we move to the stronger calculus LPC , and ask again which types admit of Boolean coordination in the above sense. The answer is as follows:

The sequent $a \quad (t, (t, t)) \quad a \Rightarrow a$ is derivable in LPC if and only if a is a *Boolean type* , of the form $a = (a_1, (\ldots (a_n, t)\ldots))$.

In one direction, this follows from the semantic recipe for generalized conjunction in arbitrary Boolean types, being

$\lambda P_a \bullet \lambda Q_a \bullet \lambda x_{a1} \bullet \ldots \lambda x_{an} \bullet \wedge_{(t, (t, t))}(P_a(x_{a1}) \ldots (x_{an}))(Q_a(x_{a1}) \ldots (x_{an}))$.

In the other direction, it suffices to observe that for non-Boolean types, ending in a final type e : $(a_1, (\ldots (a_n, e) \ldots))$, the following sequent is not derivable in LPC :

$$(a_1, (\ldots (a_n, e) \ldots)) \quad (t, (t, t)) \quad (a_1, (\ldots (a_n, e) \ldots)) \quad a_1 \quad \ldots \quad a_n \quad \Rightarrow \quad e \, .$$

Again, this must be established by actual proof analysis in the presence of Contraction.

Next, for non-Boolean conjunction in type $(e, (e, e))$, similar questions arise. Since the two cases are formally identical, some outcomes may be listed straightaway:

- In LP , non-Boolean conjunction is possible,
  for instance, in the cases $a = e$ or $a = ((e, t), t)$
  (this was observed already in Chapter 5)
- In LPC , non-Boolean conjunction also becomes possible
  for such cases as $a = (e, e)$ or $((e, t), (e, t))$.

Krifka 1989A sees further relevant conjunctions occurring in types like $(e, (e, t))$ (transitive verbs) or $((e, t), ((e, t), t))$ (determiners) , where not even an LPC derivation is possible. For instance,

$$(e, (e, t)) \quad (e, (e, e)) \quad (e, (e, t)) \quad \text{does not LPC-derive} \quad (e, (e, t)) \, .$$

In such cases, non-trivial semantic recipes for non-Boolean conjunction will have to involve type-theoretic apparatus beyond the pure lambda calculus. (Krifka himself uses a set-theoretic 'image function' over an original non-Boolean addition of individuals.)

Remark. A Further Option.

The above has by no means exhausted the combinatorial interest of conjunction in natural language. For instance, the reader might wish to ponder the following three sentences:

"Julius came and conquered"

"Marc and Julius came"

"Marc and Julius came and conquered".

What kinds of conjunction are involved where? At least, in the "respectively" reading of the third sentence, one could also make a case for *product* types now, letting the first "and" have type $(e, (e, e \bullet e))$, forming the ordered pair of Marc and Julius. Then, the verb phrase "came and conquered" may be read with Boolean conjunction, *not* in LPC as above, but merely in LP :

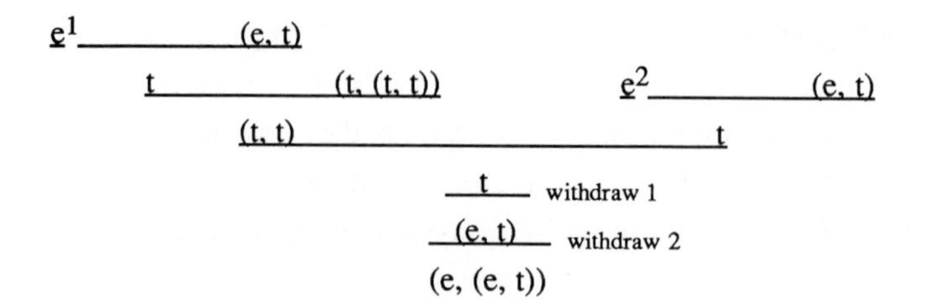

whose semantic meaning is the Λ-term

$$\lambda x_{e} \bullet \lambda y_{e} \bullet \wedge_{(t,(t,t))} (\mathrm{CAME}(x))(\mathrm{CONQUERED}(y)) .$$

The latter can then combine with the ordered pair <MARC, JULIUS> living in the product type $e \bullet e$, to the truth value type t , in the manner already demonstrated in Chapters 4 and 5.

Thus, the combinatorics of linguistic expressions may be related directly to options between various calculi in the Categorial Hierarchy. And the mathematical task broached in this Chapter is then to acquire a good general grasp of the (shifts in) proof-theoretic meta-properties that are involved in such choices.

To conclude, it should be stressed that there are many other interesting proof-theoretical aspects of categorial calculi that have not been considered here. Examples are Interpolation or Beth Definability properties. Indeed, the presentation so far does not even exhaust the topic of structural rules in our systems of deduction. For instance, we do not have a full grasp yet of all the *derived rules of inference* that hold for L . Rybakov's Theorem (Rybakov 1985) tells us that this notion is *decidable* for the full constructive implicational logic: does the same hold for the Lambek Calculus?

8 Recognizing Power

In this chapter, we turn to a more traditional question, which has been central in Mathematical Linguistics, namely that of recognizing power of our categorial systems.

Here, 'recognition' means the following. In the background, there is a finite alphabet Σ, assigning one or more initial categories (though always finitely many) to each symbol. One category shall be distinguished: say t. Then all those sequences of symbols constitute the 'language' recognized, for which there exists a matching sequence of categories, via the initial assignment, which derives t in the relevant categorial calculus.

8.1 The Directed Case

Example. Polish Propositions.

Let Σ be the alphabet $\{p, q, \neg, \wedge\}$, and let the initial assignment be

$p, q: \quad t \qquad\qquad \neg: \quad (t/t) \qquad\qquad \wedge: \quad ((t/t)/t)$.

Then the language recognized is precisely the set of well-formed Polish propositional formulas over these two atoms.

That all such formulas are recognized may be shown by induction on their construction. For instance, if A, B have been recognized already, then $\wedge AB$ will be recognized via the following pattern

$$\dfrac{\dfrac{\overset{\wedge}{((t/t)/t)} \qquad \overset{\text{'A-derivation'}}{t}}{(t/t)} \qquad \overset{\text{'B-derivation'}}{t}}{t}\ .$$

Conversely, that each recognized expression is Polish may be seen by induction on categorial derivations. For instance, given the initial assignment and the working of Modus Ponens, such derivations can only contain nodes with types t, (t/t) or $((t/t)/t)$. Therefore, the final step must have been an application of (t/t) to t: where the latter already represents some final segment which is Polish, by the inductive hypothesis. The former can come either directly from a symbol $\neg$: and we are done, or, it is the result of an application of $(t/((t/t))$ to t. In the latter case, there is a Polish formula again, preceded by a symbol $\wedge$: as $((t/t)/t)$ cannot have been the result of a function application itself.

In general, arguments for identifying the language recognized may be more complex than the preceding one.

The above example concerned recognition in the original Ajdukiewicz system. Outcomes may change, however, when the categorial engine is varied. For instance, the assignment

a : (t/t) b : ((t/t)\t)

recognizes the language { ab } in the Ajdukiewicz system, but { ab, aab, aaab, ... }, i.e., aa^*b , in the Lambek calculus. Indeed, with the above notion of language recognition, it would be a fallacy to expect that a stronger categorial calculus will automatically be able to recognize all languages recognized by some weaker calculus. Nevertheless, we have the following

Proposition. Each language recognizable in the Ajdukiewicz system can also be recognized in the Lambek calculus.

Proof. If a language is recognized in the former system, then we can change the initial assignment to one in which the distinguished type is atomic, and in which all other types are 'first-order' : i.e., having (iterated) *atomic* arguments only. The main trick here is to introduce new atomic types for each subtype occurring in the initial types, and then switch to the 'first-order substitution instances' of the latter. (This can be shown to be adequate by a simple inspection of the relevant categorial deductions.)

But then, the same assignment will work for L as well, since a simple induction on cut-free L-derivations shows that

Sequents with first-order premises and atomic conclusion are derivable in the Ajdukiewicz system if and only if they are derivable in Lambek's.

Remark. First-Order Recognition.

For an illustration, here is a first-order assignment recognizing the language aa^*b both in Ajd and in L :

b : (s\t) , a : { s, (s\s) } .

Remark. The syntactic reduction to non-iterated 'first-order' types in categorial recognition of context-free languages might also have some semantic significance. What it suggests is that, in linguistic description, one can get by with 'flat many-sorted models', much as happens in the many-sorted first-order version of higher-order logic proposed in Henkin 1950 (cf. Chapter 2). It remains to be seen whether there is a deeper connection between these two perspectives.

A well-known historical result of Bar-Hillel, Gaifman & Shamir 1960 says that the languages recognized by the original categorial grammars are precisely the *context-free* ones. This result was extended to the Lambek calculus in Cohen 1966, but unfortunately, the proof turned out to contain an error. Right now, what we know is only this:

- The languages recognized by a one-directional Lambek calculus are precisely the context-free ones (Buszkowski 1982)
- The question as to the recognizing power of the full directional system is still open, although the evidence gathered by Buszkowski seems to point toward exactly context-free recognizing power too (the survey Buszkowski 1988 has many partial results and useful techniques).

If the latter question is settled positively, then the move toward a more flexible Categorial Grammar has given us additional 'strong recognizing capacity', in terms of new constituent structures, while leaving 'weak capacity', concerning sets of flat strings recognized, at the old level.

Remark. A More Revolutionary Perspective.
The desire to fit recognition levels of the Categorial Hierarchy precisely into those supplied by its more traditional rival, being the Hierarchy of Rewrite Grammars, will not be all that urgent, of course, for those who really believe in the new paradigm. The burden of proof with the above result, on their view, would rather lie with the Establishment in Mathematical Linguistics.

Finally, questions of recognition by categorial calculi are intertwined with questions concerning their proof theory. For instance, we can distinguish various natural relations between such calculi now:

a L_2 is a deductive extension of L_1

b L_2 provides a faithful embedding for L_1

c L_2 recognizes every language which L_1 recognizes .

Can we identify the notion $\mathbb{c}$ with some more standard proof-theoretic relation between categorial calculi? No satisfactory answer exists right now. Here are only some relevant observations:

Fact. 1 $\mathbb{b}$ implies $\mathbb{c}$, but not conversely.

2 $\mathbb{a}$ does not imply $\mathbb{c}$, and neither conversely.

Proof. Ad 1 . Translating the recognizing L_1-assignment via the embedding will work for L_2 as well. A counter-example to the converse is provided by the case of $L_2 = L$, $L_1 = Ajd$.

Ad 2 . As we shall see later on in this Chapter, $\mathbb{a}$ holds when L_2 is taken to be Intuitionistic conditional logic, $L_1 = L$ - while $\mathbb{c}$ does not. A counter-example to the converse is provided by L_2 consisting of Ajd plus additional rules of Lifting, $L_1 = Ajd$ (cf. Koenig 1989).

8.2 The Undirected Case

From now on, we shall study our earlier undirected calculi, starting from LP , admitting a rule of Permutation. These may not be serious candidates for linguistic description, since they only recognize languages with extremely free word order, being closed under permutations of their expressions (but see van Benthem 1988C for a possible defense). All the same, the study of their behaviour seems illuminating, as well as somewhat more accessible from a technical point of view.

First, here is a simple illustration of a useful pattern of argument concerning Lambek recognition.

Example. Unordered Polish Propositions.

Let $\Sigma = \{ p, q, \neg, \wedge \}$, and assign

p, q : t $\qquad$ $\neg$: (t, t) $\qquad$ $\wedge$: $(t, (t, t))$.

As in an earlier example, this will recognize all well-formed Polish formulas in the Ajdukiewicz system already. Hence, these strings will also be recognized in its extension LP . Moreover, so will all of their permutations, because of the presence of the Permutation rule.

But, what about the converse? Could not LP recognize further strings: and, how is one to find out? Here is where the earlier notion of 'count' may be used again. Any LP-derivable sequent $X \Rightarrow t$ must have equal t-counts on both sides. That is, the sum of all t-counts on the left-hand side must equal 1 . Now, this sum is composed out of contributions for the three types figuring in the intial assignment, being

$\#_t(t) = 1$, $\#_t((t,t)) = 0$, $\#_t((t, (t, t))) = -1$.

Therefore, the only sequences reducing to t must have, say,

n occurrences of $\wedge$, n+1 occurrences of atoms p , q
and an arbitrary number of occurrences of $\neg$.

But, any such sequence is in fact a permutation of a well-formed Polish formula. ◆

Our next illustration shows that LP may recognize languages that are more complex than context-free.

Example. Equal Numbers of Three Symbols.
The language consisting of all strings having equal numbers of symbols a, b, c may be LP-recognized via the following assignment:

a : e   b : (e, s)   c : { (s, t), (s, (t, t)) } .

Again, the crucial argument involves counting. Here are the count values for the relevant types e , s , t :

type	e-count	t-count	s-count
e	1	0	0
(e, s)	–1	0	1
(s, t)	0	1	–1
(s, (t, t))	0	0	–1
t	0	1	0

Now, to derive any sequence of initial types to t , there might be n occurrences of e (i.e., symbol a), which have to be compensated for by an equal number of (e, s) (i.e., symbol b). But, the latter introduce value +n for the s-count, which has to be compensated for again by means of n occurrences of types corresponding to the third symbol c . Therefore, all three symbols occur with the same multiplicity.

That the language recognized is not context-free may be seen by intersecting it with the regular language $a^*b^*c^*$: the result is the well-known non-context-free language $\{a^nb^nc^n\}$. ◆

Nevertheless, the preceding example may still be described quite simply as the permutation closure of a context-free, in fact even a regular, language, namely $(abc)^*$.

Now, we are ready for our first main result.

Theorem. All permutation closures of context-free languages are recognizable in LP.

Proof. Fix some alphabet $\Sigma = \{ s_1, ..., s_n \}$. Let T_1 be a permutation closure of some context-free language T. *Parikh's Theorem* (see Ginsburg 1966 for a proof), says that the set of numerical 'occurrence vectors' for context-free languages T, i.e.,

$$\{ (\text{number of } s_1 \text{ in } \varepsilon, ..., \text{number of } s_n \text{ in } \varepsilon) \mid \varepsilon \text{ a string in } T \}$$

is *semi-linear*. That is, the latter set is a finite union of 'linear' sets of vectors, being those generated by one fixed k-tuple of natural numbers (the 'base') and addition of arbitrary multiples of some finite set of k-tuples (the 'periods').

Example. Propositional Formulas.

The occurrence set for the earlier undirected propositional logic is itself linear, being the union of two linear forms [with vectors '(#(p), #(q), #(¬), #(∧))']:

$$(1, 0, 0, 0) + \lambda(0, 0, 1, 0) + \mu(1, 0, 0, 1) + \nu(0, 1, 0, 1),$$

$$(0, 1, 0, 0) + \lambda(0, 0, 1, 0) + \mu(1, 0, 0, 1) + \nu(0, 1, 0, 1).$$

●

Now, for any linear set, there already exists a regular language with precisely that occurrence set. The procedure will be clear from the following illustration.

Example. Propositional Formulas, Continued.

A regular language producing the occurrence set given above is

$$p ; \neg^* ; (p\wedge)^* ; (q\wedge)^* \ \cup \ q ; \neg^* ; (p\wedge)^* ; (q\wedge)^* .$$

●

Hence, any permutation closure of a context-free language is already a permutation closure of some *regular* language. And to show that the latter can all be LP-recognized, it suffices to follow their inductive definition via regular expressions, and prove the following four assertions:

1 every singleton language { a } is LP-recognized
2 every union of LP-recognized languages is itself LP-recognizable. (It is automatically permutation-closed if its components were.)
3 every permutation closure of the concatenation of two LP-languages is LP-recognizable.
4 every permutation closure of the Kleene iteration of an LP-language is LP-recognizable.

And, all these assertions can indeed be proved, relying steadily on proof-theoretic properties of the cut-free sequent calculus format for LP (cf. van Benthem 1988C).

Singletons

Let a be assigned t , all other symbols in Σ going to some other type e . Among sequences of basic types, only <t> itself reduces to t .

Union

Let T_1 be recognized via the assignment G_1 , and T_2 via G_2 , with distinguished types t_1, t_2 , respectively. Without changes in recognition, it may be assumed that all types used in the two grammars are distinct. Now, choose some new basic type $t^\#$. Then, assign types to symbols as follows. Retain all old types in the assignment. If a symbol s has a type of the form $(a_1, ..., (a_n, t_1)...)$, with a final type t_1 , then also give it a new type $(a_1, ..., (a_n, t^\#)...)$. And do likewise for types ending in t_2 .

Claim. This assignment recognizes exactly $T_1 \cup T_2$.

Proof. If a sequence σ was in T_1 , then it had an LP-derivation $b_1 \ ... \ b_m \Rightarrow t_1$ for some distribution $b_1 \ ... \ b_m$ of G_1-basic types over its symbols. Now, consider the form of this proof, in the cut-free version of LP . Continuing upward in the proof tree, there will be some branch with always t_1 on the right-hand side (as basic types cannot be introduced by a rule). Such a branch must end in an axiom $t_1 \Rightarrow t_1$. Now, replace its two occurrences of t_1 by $t^\#$, and then, going down again, likewise with all their 'offspring' in the derivation. In this fashion, a derivation is obtained for $t^\#$ from a sequence of types exactly one of which has a final $t^\#$. (To see this, observe that the left-hand type $t^\#$ on top can only end up on the *right-hand* side of comma-introductions.) And such a sequence can be recognized in the new grammar.

The case of T_2 is symmetric, of course.

Conversely, it is to be shown that any recognized sequence already belongs to T_1 or T_2. So, let $b_1 \dots b_m \Rightarrow t^\#$ be derivable in LP. The $t^\#$-count of $t^\#$ itself is 1. Now, all types with $t^\#$ inserted in final position also have $t^\#$-count 1. Therefore, exactly one such type must occur among $b_1 \dots b_m$.

Case 1. That type is $t^\#$ itself.

Then, the following assertion may be used, which is based on a simple induction.

Lemma. If $B\ t^\# \Rightarrow t^\#$ is LP-derivable, with $t^\#$ not occurring in B, then B must be empty.

So, in this case, there is just one symbol, which belonged already to T_1 (if it came from the type t_1) or to T_2 (if it came from t_2).

Case 2. That type also contains subtypes from G_1.

Then, the whole sequence must consist of G_1-types, in addition to the manipulated one, because of the following property of 'Inseparability':

Lemma. If $A, B \Rightarrow c$ is derivable in LP, then either A, B share some occurrence of at least one basic type, or B, c do.

The proof of this result is again by a simple induction on cut-free proof trees.

Then, finally, replacing $t^\#$ by t_1 in the derivation, the sequence is recognized as being in T_1.

Case 3. This is symmetric, with G_2 instead of G_1.

Concatenation

Let T_1 be recognized by G_1 (with t_1) and T_2 by G_2 (with t_2), disjoint as before. Again, choose a new distinguished type $t^\#$. This time, admit additional G_1-types with final t_1 replaced by the type $(t_2, t^\#)$.

Claim. This assignment LP-recognizes the permutation closure of $T_1 ; T_2$.

Proof. First, to recognize a sequence $\sigma_1\sigma_2$ with σ_1 in T_1, σ_2 in T_2, recognize σ_1 as in G_1, but with $(t_2, t^\#)$ inserted as in the earlier argument for unions $T_1 \cup T_2$. Then, recognize σ_2 as before in G_2, and combine to obtain $t^\#$.

This shows one inclusion - further permutation closure being automatic for LP.

Next, let some sequence σ be recognized to $t^{\#}$. As before, $t^{\#}$-count tells us that exactly one type will occur in σ with $(t_2, t^{\#})$ in final position. Then, an analysis of the proof tree, using the form of the admissible rules in the cut-free format, reveals the following shape:

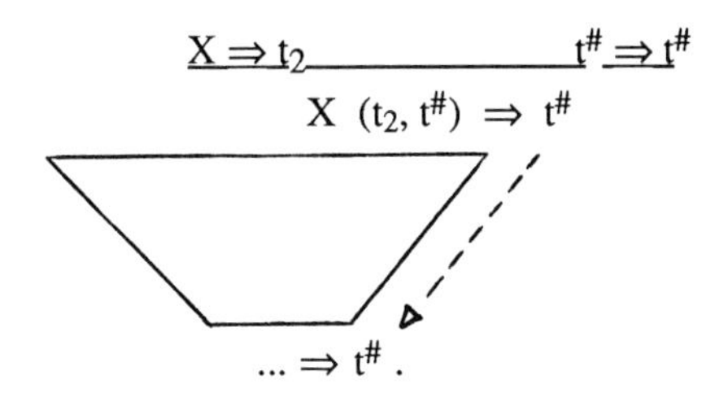

Here, X consists of G_2-types only, by the earlier Inseparability lemma. In the final sequent, the part $(t_2, t^{\#})$ occurs inside a type, say, $(x_1, ...(t_2, t^{\#})...)$. This must have been brought about by a number of comma-introductions on the left, of the form

$$\frac{B_i \Rightarrow x_i \qquad\qquad C \;...\; (t_2, t^{\#}) \Rightarrow t^{\#}}{C\ B_i\ (x_i, (\ldots (t_2, t^{\#})) \Rightarrow t^{\#}}$$

Again by Inseparability, all B_i-types must belong to G_1 (since x_i does).

Now, consider the general form of the '$t^{\#}$-branch'. Its steps are either of the form just indicated, or they introduce some antecedent for another type on the left-hand side, together with enough types to prove its antecedent. Throughout this process, the following holds, however, at each stage $A \Rightarrow t^{\#}$:

- The totality of all G_2-types in A derives t_2
- The totality of all pure G_1-types in A , being all the non-G_2 types except for the single one in which $t^{\#}$ occurs, derives all types in the prefix of the latter (preceding its final part $(t_2, t^{\#})$).

Applying this to the final sequent in the derivation, we find that the symbols contributing the G_2-types form a T_2-expression, the remaining ones an expression in T_1 (remember the original form of the special type). Hence, the sequence recognized is a permutation of one in $T_1 ; T_2$.

Iteration

Let T be recognized by G (with distinguished type t). Again, we choose a new type marker $t^{\#}$, and add both $t^{\#}$ and $(t^{\#}, t^{\#})$ in final positions t . This allows us to recognize all of T^* , by suitable $t^{\#}$ - insertion (for one string in T) and $(t^{\#} , t^{\#})$ -

insertions (for all other strings in T), combining the resulting types $t^{\#}$ and $(t^{\#}, t^{\#})$ into one single $t^{\#}$.

The converse again requires a proof analysis. Let any sequence of types reduce to the type $t^{\#}$. Following the upward path of the conclusion $t^{\#}$, there must have been an initial situation of the form

$$\frac{A \Rightarrow x \qquad\qquad t^{\#} \Rightarrow t^{\#}}{A\ (x, t^{\#}) \Rightarrow t^{\#}}$$

Case 1. x is not $t^{\#}$.

Then, the expression recognized is in T (possibly permuted). This follows from an argument about the remaining form of the derivation, as earlier on for Concatenation, using one additional property:

Fact. No symbol $t^{\#}$ occurs in the types of A.

Proof. As x has $t^{\#}$-count 0, no T-type in A has just one $t^{\#}$ inserted. But also, insertion of $(t^{\#}, t^{\#})$ is impossible. If, for instance, $(y, (t^{\#}, t^{\#}))$ occurred among A, then, somewhere up in the tree for $A \Rightarrow x$, $(t^{\#}, t^{\#})$ must have been introduced, resulting in a premise $B \Rightarrow t^{\#}$. Consider the first time that this happened in the derivation: B contains no occurrences of $t^{\#}$, and hence there is a mismatch in $t^{\#}$-count - which is a contradiction.

Case 2. x is itself $t^{\#}$.

Then, arguing as in previous cases, the final sequent may be seen to be of the following form (modulo permutation)

$$A\ B\ (x_1, \ldots (x_k, (t^{\#}, t^{\#}))\ldots) \Rightarrow t^{\#},$$

with A a sequence of types deriving $x_1, \ldots, x_k$, and B deriving $t^{\#}$. As before, since $x_1, \ldots, x_k$ are pure G-types, the A-types cannot contain occurrences of $t^{\#}$. It follows that the symbols contributing B plus the final type form an expression in T.

Applying the same reasoning to the shorter sequence of symbols contributing A decomposes this eventually into a finite sequence of T-expressions.

This concludes the proof of the Theorem. ■

Remark. Alternative Arguments.

The above proof could also have been presented in other ways, for instance, relying more heavily on the Subformula Property of the cut-free calculus.

Moreover, there is also an advantage in working with recognition of *two* languages: one the original level of strings of symbols, the other consisting of strings of types regarded as symbols by themselves. Any initial assignment may then be viewed as establishing a substitution homomorphism from the 'type language' to the 'symbol language'. Therefore, if one can only show that some type language is context-free, or regular, the same must hold automatically for its homomorphic 'symbolic' image. This perspective may also be used in what follows.

8.3 The Effect of Contraction

Let us now move up in the Categorial Hierarchy, and see what happens with calculi having the structural rule of Contraction. This time, there will be an expected loss in recognizing power, since the latter rule can sabotage genuine recursion, already upon addition to the Ajdukiewicz system.

Example. Polish Propositions under Contraction.
With an earlier assignment

$p: \quad t \qquad \wedge: \quad ((t/t)/t)$

one could recognize precisely the Polish propositional language over the atom p with binary conjunction. But, with Contraction added, the language recognized becomes a *regular* one, viz. all strings beginning with $\wedge$ and ending with p. (Here, Contraction is taken in the suitably general formulation which it needs when no Permutation is assumed.) For instance, the ill-formed string $\wedge pp\wedge\wedge p$ will now be recognized, by contraction on the types used in recognizing the well-formed formula $\wedge(\wedge pp\wedge\wedge p)pp$.

In fact, the set of correct propositional formulas cannot be recognized at all in the presence of Contraction. For, any correct string $\wedge^k p^{k+1}$ (where k is the maximum number of types involved in the initial assignment) will also have a non-Polish substring recognized of the form $\wedge^k p^i$ with $i \leq k$.

In order to state more precise results, some notation is needed.
Fix some alphabet Σ with n symbols. First, we work in the presence of Permutation. Then, languages T can be described completely as sets of n-tuples $X = (X_1, ..., X_n)$, being the 'occurrence vectors' for symbols in the strings of T. Next, when considering grammars, let $\Bbbk$ be the maximum number of initial types assigned to any one symbol.

Here are some further numerical effects of structural rules higher up in the Categorial Hierarchy (cf. Chapter 4):

Contraction
If $X \in T$ and $X_i > \Bbbk$, then $(X_1, ..., X_i', ..., X_n) \in T$ for some $X_i' \leq \Bbbk$.
That is, for the n_i symbols a_i involved in the recognition of X,
one can use at most $\Bbbk$ different types: to which all others can be contracted.
Expansion
If $X \in T$ and $X_i > 0$, then $(X_1, ..., Y, .., X_n) \in T$ for all $Y > X_i$.

Using these observations, it is easy to establish the recognizing power of some strong implicational logics above the basic Lambek calculus.

We start with the calculus LPCE, which treated its premises like *sets*. Consider n-tuples X of natural numbers, with a symbol \$ added. The language $T(X)$ consists of all tuples Y such that

$$Y_i \geq X_i \qquad \text{if } X_i \in N, \qquad\qquad Y_i = 0 \qquad \text{if } X_i = \$.$$

Call such languages *existential*. Note that all of them are regular.

Theorem. LPCE grammars recognize precisely the class of all finite unions of existential languages.

Proof. First, let T have a PCE grammar. Consider the set $T^{£}$ of all tuples $X \in T$ none of whose members exceeds $\Bbbk$. $T^{£}$ must be finite (there are only $\Bbbk^n$ candidates) - and, using the above two observations concerning Contraction and Expansion, T can be described as the finite union of all languages $T(X)$ for $X \in T^{£}$.

For the converse, here is a representative example. Let $n = 2$. T is the union of T_1, T_2 with

T_1: all strings with at least two symbols a_1, and no a_2,

T_2: all strings having at least one a_1, and at least one a_2.

Now, T_1 is PCE-recognized by the following assignment:

a_1: $\{ x, (x, t), (t, t) \}$ a_2: $\{ y \}$

And for T_2, one can use

a_1: $\{ u, (u, u) \}$ a_2: $\{ (u, u), (u, t) \}$.

Moreover, the union of these assignments will recognize $T_1 \cup T_2$. (Note how all basic types are distinct in the two assignments, except for t .) That it recognizes all of $T_1 \cup T_2$ is immediate. The converse requires some further proof analysis.

Suppose that some set A of the above types derives t . As the inference must be logically valid, at least one of (u, t), (x, t) must occur in A (since the remaining types do not imply t , as a matter of implicational logic). If (x, t) occurs, then x must occur too (as the remaining types do not imply x - and some occurrence of (x, t) must be used at some stage). In that case, x and (x, t) already produce t , and other types cannot have been used in the derivation, except for (t, t) . Thus, we are in T_1 . Likewise, if (u, t) occurs, we must be in T_2 . ♠

From this result, it is easy to derive the recognizing capacity of the full Intuitionistic implicational logic I . It will only recognize a further subset of the existential languages, losing even the power to forbid the occurrence of a symbol in its expressions. (That is, \$ drops out of the description altogether.)

Next, we consider the logic LPC by itself: which after all played an independent semantic role in earlier Chapters. Without the Expansion rule, the earlier argument breaks down - and we have to use a more complex method of proof. Even so, the same collapse occurs in recognition.

Theorem. LPC grammars recognize only regular languages.

Proof. Let T be recognized by some PC assignment. Define the following relation $\leq_{\mathbb{k}}$ on natural numbers:

$$i \leq_{\mathbb{k}} j \quad \text{if} \quad i = j \leq \mathbb{k} \quad \text{or} \quad \mathbb{k} < i \leq j .$$

Then, extend this to n-tuples X , Y by requiring $X_i \leq_{\mathbb{k}} Y_i$ for each coordinate i ($1 \leq i \leq n$). The reason for this definition lies in the following consequence of Contraction. Let T^c be the complement of T . Reasoning as in earlier observations, we see that, in LPC-recognition:

If $X \in T^c$ and $X \leq_{\mathbb{k}} Y$, then $Y \in T^c$.

Now, we need an earlier-mentioned technical result. On the natural numbers, $\leq_{\mathbb{k}}$ is what is called a *well-quasi-order*, having only *finite antichains* (i.e., subsets without any $\leq_{\mathbb{k}}$-comparable pair). By a well-known result on well-quasi-orders, the same then holds for this order lifted to n-tuples as above (see Kruskal 1972, Nash-Williams 1963).

So, let A be a *maximal* antichain in T^c. (These exist, by Zorn's Lemma.) A must be finite, by Kruskal's result. By maximality, each tuple in T^c must be comparable in the ordering $\leq_k$ with at least one tuple in A. Now, let A^+ be the *finite* set consisting of A together with all $\leq_k$-smaller tuples in T^c. A^+ is a 'basis' for the language T^c, in the sense that a vector X belongs to T^c if and only if some tuple in A^+ is $\leq_k$-included in it. Hence, T^c is a finite union of languages of the form

$$T^{\S}(X) \quad = \quad \{ Y \mid X \leq_k Y \}, \qquad \text{where X is any tuple in } A^+.$$

It is easy to see that all languages $T^{\S}(X)$ are regular - and therefore, so is their finite union.

Finally, when the complement T^c is regular, the original language T itself must be regular too. ◆

The preceding argument in fact supplies further information about which regular languages are recognized in LPC. All languages $T^{\S}(X)$ are defined by numerical conditions of the forms

$$X_i \geq m, \qquad X_i = m.$$

Finite unions of such languages were called *first-order* regular languages in van Benthem 1988D, because these numerical conditions may be described using precisely first-order definable generalized quantifiers. The latter also have an independent interest among the regular languages because of their recognition behaviour:

The first-order regular languages are precisely those which can be recognized by some *permutation-invariant acyclic* finite state automaton.

Corollary. LPC grammars recognize only first-order regular languages.

Proof. T^c is clearly a first-order regular language. Hence, so is T itself. For, by reversing accepting and rejecting states, complement languages can be recognized in the above acyclic finite-state format too. ◆

Again, there is a converse.

Theorem. All first-order regular languages are recognized by LPC grammars.

Proof. The theorem follows from two auxiliary results.

Lemma 1. Each basic first-order regular language is LPC-recognizable.

Proof. To avoid cumbersome notation, here is one representative example.

Let $n = 2$, with numerical conditions $X_1=1$, $X_2 \geq 3$.

The following assignment recognizes the corresponding language:

a_1: { u }, a_2: { (u, v), (v, s), (s, t), (s, s) }.

First, every string satisfying the occurrence conditions on a_1, a_2 is obviously recognizable. Conversely, suppose that some set A of these five initial types derives t in LPC. As this inference must be logically valid, at least (s, t), (v, s), (u, v) and u must be in A. (Without any of these, the conclusion t does not follow.) So, there is at least one a_1 and at least three a_2. On the other hand, there can be no more than one symbol a_1, since two occurrences of u would produce at least two occurrences of t, which cannot be contracted. ∎

Lemma 2. The LPC-recognizable languages are closed under unions.

Proof. The proof of this assertion involves techniques similar to those employed before, in studying the recognizing power of LP. For further details, see van Benthem 1989F. ∎

The desired result is then immediate from the Lemmas, given the definition of first-order regular languages. ∎

The two preceding results completely determine the recognizing power of the undirected Lambek Calculus with Contraction:

Theorem. LPC recognizes precisely the first-order regular languages.

Still, for all we know at this stage, the recurrent collapse in recognizing power found here might be due to a combination of two factors: Contraction and Permutation. That the former is really the culprit, however, may be seen in the following result, whose full proof involves a generalized form of Kruskal's Theorem, for sequences of arbitrary finite length (see again the above reference):

Theorem. LC grammars recognize only regular languages.

The general picture to come out of this investigation is as follows. The Categorial Hierarchy starts with recognition of context-free languages. Somewhere in the middle, recognition of non-context-free languages becomes possible. But, in the upper reaches, only regular languages can be recognized. In a picture:

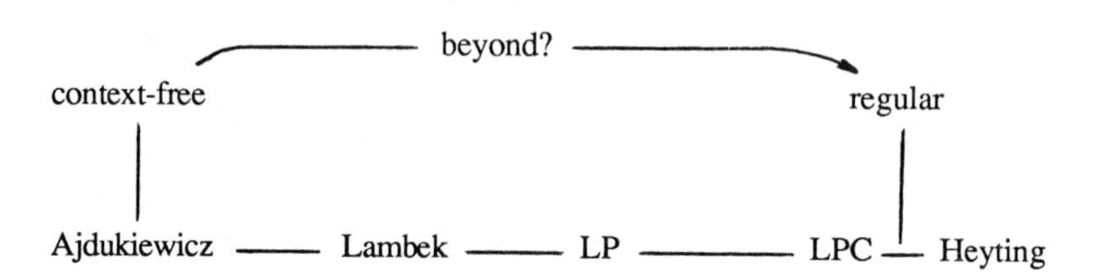

There is much to be pondered over in this scheme.

Moreover, there are many further questions of a systematic linguistic nature, which have been left out of consideration here.

Example. Fine-Structure of Initial Assignments.
There are various natural restrictions that might be imposed on initial assignments.

First, now that so much type change has become posssible in the mechanism of categorial combination, there are fewer reasons for having multiple assignment at all: and one obvious question is what remains recognizable if we restrict attention to *single-valued* symbol-to-type mappings.

Another, more semantically oriented, restriction would be to order assignments by the number of basic types involved in them. For instance, much can be done already with just the basic type t itself:

The assignment

a : { (t, t) } , b : { t , (t, t) }

recognizes all strings in the alphabet { a , b } having at least one symbol b (this is the quantifier "some" in the terminology of van Benthem (1986, chapter 7)). Changing to

a : { (t, (t, t)) } , b : { t , (t, t) }

recognizes the strings having more occurrences of b than of a ("most").

Moreover, adding basic types often turns out redundant for the purpose of recognition. For instance, the language { ab , ba } has a prima facie categorial description involving two basic types:

a : { e } , b : { (e, t) } .

But in fact, the following pure t-type asignment will do as well:

a : { ((t, t), t) } b : { (((t, t), t), t) } .

And indeed, Ponse 1988 has shown that there is no genuine hierarchy here:

Every language which can be categorially recognized at all
can be recognized using only (very complex) types involving t only.

To conclude, this has been a rather long chapter: partly because of the intrinsic attraction of the topic, but partly also as a demonstration of a more general moral. Those who are willing to accept the principle that

"Parsing is Deduction"

(as expounded in Chapters 3 and 4) should also be willing to draw a further corollary from that thesis, namely that

"Theoretical Grammar is Proof Theory" .

And the point of many of the preceding results is that, indeed, linguistic insights can be achieved by logical proof-theoretic methods. It is our hope that the currently separate subjects of Mathematical Linguistics and Proof Theory will become more intertwined in the process.

Post-Script. Logical Syntax.

Linguistic recognition as studied here makes sense, not just for natural languages, but also for logical formalisms themselves. This was already shown by the example of formulas in propositional logic at the beginning of this Chapter. Perhaps more strikingly, similar considerations may be brought to bear upon the type-theoretic logical formalisms behind our very categorial calculi (cf. van Benthem 1987$), with the following outcomes:

- the set of terms of the full typed lambda calculus is *context-free*
- the Lambek fragment Λ is *not context-free.*

Thus, with this formal mode of description, our semantic restriction to single occurrences increases the grammatical complexity of the class of special terms describing the intended denotations.

IV MODEL THEORY

There already exists a well-developed area of logical semantics for natural language. In this part, we shall consider further semantic aspects of the type-theoretic apparatus studied so far. In more technical terms, we are going to investigate the applied model theory of a lambda calculus having a distinguished Boolean truth value type (though lacking the richer structure of a full theory of types).

9 Enumerating Readings

What we have seen in Part II is the great diversity of possible 'outcomes' derivable from a given sequence of types in our categorial calculi - although there were also some systematic restrictions (such as those induced by the count invariants). But, there is also a possible variety for type sequences with already given conclusions, namely, in the manner of their proofs, recording possible semantic interpretations of the transition. Therefore, we shall look into numbers of different proofs for an inference, as a matter of logical interest, or different 'readings' of grammatical expressions in the underlying linguistic application. (From a more mathematical point of view, a category theorist would recognize this concern as one of determining equalities and differences between arrows in a category.)

9.1 Variety of Readings

For a start, the basic categorial pattern itself is unambiguous.

Example. Function Application.
The categorial transition

$a \quad (a, b) \Rightarrow b$

has a reading expressible as follows:

$u_{(a,\,b)}(v_a)$,

which is associated with the derivation

$$\frac{a \qquad\qquad (a, b)}{b}\ \text{MP}$$

To be sure, there are other possible derivations here, such as

$$\dfrac{\dfrac{\dfrac{a \qquad (a, b)^1}{b}\ \text{MP}}{((a, b), b)}\ \text{COND, withdrawing 1} \qquad (a, b)}{b}\ \text{MP}$$

whose associated lambda term is

$(\lambda x_{(a, b)} \bullet x(v_a)) (u_{(a, b)})$.

But, the latter term is lambda-convertible to the former. And indeed, no essentially different derivations exist for this transition. ♠

Also, many principles of type change proposed in Chapter 3 are in fact unambiguous, having only one reading.

Example. Montague and Geach.

The Montague transition

$a \Rightarrow ((a, b), b)$

had the following semantic recipe computed in Chapter 5:

$\lambda x_{(a, b)} \bullet x(u_a)$.

And this is in fact the only possible lambda term of type $((a, b), b)$ having one parameter of type a , modulo normalization.

For, consider the shape of any normal form τ of this kind. Note that the types of all subterms occurring in it must be subtypes of a, $((a, b), b)$. (Recall the 'Subformula Property' of normalized, cut-free derivations stated in Chapter 7.) Therefore, τ must begin with a prefix $\lambda x_{(a, b)} \bullet$ (the parameter u_a does not qualify here, and neither does a function application: as its head would involve a more complex type than $((a, b), b)$). After that, there must be a normal form of type b , involving the two parameters u_a and $x_{(a, b)}$. Again, the latter cannot be a single parameters, as neither type fits. Moreover, evidently, it cannot be a lambda form. Therefore, it must be an application: for which there is only one candidate, namely a term of type (a, b) applied to one of type a . As the whole term is a normal form, the former cannot be a lambda form - while it cannot be an application either: none of the relevant types has (a, b) for a value. So, it must be a parameter, and the only candidate for that is $x_{(a, b)}$. Next, by similar reasoning, the only admissible shape for the argument term of type a is the parameter u_a itself. And so, we have precisely the Montagovian recipe of 'lifting'.

Repeating the considerations on lambda normal forms spelt out here, it may be shown that also the Geach transition (a, b) (b, c) ⇒ (a, c) has only one meaning, modulo logical equivalence, namely, the familiar function composition

$\lambda x_a \bullet u_{(b,c)}(v_{(a,b)}(x_a))$.

■

Remark. Checks of non-ambiguity may be performed much faster in specialized formats of notation. A case in point are the proof nets of Roorda 1989A. ■

Nevertheless, quite familiar sentential patterns may be ambiguous.

Example. Complex Transitive Sentences.
In a sentence with a transitive verb whose subject and object are complex noun phrases, the type pattern was as follows

((e, t), t) (e, (e, t)) ((e, t), t) ⇒ t .

Here, no categorial combination was possible in the Ajdukiewicz system, while the Lambek calculus did work. In fact, the latter system provides exactly *four* semantically distinct readings here, corresponding to two different scope orders for the noun phrases, and two different argument orders for the binary predicate. Moreover, when analyzed in the full intuitionistic system, this transition even acquires *infinitely* many readings: as will be shown in more detail below. ■

Thus, numbers of readings will also depend on the particular categorial calculus employed. Here is a table showing some of the numerical phenomena that may occur.

number of readings calculus / transition	Ajd	Lambek	Int
e (e, t) ⇒ t	1	1	1
e (e, (e, t)) ⇒ (e, t)	1	2	4
e ⇒ t	0	0	0
t ⇒ (e, t)	0	0	1
e (e, (e, (t, t))) ⇒ (t, t)	0	0	∞
t ⇒ ((t, t), t)	0	1	∞

Several hypotheses come to mind here. In particular, numbers in the middle column are always *finite*. And indeed, we have the following result.

Theorem. If b is Lambek derivable from $a_1, ..., a_n$, then there are only finitely many readings corresponding to its derivation, modulo logical equivalence.

Proof. Recall that all these readings must fall within the lambda calculus fragment Λ. Now, the argument is by induction on the number of comma operators in the set of type occurrences $\{ b, a_1, ..., a_n \}$. Lambda terms $\tau_b [\{ u_{a1}, ..., u_{an} \}]$ in normal form can only have one of the following three shapes, each of which allows only a finite variety:

- a variable u_{ai}
  (in which case $b = a_i$, $1 = i = n$),
- an application $\tau^1{}_{(c, b)}(\tau^2{}_c)$,
  where the occurrences of $u_{a1}, ..., u_{an}$ are distributed over τ^1, τ^2, and $(c, b), c$ are distinct subtypes of some a_i. The latter is the type of what may be called the *leading variable* in τ^1, which term must have the form of an application
  $$u_{ai}(\sigma^1)...(\sigma^m) ,$$
  by a simple argument about normal forms. Now, one appeals to the inductive hypothesis with respect to $\sigma^1, ..., \sigma^m, \tau^2$, noting that the 'comma complexity' has gone down in each case.
- an abstraction $\lambda x_c \bullet \tau^1{}_d$,
  where $b = (c, d)$ and τ^1 has parameters $x_c, x_{a1}, ..., x_{an}$. There are only finitely many such candidates, again by the inductive hypothesis.

■

Note that this argument typically breaks down for the full lambda calculus in the application step, when occurrences are to be 'distributed'. Compare, for instance, the infinitely many readings for a transition like

$$t \Rightarrow ((t, t), t) :$$

which are of the forms

$$\lambda x_{(t, t)} \bullet x^n(u_t) \qquad (n = 1, 2, 3, ...) ,$$

when multiple binding is allowed.

Remark. Extensions and Variations.
The above Finiteness Theorem can be generalized, at least, to the case of *product* types. This may be shown by an extension of the above argument, using the earlier-mentioned characterization of the Lambek fragment for implications and products. Perhaps an easier route is provided by an alternative argument. Since proof-theoretic normalization does not affect meaning, all possible readings for a derivable transition are also encoded in its *cut-free sequent-calculus* derivations. But of the latter, there are obviously finitely many, also in the presence of products.

Next, we must consider a subtlety which has escaped discussion so far.
A tacit presupposition during our calculations has been this:

Each parameter for a lambda term computed
may occur *only once*.

This requirement reflects one reasonable use of such terms, as schemata of definition for some given grammatical form. Note that, then, a transition like

$e \ (e, e) \Rightarrow e$

will be unambiguous, even in the full intuitionistic calculus. (But of course, its conditionalized version $e \Rightarrow ((e, e), e)$ will acquire infinitely many readings through multiple binding.)

Nevertheless, there is also a plausible semantic perspective where an emphasis on single parameter occurrences seems quite inappropriate.

Digression. Semantic Definability.
Within the setting of some specific model, there is a natural question of *definability* from given objects:

Let objects $u_1, ..., u_n$ be given, in domains of types $a_1, ..., a_n$.
Which objects in some target domain D_b are lambda-definable from these?

In this case, unlike the previous one, we are not concerned with any particular piece of syntax awaiting interpretation. And in fact, no constraints on the occurrences of parameters in definitions for these objects seem implied.

Therefore, the question can be of high complexity. For instance, given some function u_1 on the natural numbers and some natural number u_2 , the question whether some other natural number v is definable from these two is undecidable in general. On *finite* base domains, however, the general question may still be decidable. The statement that it does, is in fact a general form of what is known in the literature as 'Plotkin's Conjecture' (cf. Statman 1982).

This deviant kind of question has a more linguistic counterpart after all - be it not in terms of interpreting expressions, but rather in terms of language recognition (cf. Chapter 8) . Let the above parameters correspond to items in some symbolic alphabet. Then, problems like the above arise, once we enlarge the scope of traditional questions in Mathematical Linguistics, and no longer talk about recognizing strings themselves, but rather about objects denoted by strings. For instance, given a context-free grammar and some map from symbols in its alphabet to semantic objects, as well as some interpretation schema for its rewrite rules, we can ask of an arbitrary object in the relevant domain whether it belongs to the obvious 'denotation class' of this grammar. (Here, the original language recognition problem is the special case where the interpretation map is the syntactic identity function, with mere concatenation for the rules.) The resulting 'semantic recognition problem' for context-free grammars is undecidable in general, witness the above example in the natural numbers.
But again, it might be decidable on *finite* structures.

Thus, the concerns of Mathematical Linguistics may be transferred from syntax to semantics. ●

Finally, there is a new question concerning the earlier categorial calculi which arises in this context. If we liberalize our occurrence policy concerning parameters (though not with respect to their categorial combination!), the relevant question of derivability remains no longer

" Is $a_1, ..., a_n \Rightarrow b$ derivable? " Derivability

but rather becomes one involving Kleene iteration of premises:

" Is $a_1^*, ..., a_n^* \Rightarrow b$ derivable? " *-Derivability

Prima facie, this adds an *infinitary* element to the situation, even for simply decidable calculi like Ajdukiewicz's or Lambek's: an infinite number of premise patterns should be tried in principle. Nevertheless, it is easy to show that the new problem of *-derivability for the Ajdukiewicz system is decidable (cf. van Benthem 1989*). But the following natural issue is still open.

Question: Is the *-derivability problem for the Lambek Calculus decidable?

(Note that, contrary to first expectations, there is no *obvious* reduction to derivability in the decidable calculus L plus Contraction here, since the latter also allows multiplicity 'inside proofs', so to speak.) For the full Intuitionistic conditional logic, again, decidability is immediate, since the *-derivability problem is equivalent there to the ordinary derivability problem, given its structural rules. ◆

9.2 Grammars and Automata for Semantic Description

It is one thing to determine numbers of readings for a given derivable transition, and another to find a way of systematically enumerating all possibilities. Here, we shall present one method for the latter purpose, based on the use of formal grammars and automata. Its main features will come out in two extensive illustrations.

Multiple Quantification in Transitive Sentences

The first illustration concerns the earlier-mentioned case of transitive sentences with complex subjects and objects:

What is the set of all possible lambda readings for the derivable transition
((e, t), t) (e, (e, t)) ((e, t), t) $\Rightarrow$ t ?

As it turns out, we can describe all these lambda forms using a *context-free grammar* having symbols

X_a , $X_a{}^{\#}$, C_a , V_a

for each of the relevant types here: being those occurring explicitly in the given transition as well as their subtypes. (Recall the earlier 'Subformula Property' of lambda normal forms.) Here, V_a stands for a variable of type a , C_a for a constant (parameter), X_a for any term of type a , and $X_a{}^{\#}$ for any such term which does not start with a lambda symbol. The point of this division will become clear from the rewrite rules for terms in normal form to be presented now.

First, we have some obvious connections, for all types a :

$$X_a \Rightarrow X_a^{\#}$$
$$X_a^{\#} \Rightarrow C_a$$
$$X_a^{\#} \Rightarrow V_a \ .$$

Then, there are rules for application or abstraction terms, depending on the actual types being present. In understanding the following set of arrows, it should be borne in mind that lambda normal forms contain no more so-called 'redexes' of the unreduced shape " $(\lambda x\bullet \alpha)(\beta)$ " , while the types of all variables occurring in them must be subtypes of either the resulting type or one of the parameter types:

$$X^{\#}_t \Rightarrow X^{\#}_{(e,\, t)}(X_e)$$
$$X^{\#}_t \Rightarrow X^{\#}_{((e,\, t),\, t)}(X_{(e,\, t)})$$
$$X^{\#}_{(e,\, t)} \Rightarrow X^{\#}_{(e,\, (e,\, t))}(X_e)$$
$$X_{(e,\, t)} \Rightarrow \lambda V_e \bullet X_t$$
$$X_{(e,\, (e,\, t))} \Rightarrow \lambda V_e \bullet X_{(e,\, t)}$$
$$X_{((e,\, t),\, t)} \Rightarrow \lambda V_{(e,\, t)} \bullet X_t \ .$$

Now, this grammar will be used in a somewhat unorthodox 'procedural' fashion. Each possible reading may be obtained by following some grammatical derivation from the symbol X_t to some string of terminal symbols - under the following conventions:

- symbols V_a can only be replaced by variables of type a ,
- constants C_a can only be replaced by the appropriate original parameters,
- successive variables substituted become available for later substitution in X_a positions of the appropriate type.

Then, a derivation is called *admissible* if it can be completed to a correct lambda term in this fashion. For instance, here is an admissible derivation:

$$X_t\,, \quad X_{((e,\, t),\, t)}(\,X_{(e,\, t)}\,)\,, \quad X_{((e,\, t),\, t)}(\,\lambda V_e \bullet X_t\,)\,,$$
$$X_{((e,\, t),\, t)}(\,\lambda V_e \bullet X^{\#}_{((e,\, t),\, t)}(\,X_{(e,\, t)}\,)\,)\,,$$
$$X_{((e,\, t),\, t)}(\,\lambda V_e \bullet X^{\#}_{((e,\, t),\, t)}(\,X^{\#}_{(e,\, t)}\,)\,)\,,$$
$$X_{((e,\, t),\, t)}(\,\lambda V_e \bullet X^{\#}_{((e,\, t),\, t)}(\,X^{\#}_{(e,\, (e,\, t))}(\,X_e\,)\,)\,)$$
$$C_{((e,\, t),\, t)}(\,\lambda V_e \bullet C_{((e,\, t),\, t)}(C_{(e,\, (e,\, t))}(V_e))\,)\,,$$

which can be processed to the lambda term

$$u_{((e,\, t),\, t)}(\lambda x_e \bullet v_{((e,\, t),\, t)}(w_{(e,\, (e,\, t))}(x_e))\,.$$

And here is an inadmissible derivation:

$$X_t\,, \qquad X^{\#}_t\,, \qquad X^{\#}_{(e,\,t)}(\,X_e\,)\,, \qquad C_{(e,\,t)}(\,V_e\,)\,.$$

In this particular case, the description becomes much facilitated, because we can make this grammar *regular*. Therefore, all its productions may be displayed graphically in the following visualization.

Proposition. All readings for multiply quantified transitive sentences arise by traversal of the following diagram of a finite state machine:

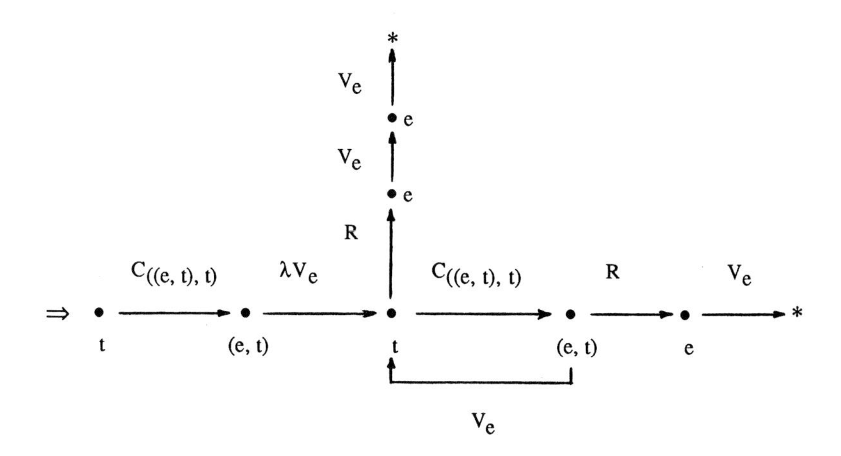

Traversing this diagram until an exit * is reached, putting in appropriate parameters or variables stored earlier 'en route', yields an infinity of readings for the above sentence pattern. These include the four cases obtained in the Lambek calculus, but also such candidates as the reflexive form

NP1 $(\lambda x_e \bullet \mathrm{TV}(x)(x))$,

as well as various stackings of noun phrases. ◆

Semantic Reducibility of Determiners

Now, let us consider another example of enumeration by the grammar method. This time, our question will be slightly different, concerning possible definability of an important layer in our type hierarchy 'from below':

When is a semantic determiner, living in type $((e, t), ((e, t), t))$, lambda-definable from items in lower types, i.e., $\{ e, t, (e, t), ((e, t), t) \}$?

Here is the context-free grammar for the relevant types:

$$X_{((e, t), ((e, t), t))} \Rightarrow \lambda V_{(e, t)} \bullet X_{((e, t), t)}$$
$$X_{((e, t), t)} \Rightarrow \lambda V_{(e, t)} \bullet X_t$$
$$X_t \Rightarrow X^{\#}_{(e, t)}(X_e)$$
$$X_t \Rightarrow X^{\#}_{((e, t), t)}(X_{(e, t)})$$
$$X_{(e, t)} \Rightarrow \lambda V_e \bullet X_t \ .$$

And again, there is an equivalent regular grammar, whose diagram looks as follows:

Proposition. All possible definitions for determiner denotations from lower type domains arise by traversal of the following finite state machine [with 'D' standing for 'C' or 'V']

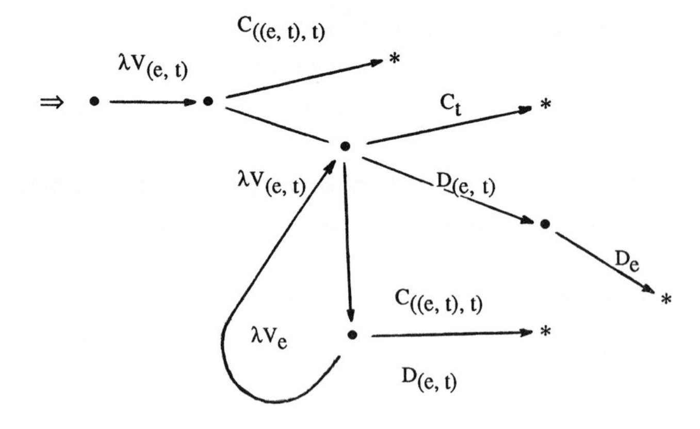

This scheme produces determiner denotations of forms such as the following:

1 $\lambda x_{(e, t)} \bullet \ A_{((e, t), t)}$

2 $\lambda x_{(e, t)} \bullet \lambda y_{(e, t)} \bullet A_{((e, t), t)}(x)$

3 $\lambda x_{(e, t)} \bullet \lambda y_{(e, t)} \bullet x(B_e)$

4 $\lambda x_{(e, t)} \bullet \lambda y_{(e, t)} \bullet A_{((e, t), t)}(\lambda z_e \bullet A'_{((e, t), t)}(\lambda u_e \bullet x(z)))$.

Here, the latter format is 'iterative', producing infinitely many forms by repeating the subroutine " $A'_{((e, t), t)}(\lambda u_e \bullet$ " . Thus, globally, infinitely many distinct possibilities exist for defining determiners from objects in lower types. ◆

In both cases considered here, an *infinite* number of possible readings emerged, due to the existence of at least one derivational *loop* in their context-free grammar (or finite state machine, in the simplified version). Here is a

Conjecture

A derivable type transition has finitely many readings in the full Lambda Calculus if and only if its associated context-free grammar has no loops.

Statman 1980 characterizes the special case of those derivable transitions in a typed lambda calculus with one basic type which have finitely many readings.

Remark. Readings in the Lambek Calculus.

With the earlier calculi L or LP , however, there was finiteness in every case, even if the grammar contained loops. The reason for that phenomenon, however, is a different one. In the associated Λ-fragment, only single binding occurs, and hence the above procedure changes. Now, each passage through an arrow λV_a introduces a variable which must be used *exactly once* in the sequel. And, since the number of positions where these variables can be put in order to effect an exit from the diagram is finite, the number of possible succesful traversals is finite too. This sketch may be worked into an alternative proof for the Finiteness Theorem established above.

This finiteness also implies that all the above questions concerning numbers of readings must be *decidable* for Lambek calculi. ◆

In practice, however, there are various reasons why a prima facie infinity of possible outcomes may reduce to only a finite number after all, even in the setting of a full typed lambda calculus. Here are two possibilities, continuing the previous topics.

Reducing Determiners

Locally, within a fixed model, there are only finitely many ways of defining a determiner from below.

The reason is as follows. Any scheme of definition found in the above analysis will start with a prefix $\lambda x_{(e, t)}\bullet$, and then some further term τ of type $((e, t), t)$. Now, if the latter contains no free occurrences of the variable $x_{(e, t)}$, then it defines some fixed object: for which there is already one parameter $A_{((e, t), t)}$ denoting it. Hence, we are in the above case 1 . Next, if the variable $x_{(e, t)}$ does occur in t , then, analyzing the latter term one step further via the state diagram, we can rewrite the whole scheme of definition to

$$\lambda x_{(e, t)}\bullet \lambda y_{(e, t)}\bullet [(\lambda z_{(e, t)}\bullet \tau[z/x]) (x)] ,$$

where the subterm $\lambda z_{(e, t)}\bullet \tau[z/x]$ does not contain any free occurrence of the variable $y_{(e, t)}$. (To see this, check the available 'exit routes' in the diagram.) Now, the latter subterm again denotes some fixed object in the $((e, t), t)$ type domain: and hence, we arrive at the above form 2 .

Therefore, the general outcome becomes even this:

Proposition. Within any fixed model, the two patterns

1 $\quad \lambda x_{(e, t)}\bullet\ A_{((e, t), t)}$

2 $\quad \lambda x_{(e, t)}\bullet \lambda y_{(e, t)}\bullet A_{((e, t), t)}(x)$

suffice when defining determiners from below.

For a check, one may reduce the earlier pattern 3 to the pattern 2 as follows:

$$\lambda x_{(e, t)}\bullet\ \lambda y_{(e, t)}\bullet\ (\lambda z_{(e, t)}\bullet\ z(B_e)) (x) .$$

What this result says is that determiners admit of no non-trivial reductions to lower types: they are a genuinely new category of expression.

Evidently, this is just one of a number of general questions which may be investigated about reducibility in type domains. For instance, can we prove general 'hierarchy results' to this effect?

Multiple Quantification

With the above double quantifier patterns too, there is a collapse to finitely many readings within any fixed model.

This time, however, the reason is one having to do with the special structure of the Boolean base domain D_t . The local reduction obtained is as follows:

Proposition. Let A, B be any two objects in the type domain ((e, t), t) of some model, and R one in the domain of type (e, (e, t)). The items in type t definable from these must be of one of the following forms

- $(\neg)Q(\lambda x_e \bullet (\neg)R(x)(x))$
- $(\neg)Q(\lambda x_e \bullet (\neg)Q((\neg)R(x)))$
- $(\neg)Q(\lambda x_e \bullet (\neg)Q(\lambda y_e \bullet (\neg)R(y)(x)))$.

Here, 'Q' indicates either A or B , and '$(\neg)$' denotes an optional negation.

Proof. To prove this assertion, it suffices to show how the single iterative pattern in the finite state diagram found above produces readings which all fall under this description. The reason for this will become clear from the treatment of one typical example:

$$B(\lambda k \bullet A(\lambda y \bullet B(\lambda s \bullet A(\lambda x \bullet B(\lambda u \bullet A(\lambda z \bullet R(x)(y))))))) .$$

The following sequence of steps demonstrates the algorithm which effects the desired reduction:

1 $A(\lambda z \bullet R(x)(y))$ is equivalent by Boolean reasoning to
$(R(x)(y) \wedge A(\lambda z \bullet \mathrm{TRUE})) \vee (\neg R(x)(y) \wedge A(\lambda z \bullet \mathrm{FALSE}))$.
Now, since both $A(\lambda z \bullet \mathrm{TRUE})$ and $A(\lambda z \bullet \mathrm{FALSE})$ have fixed truth values in our model, the above disjunction reduces to one of the following, *uniformly* in the relation R :
$R(x)(y)$, $\neg R(x)(y)$, TRUE or FALSE .

For the sake of illustration, assume that the outcome is $R(x)(y)$. Now, we move one step to the left in our analysis of the above lambda term.

2 $B(\lambda u \bullet R(x)(y))$ again reduces to one of the forms
$R(x)(y)$, $(\neg)R(x)(y)$, TRUE or FALSE ,
by a similar Boolean argument.

3 One step further to the left, the operator λx can bind non-vacuously in the first two terms, yielding $\lambda x \bullet R(x)(y)$ $(= R^{\vee}(y))$ or $\lambda x \bullet \neg R(x)(y)$ $(= \neg R^{\vee}(y))$, the remainder being again two constant cases.

4 Inserted into A , this gives one of
$A(R^{\vee}(y))$, $A(\neg R^{\vee}(y))$, TRUE or FALSE .

Now, apply similar reductions as above, distinguishing cases for $A(R^{\vee}(y))$, etcetera, to arrive at one of the forms listed in the above enumeration. ∎

Again, there are more general questions in the background here, in addition to those already posed about 'global' finiteness of readings. In the meantime, we have also found natural 'local' forms of finiteness. In particular, then, what would be an effective characterization of those cases where a derivable transition has only finitely many readings in the local sense?

Another general question concerns the existence of *algorithms* for computing the actual *numbers of readings* for derivable transitions. This number is obviously subject to exponential explosion, witness the case when some finite string of extractable scope-bearing operators occurs. There are some numerical formulas describing all readings in special cases, but the general function remains unknown. These phenomena involve a more empirically oriented discussion.

At least for ordinary linguistic purposes, our undirected categorial calculi produce too many readings. Admittedly, some sources of combinatorial explosion may be facts of natural language use too, such as the scope permutations just mentioned. But, there are other permutations which are simply implausible. For instance, on an LP analysis, an unambiguous sentence like "Every dinosaur died" would still obtain *two* readings, namely both the intended inclusion of DINOSAUR in DIED , but also its converse.

There are various strategies for coping with this over-generation. One is to impose restrictions on categorial calculi: either on their axioms and rules of inference, or by imposing some 'filter' on admissible derivations. (For instance, proofs with an overly high 'computational cost' might be banned.) Another strategy is to retreat to the earlier directed calculi after all, which produce fewer readings, due to their encoded syntactic constraints (see Moortgat 1988 for an up-to-date presentation). Then, Hendriks 1987 is an interesting proposal for setting up a special purpose *subcalculus* of the full LP for the proper treatment of admissible quantifier readings. Moortgat 1989B has a re-analysis of the resulting system via a tactic inspired by Linear Logic (cf Chapters 6, 14): viz. adding a suitable new 'exponential' connective to the basic Lambek Calculus encoding a certain amount of permutation for quantifiers.

But in our view, it is quite attractive to have precisely some 'strategic depth'. The type-theoretic semantics for categorial derivation maps out a *logical space* of a priori possibilities for semantic interpretation: out of which natural language then realizes only those satisfying certain additional *denotational constraints*. In the latter vein, Keenan 1988 has some interesting empirically motivated semantic universals constraining the logical latitude of quantifier scoping. Perhaps the archetypal example of such a denotational constraint, however, which has been studied extensively in the literature (cf. Keenan & Stavi 1986, van Benthem 1986, Westerstahl 1989) is that of *Conservativity* for determiners, being the equivalence:

$$D\,AB \quad \text{if and only if} \quad D\,A(B\cap A)\,.$$

That is, the first argument provides the context for the action of the second. And this phenomenon too may be viewed as constraining possible readings. In particular, the above inversion problem disappears. The sentence "Every dinosaur died" should be equivalent to "Every dinosaur was a dinosaur who died" - and this will hold only on the standard construal: not on the permuted reading, where the determiner "all" becomes converted to mean "only" , an expression which is not conservative.

To summarize, the purpose of this Chapter has been to develop some general tools for systematic description of all possible denotations for given expressions, insofar as definable in the lambda calculus formalism. On top of this, various special denotational properties of such expressions must come into play, which is the topic of the next Chapter.

10 Computing Denotational Constraints

Now, we want to consider the more general semantic effects of categorial combination and type change. There are various ways of focussing on the relevant phenomena here. One is to view the lambda machinery as providing the 'abstract glue' for composition of meanings in natural language according to Frege's Principle (cf. Chapter 5): and then, of course, we want to know the semantic peculiarities of this glue. For instance, how does it affect basic semantic properties which have been discovered in logical semantics concerning the behaviour of various categories of linguistic expression? Another angle is that of 'polymorphism': many expressions can live in different categories, as we have seen, and one wants to know about invariants and changes in their semantic properties when moving from one to the other. And finally, the whole combinatorics of typed denotations is also one means of bringing out the 'coherence' of the hierarchy of type domains, whose various levels are internally related.

But for a start, one immediate attraction of the categorial mechanism developed so far is that it helps us in attaining maximum generality across natural language. That is, a good semantic account found for an expression occurring in one syntactic context should preferably generalize automatically to other contexts where it occurs. And indeed, in many cases, the present apparatus allows us to compute suitable further denotations from the original one.

Example. Reflexives.
The basic use of the reflexivizer "self" is to turn binary relations into unary properties by identifying their two argument positions, thus acting in type

$$((e, (e, t)), (e, t)) .$$

This reflects constructions with transitive verbs, such as "know oneself" . But, there are also further syntactic contexts where reflexives can appear, such as prepositional phrases:

" [Mary kept the Bible] to herself " .

Now, we do not want a new ad-hoc account of denotations here. Our semantics should automatically derive the correct meaning for reflexivized prepositions. And in fact, it does.

For the sake of simplicity, let prepositions have type $(e, ((e, t), (e, t)))$: van Benthem 1986 has a motivation for this assignment, denoting maps from individuals to adverbial phrases. (Incidentally, the latter type should not be confused with that of the reflexivizer itself.) The relevant sequent will then be

$$(e, ((e, t), (e, t))) \quad ((e, (e, t)), (e, t)) \quad \Rightarrow \quad ((e, t), (e, t)) .$$

A straightforward categorial derivation is as follows:

$$\underline{e^1 \qquad\qquad\qquad\qquad (e, ((e, t), (e, t)))} \; \text{MP}$$
$$\underline{(e, t)^2 \qquad\qquad\qquad\qquad ((e, t), (e, t))} \; \text{MP}$$
$$\underline{(e, t)} \; \text{COND, withdrawing 1}$$
$$\underline{(e, (e, t)) \qquad\qquad\qquad\qquad ((e, (e, t)), (e, t))} \; \text{MP}$$
$$\underline{(e, t)} \; \text{COND, withdrawing 2}$$
$$((e, t), (e, t))$$

Its corresponding lambda term is constructed thus:

$$\underline{x_e \qquad\qquad\qquad\qquad TO_{(e, ((e, t), (e, t)))}}$$
$$\underline{P_{(e, t)} \qquad\qquad\qquad\qquad TO(x)}$$
$$\underline{TO(x)(P)}$$
$$\underline{\lambda x \bullet TO(x)(P) \qquad\qquad\qquad\qquad REF_{((e, (e, t)), (e, t))}}$$
$$\underline{REF(\lambda x \bullet TO(x)(P))}$$
$$\lambda P \bullet (REF(\lambda x \bullet TO(x)(P)))$$

Finally, by the definition of REF , being

$$\lambda R_{(e, (e, t))} \bullet \lambda y_e \bullet R(y)(y) \,,$$

the latter term computed reduces to the desired reading

$$\lambda P \bullet \lambda x \bullet TO(x)(P)(x) \,.$$

●

Remark. Quine Operators.

The expression "self" is of a more general semantic interest, because its linguistic function is to contract arguments. Thus, it serves as an explicitly lexicalized instruction for performing the function of the earlier structural rule of Contraction (cf. Chapters 4 and 5). If the latter principle were freely available in natural language, then this facility would have been superfluous.

In fact, most of the structural rules beyond the Lambek calculus seem to have corresponding lexicalized instructions, witness the use of *passives* for Permutation of arguments, and various forms of Projection deleting arguments ('argument drop'). There is a connection here with the well-known operators of Quine's variable-free notation for predicate logic (cf. Quine 1971, as well as the survey in van Benthem 1989G of variable-free formalisms in Logic and natural language), which does away with all explicit variable binding by means of operations of 'identification', 'inversion' and 'projection' . The issue of hidden versus explicitly lexicalized structural rules will come up again with the Linear Logic of Chapter 14. ●

10.1 Denotational Constraints

In the study of various special categories of expression, one encounters so-called *denotational constraints*, being general structural properties which all, or at least, many important denotations for linguistic expressions in that category turn out to possess. Prominent examples are found in the study of *generalized quantifiers* (cf. van Benthem 1986A): where all determiners turn out to be 'conservative' (cf. Chapter 9), while many important ones are also 'logical' or 'monotone', two notions to be introduced below. Now, through our categorial apparatus, a phenomenon like Conservativity proliferates across natural language, starting from the initial requirement on determiners D that

$$D\,AB \quad \text{if and only if} \quad D\,A\,(B \cap A)\,.$$

Example. Conservativity in Complex Transitive Sentences.
In transitive sentences like "Every boy loves a girl" , there is no Conservativity in the naive sense of having an equivalence with "Every boy loves a girl who is a boy" .
So, what is the proper form of this constraint here? The denotation of a sentence pattern

$$Q1\,A \quad R \quad Q2\,B$$

may be computed as follows, in accordance with the simplest categorial derivation of the type transition

$$((e, t), ((e, t), t)) \quad (e, t) \quad (e, (e, t)) \quad ((e, t), ((e, t), t)) \quad (e, t) \quad \Rightarrow \quad t :$$

$$Q1\,(\,A, \lambda x\text{•}\,Q2\,(\,B, R(x)\,)\,) \qquad \text{iff}$$

(by two applications of ordinary Conservativity)

$$Q1\,(\,A, \lambda x\text{•}\,Q2\,(\,B, \lambda y\text{•}\,(R(x)(y) \wedge B(y)) \wedge A(x))\,) \qquad \text{iff}$$

$$Q1\,(\,A, \lambda x\text{•}\,Q2\,(\,B, \lambda y\text{•}\,(R(x)(y) \wedge B(y) \wedge A(x)) \wedge A(x))\,) \qquad \text{iff}$$

$$Q1\,(\,A, \lambda x\text{•}\,Q2\,(\,B, \lambda y\text{•}\,(R(x)(y) \wedge B(y) \wedge A(x)))\,)\,.$$

I.e., there is a restriction of argument roles:

$$Q1A \;\; R \;\; Q2B \quad \text{holds} \quad \text{iff} \quad Q1A \;\; R \cap (A \times B) \;\; Q2B \quad \text{does.}$$

●

Similar phenomena occur in other linguistic patterns. For instance, in evaluating an expression like "walk to every city", the common noun "city" will come to restrict the individual argument of the preposition "to". A full calculation goes as follows.

Example. Conservativity in Prepositional Phrases.
The relevant derivation tree is this:

$$\dfrac{\underline{e}^3 \quad \dfrac{\underline{(e,t)}^2 \quad \dfrac{\underline{e}^1 \quad (e,((e,t),(e,t)))}{((e,t),(e,t))}\,\text{MP}}{(e,t)}\,\text{MP}}{\dfrac{t}{(e,t)}\,\text{COND, withdrawing 1}}\,\text{MP}$$

$$\dfrac{\dfrac{\dfrac{(e,t) \quad ((e,t),t)}{t}\,\text{MP}}{(e,t)}\,\text{COND, withdrawing 3}}{((e,t),(e,t))}\,\text{COND, withdrawing 2}$$

Its corresponding lambda term reads as follows:

$$\dfrac{y_e \quad \dfrac{z_{(e,t)} \quad \dfrac{x_e \quad TO_{(e,((e,t),(e,t)))}}{TO(x)}}{TO(x)(z)}}{\dfrac{TO(x)(z)(y)}{\lambda x_e\bullet TO(x)(z)(y)}}$$

$$\dfrac{\dfrac{\dfrac{\lambda x_e\bullet TO(x)(z)(y) \quad EVERY\ CITY_{((e,t),t)}}{EVERY\ CITY_{((e,t),t)}(\lambda x_e\bullet TO(x)(z)(y))}}{\lambda y_e\bullet (EVERY\ CITY_{((e,t),t)}(\lambda x_e\bullet TO(x)(z)(y)))}}{\lambda z_{(e,t)}\bullet (\lambda y_e\bullet (EVERY\ CITY_{((e,t),t)}(\lambda x_e\bullet TO(x)(z)(y))))}$$

Here, by ordinary Conservativity, the last three lines may replace their part

$\lambda x_e\bullet TO(x)(z)(y)$ by $\lambda x_e\bullet (TO(x)(z)(y) \wedge CITY(x))$.

Remark. Alternatives.
Incidentally, there are several different semantic construals of this expression in our format. In general categorial terms, the transition

$$(e,((e,t),(e,t))) \quad ((e,t),t) \Rightarrow ((e,t),(e,t))$$

also has readings such as the following [with the parameter U of type (e, ((e, t), (e, t))) and V of type ((e, t), t)] :

$$\lambda z_{(e,\, t)} \bullet \lambda x_e \bullet V(U(x)(z)) .$$

When plugged into the above example, the latter would rather express the property of

BEING WALKED TO BY EVERY CITY .

By the above mechanism of categorial evaluation, each common noun in a quantifier phrase will come to restrict at least one individual argument role, in an entirely predictable manner. By itself then, the combination of basic Conservativity with the categorial combinatorics induces a general principle, which has also been observed on independent empirical grounds, of what may be called 'Domain Restriction' across the whole language.

Moreover, this result also provides an explanation after all for the difference between the grammatical categories of common nouns and intransitive verbs, whose semantic types were identified in Chapter 3. Although their denotations are similar, their linguistic functions are different. Verbs set up 'frames of predication', while nouns serve to restrict the range of arguments occurring in these frames.

In Chapter 11, further examples of this categorial 'spreading' of denotational constraints will be presented, having to do with Boolean forms of inference (of which, indeed, Conservativity itself is already an example).

Next, in addition to denotational properties arising from basic constraints, there is also an issue of *emergent* semantic behaviour, produced by the categorial glue itself. For instance, as Keenan & Faltz 1985 have stressed, inside the noun phrase domain, individuals behave in a very special manner, namely as 'Boolean homomorphisms' - and the same holds for intransitive verbs in their lifted third-order type (((e, t), t), t) (cf. Chapter 5). Now, here is one case where the lambda semantics itself is entirely responsible for the phenomenon. For, the relevant terms, such as that for lifted individuals

$$\lambda x_{(e,\, t)} \bullet x_{(e,\, t)}(u_e) ,$$

automatically denote homomorphisms. This is due entirely to their syntactic form, having the argument x occurring in the 'head position' of their matrix. We shall return to this phenomenon of 'denotational embellishment' below.

What this observation brings to the fore again is a more general issue, viz.

How to determine special semantic behaviour
from the form of lambda terms?

Several examples of this will be studied in what follows. What this question involves is in fact ordinary model theory of the Lambda Calculus, being concerned with the interplay between syntactic form of terms and their semantic behaviour. And what would be desired here in particular are so-called 'preservation theorems', stating some precise connection between the two - such as Los' well-known result in first-order predicate logic, equating syntactic definability by universal quantifier prefix forms with semantic preservation under submodels. This kind of issue was already raised in Chapter 2, in particular, concerning a class of lambda terms that figured quite prominently in our previous explications of type change:

Is there a fitting semantic characterization for the 'pure combinators'
definable in the form " lambda prefix, followed by pure application term " ?

10.2 Semantic Effects of Polymorphism

Besides 'pooling' of constraints, another important function of categorial combination is that of producing 'polymorphism', i.e., the ability of linguistic expressions to modify their types in different environments, as required by the needs of interpretation. The earlier semantic account of categorial derivations may be viewed as a theory of polymorphism too, explaining how items in one type can also occur in another one, with a systematic transfer of meaning:

derivation of $a \Rightarrow b$:
lambda term $\tau_b [\{u_a\}]$.

For instance, one obvious question is which polymorphic changes reproduce the old items exactly, in that they establish an *injective* map from objects in the source domain D_a into a new domain D_b. For instance, the earlier-mentioned lifting of individuals to noun phrases is indeed one-to-one, when viewed as a map from items in the domain D_e (represented by the parameter 'u_e') to values in the noun phrase domain ((e, t), t) . Put differently,

The lambda term $\lambda y_e \bullet \lambda x_{(e, t)} \bullet x(y)$ defines a one-to-one function
in the domain of type (e, ((e, t), t))) .

At present, no effective syntactic characterization seems to be known of those lambda terms having this natural semantic property.

Here is another illustration of this notion.

Example. Individual Predicates and Properties of Noun Phrases.
The earlier lifting of intransitive verbs in type (e, t) to type (((e, t), t), t) (cf. Chapters 3, 4) is an injection again. But, the converse 'lowering' map from
(((e, t), t), t) to (e, t) , computed as being

$$\lambda x_e \bullet u_{((e,t),t)}(\lambda y_{(e,t)} \bullet y(x)),$$

is not one-to-one: and indeed, the two type domains are not isomorphic.

Digression. 'Translations' of the Hierarchy.
One additional reason for interest in one-to-one definable maps is the possibility of *recreating* the hierarchy of type domains within itself at some higher level. For instance, as was observed in Chapter 5, in setting up a viable semantics of *plurality*, one possible route is to switch from individuals in type e to sets of individuals, living in type (e, t) , as the basic objects. This may be done without loss of earlier structure by mapping old individuals uniquely to their singletons in the new type:

$$\lambda x_e \bullet \lambda y_e \bullet x = y .$$

Likewise, the study of higher forms of 'polyadic quantification' sometimes requires viewing *pairs* or *sequences* of individuals as new basic objects (cf. van Benthem 1987B): which again involves an upward translation of the universe into itself.

Denotational properties may be gained or lost in the process of polymorphism. For instance, ordinary individuals became homomorphisms in their lifted type. On the other hand, in Chapter 12, we shall find loss of Boolean monotonicity in certain cases.

An important example of an invariant characteristic is *logicality* , in the sense of the *permutation invariance* defined in Chapter 2. Now, one of the results stated there was the following 'Invariance Lemma', that may be formulated in a somewhat ad-hoc notation:

For all terms τ_b with n free variable occurrences of types $a_1, ..., a_n$,
and all permutations π of the individuals, suitably lifted to higher types,

$$\pi_b ([[\tau (u_1, ..., u_n)]]) \quad = \quad [[\tau (\pi_{a1}(u_1), ..., \pi_{an}(u_n)]] .$$

As a consequence, we have this outcome.

Proposition. Logicality is preserved under polymorphism.

Proof. If u is a permutation-invariant object itself of type a , then so are all its polymorphic forms $\tau(a)$ of type b , since

$$\pi_b([[\tau (u)]]) = [[\tau (\pi_a(u))]] = [[\tau (u)]] .$$

●

Digression. Permutation Invariance and Definability.

It may be of interest to observe that permutation invariance also implies type-theoretic definability, at least, on suitable semantic structures. More precisely, on *infinite* domains, there will be uncountably many permutation invariants, outrunning the countable supply of any ordinary formal language. But on *finite* domains, we can do better. For this purpose, however, one needs the full Type Theory L_ω of Chapter 2, having application, abstraction and identity. Recall that this language had the power to define the usual logical operators

$$\neg , \wedge , \vee , \exists , \forall .$$

Now, the earlier Invariance Lemma for permutations extends to this formalism too. This allows for a model-theoretic construction which proves the following

Proposition. In function hierarchies over finite base domains, the permutation-invariant objects are precisely those definable by a formula of the type-theoretic language.

Proof. One direction is straightforward. Consider any type-theoretic model, finite or infinite. If a formula $\phi = \phi(x_a)$ of L_ω defines an object f of type a uniquely, then - since $\pi_a(f)$ satisfies ϕ in the model whenever f does (by the above observation) - $\pi_a(f) = f$, for all permutations π . So, all type-theoretic terms define invariant denotations.

As for the converse, here is a model-theoretic argument, which works on finite models (and in fact also on suitably saturated infinite general models) for L_ω . First, an auxiliary result is needed.

Lemma. If f, g are in D_a , and $(\mathbb{M}, f)$ is elementarily equivalent to $(\mathbb{M}, g)$, then there exists a permutation π of D_e such that $\pi_a(f) = g$.

Proof. This assertion is shown by a standard zigzag argument, finding increasing sequences $\mathbb{a}, \mathbb{b}$ such that $(\mathbb{M}, \mathbb{a})$ and $(\mathbb{M}, \mathbb{b})$ always remain elementarily equivalent.

When the relevant portions of the hierarchy (i.e., the transitive closure of D_a) have been exhausted, one can read off the desired permutation on the ground domain D_e , and show that its canonical lifting respects the correspondence between all items in the matching sequences: in particular, that between f and g . ◆

Then, a logical f in D_a can be defined uniquely as follows. For every object $g \neq f$ in D_a , there must exist some formula $\alpha_g(x_a)$ with $(\mathbb{M}, f) \models \alpha_g$, $(\mathbb{M}, g) \not\models \alpha_g$. Otherwise, $(\mathbb{M}, f)$ would be elementarily equivalent to $(\mathbb{M}, g)$: and hence, by the preceding lemma, some π_a would map f onto g : which contradicts the permutation invariance of f . But then, f is uniquely defined by the conjunction

$$\bigwedge \{ \alpha_g \mid g \text{ in } D_a, g \neq f \} .$$

◆

Remark. Stronger Invariances.

Stronger demands on logicality than mere permutation invariance are possible too.
For instance, in van Benthem 1987C, so-called 'polyadic quantifiers' are studied, being generalized quantifiers of no longer the earlier type ((e, t), t) , but rather

((e, (e, t)), t) .

In this case, one can also consider invariance under permutations of *ordered pairs* of individuals, which is a much stronger requirement, while there are also several intermediate possibilities. ◆

Finally, in Chapter 11, we shall also consider some effects of polymorphism on the process of logical inference.

10.3 Coherence of Type Domains

Yet another way of looking at type-theoretically definable connections is that of 'coherence' of type domains: a topic that was already introduced at the end of Chapter 5, when further possible kinds of type change were considered. In particular, it was shown there how certain corners of the type hierarchy are related in various mathematical ways, witness the discussion of the so-called 'Partee Triangle'.

Not all such connections were lambda-definable, however: many required some additional set theory, encoded in L_ω . Nevertheless, such transitions usually satisfy certain mathematical constraints, which may make them simply definable after all. Here are two illustrations of this inverse perspective, which diverges from the main line of our study so far: as they start from the denotational properties of certain transformations, and then ask for some suitable linguistic schema of definition.

Example. Continuous Transformations.
Many type transformations proposed in practice are *continuous* , in the sense of respecting arbitrary unions of their arguments:

$$f(\cup \{A_j \mid j \in J \}) \quad = \quad \cup \{ f(A_j) \mid j \in J \} .$$

In particular, since any argument set is the union of its singleton subsets, values of the function f are determined completely by its action on singleton arguments: a fact which expresses a certain form of 'local computability'. Examples include various 'projections', such as taking the domain or range of a binary relation, and more generally, 'existentially quantified' operations like composition of binary relations.

Here is a classification of an important type of continuous operator, which works in all models, finitely based or not. For perspicuity, we choose a simple case, inspired by the theory of relation algebras (cf. Chapter 16).

Proposition. In any model, each continuous logical binary operation on binary relations is definable in the following format:

$$\lambda R_{(e, (e, t))} \bullet \lambda S_{(e, (e, t))} \bullet \lambda x_e \bullet \lambda y_e \bullet$$
$$\exists u_e \bullet \exists v_e \bullet R(u)(v) \wedge \exists w_e \bullet \exists r_e \bullet S(w)(r) \wedge$$
" Boolean condition on identities in $\{x, y, u, v, w, r\}$ " .

Proof. All forms listed define logical operations, because they involve the identity predicate only. They are also continuous, because of the special existential context for the argument predicates.

Conversely, the above observation about singleton arguments already explains the existentially quantified form: since $f(R, S)$ must equal the union of all values

$$f (\{ (u, v) \}, \{ (w, r) \})$$

with Ruv, Swr . On top of this, permutation invariance induces the given Boolean matrix of identity statements, by imposing a certain 'uniformity'. For instance, if a value (x, y) is admissible with x distinct from y, u, v, w, r , then *any* such distinct object x' must be admissible in this position. This may be seen by considering a permutation of individuals interchanging x and x' , while leaving all other objects fixed. More precisely, the Boolean matrix may be constructed as follows:

The argument objects u, v, w, r can exhibit only one of a finite number of (non-)identity patterns. For each of these, choose some pair $\{(u, v)\}$, $\{(w, r)\}$ exemplifying the pattern, and consider the set $f(\{(u, v)\}, \{(w, r)\})$. Then, for each pair (x, y) occurring inside the latter, write down the conjunction of all (non-)identity statements that are true for the objects x, y with respect to u, v, w, r. The total result is a finite disjunction of conjunctions of identities and their negations.

That this is a correct definition requires a double uniformity. First, for any fixed tuple of argument objects, the value of f may be correctly described as the set of all pairs satisfying the stated Boolean condition, by the earlier observation. But also, the descriptions obtained in this way will be uniformly valid across all argument tuples satisfying the same pattern of (non-)identities: since any two instances of the same pattern can be mapped onto each other by a permutation of individuals.

Further concrete examples of semantic effects of permutation invariance may be found in van Benthem 1986A, 1988B.

Example. New Modes of Categorial Combination.

In the linguistic literature, there has been a distinction between two basic compositional mechanisms in syntax and semantics, namely one of 'subordination', as exemplified in function-argument patterns, versus another, more conjunctive 'coordination'. For instance, many relative clauses and adverbial or adjectival phrases might involve the latter mechanism, rather than the categorial applicative mode.

This is indeed feasible, and one could envisage an extension of our framework allowing a conjunctive mode of type combination as well (cf. Chapter 14). Nevertheless, there is also an alternative analysis, in line with the main topic of this chapter - which is to employ the old functional apparatus, provided with suitable *denotational constraints*. For instance, an extensional conjunctive adjective like "blonde" is a function f in type ((e, t), (e, t)) which satisfies the following conditions:

- it is *continuous* in the above sense ,
- it is *introspective*, in that always $f(X) \subseteq X$.

Now, all such functions can in fact be *represented* in the following form:

$f(X) \;=\; X \cap F$,

where F is the set of all individuals x for which $f(\{x\}) = \{x\}$.

(Cf. van Benthem 1986 for further details.) Thus, the conjunctive behaviour via an underlying predicate becomes explained automatically.

In fact, this analysis of absolute adjectives was encountered already at the end of Chapter 5, in connection with the meaning of predicative "is" . ◆

To conclude this Chapter, here is a final point of interest.

As was observed in Chapter 5, the lambda semantics for categorial combination gives us a hierarchy of fragments of the Lambda Calculus, possessing various natural levels of variable binding, in which we can try to 'position' denotations of linguistic expressions. This may in fact be tied with many of the earlier topics. For instance, transformations going from one type to another may be classified by the complexity of lambda binding involved. Thus, an operation on binary relations like Conversion requires only single binding, while Reflexivization essentially requires double binding, etcetera. For further illustrations, see van Benthem 1989G.

This perspective again raises the issue which special semantic properties would characterize, say, the single-bond fragment Λ corresponding to the Lambek Calculus, lying low in the Hierarchy. One semantic suggestion to this effect, in terms of a special notion of Continuity, was already given in Appendix 1 to Chapter 2. (The paper last cited has another suggestion, in terms of 'Ehrenfeucht Games' with pebbling, which are subjected to a strong restriction on their final phase of comparison between partial isomorphisms constructed by the successive moves of the two players.)

11 Boolean Structure

One pervasive kind of polymorphic structure in natural language is that of Boolean expressions "not", "and", "or" . For instance, in addition to sentential negation, there is also negation of verbs, adjectives or even whole prepositional phrases ("not with a cudgel, but with an axe, the evil deed was done") . And a similar freedom of movement occurs with conjunction as a coordinator of linguistic expressions. This polymorphic behaviour has been studied in Chapters 5, 8 already, with the outcome that the behaviour of negation is accounted for by the basic Lambek Calculus LP , whereas that of conjunction required Contraction, in the calculus LPC .

Boolean structure in natural language has been studied extensively from a linguistic point of view in Keenan & Faltz 1985. What the present Chapter has to add are some general type-theoretic considerations, in line with the main perspective of this Book.

11.1 Functional Completeness

Perhaps the most basic result in elementary logic is the *functional completeness* of the standard Boolean operators. That is,

$\neg , \wedge , \vee$

suffice for defining all binary, and indeed all finitary, truth-functional connectives. When stated in terms of the preceding Chapters, these three first-order objects define all first-order pure t-type objects, using application only. Can this result be extended to higher types of objects in the pure t-hierarchy? For instance, it is an interesting exercise to try and define all objects in the second-order type ((t, t), t) (being 'properties of truth functions') in the same manner. Of course, the minimal 'logical glue' needed here is the full lambda calculus.

As it happens, there is indeed a general positive result.

Theorem. All pure t-type objects are definable by means of application / lambda terms using only the standard Booleans as parameters.

Proof. By induction on t-types, we construct terms τ_u for every object u , having the latter as their constant denotation (under all assignments). In the base domain D_t,

0 is defined by $x_t \wedge \neg x_t$, 1 is defined by $\neg(x_t \wedge \neg x_t)$.

Next, a compound type $a = (a_1, (a_2, \ldots (a_n, t)\ldots))$ may be viewed as being the set of relations on $D_{a1} \times \ldots \times D_{an}$. Now, any such relation is a finite union of singleton cases. Therefore, using *disjunction*, it suffices to define only the latter kind. For this purpose, we need a term of the form

$$\lambda x_{a1} \bullet \ldots \lambda x_{an} \bullet \text{ '}x_{a1} = u_1\text{'} \wedge \ldots \wedge \text{'}x_{an} = u_n\text{'}$$

for any tuple $(u_1, \ldots, u_n)$ in $D_{a1} \times \ldots \times D_{an}$. Thus, it remains to find formulas with one free variable

'$x_{ai} = u_i$' of type t

which are true, under any assignment A, if and only if $A(x_{ai}) = u_i$.

This may in fact be done quite easily using *identity*: but, the point of our theorem is, of course, that we can even do without the latter notion.

- First, if $a_i = t$, then take

  x_t if $u_i = 1$,

  $\neg x_t$ if $u_i = 0$.

- Next, if a_i is itself complex, then proceed as in the following illustration. Let $a_i = (b_1, (b_2, t))$. Now, take the *conjunction* of all terms

  $x_{ai}(\tau_{v1})(\tau_{v2})$ for all v_1, v_2 in the 'argument domains' of a_i such that $u_i(v_1)(v_2)$ holds

  $\neg x_{ai}(\tau_{v1})(\tau_{v2})$ for the other couples in these domains.

  The relevant terms τ exist here by the main inductive hypothesis.

It is easy to check that this procedure produces the correct outcomes in all types. ■

This argument can easily be extended to the 'many-valued' case, where the ground domain D_t has an arbitrary finite number of truth values, provided that suitable first-order parameters are added.

Remark. Monotone Functions.

An important special case, to be introduced more precisely below, is that of *monotone* objects in the pure t-hierarchy. All these can be defined by the above method too, using a simplified construction involving terms of the form

$$\lambda x_{a1} \bullet \ldots \lambda x_{an} \bullet \text{ '}x_{a1} \supseteq u_1\text{'} \wedge \ldots \wedge \text{'}x_{an} \supseteq u_n\text{'}.$$

In the further procedure for defining these inclusion statements, only conjunction and disjunction are needed in combining arguments. Hence we have the following result:

Proposition. The monotone objects in pure t-type domains are precisely those which are lambda / application definable using only $0, 1, \wedge, \vee$ in combination with closed terms denoting objects in lower types.

11.2 Lifting and Lowering

Boolean structure may be created in the process of categorial combination. Probably the most important mathematical notion involved here is that of a *homomorphism* between Boolean algebras, respecting all Boolean operations. And, as was shown in Chapter 10, lifting of individuals u , in type e , created homomorphisms in the algebra of noun phrase denotations, in type ((e, t), t) , via the Montagovian recipe

$\lambda x_{(e, t)} \cdot x(u_e)$.

In fact, the same holds for arbitrary liftings according to the pattern $a \Rightarrow ((a, b), b)$.

But, there is a reverse side to this coin too. Homomorphisms in one type will actually be just ordinary citizens in a lower class.

Proposition. Boolean homomorphisms in type ((a, t), (b, t)) correspond canonically, and uniquely, to arbitrary objects in type (b, a) .

Proof. Let f be a homomorphism in the category ((a, t), (b, t)) . As f respects arbitrary Boolean sums of its arguments, its image for any value u in $D_{(a, t)}$ (viewed, for convenience, as a subset of D_a) is equal to $\cup \{ f(\{x\}) \mid x \in X \}$. Moreover, since homomorphisms send Boolean units to units, $f(D_a) = D_b$. Therefore, D_b must equal $\cup \{ f(\{x\}) \mid x \in D_a \}$. Now, all these images of atoms are disjoint subsets of D_b : 'disjointness' being another relation preserved by Boolean homomorphisms. (Some $f(\{x\})$ may actually be empty.) Hence, a reverse function f^* from D_b to D_a may be defined unequivocally, by setting

$f^*(y)$ is that x in D_a such that $y \in f(\{x\})$.

Then, in general, the following reduction holds, for all $X \subseteq D_a$:

$$f(X) = \{ y \in D_b \mid f^*(y) \in X \} \qquad (= f^{*-1}[X]) .$$

So, f can be retrieved from f^* .

Conversely, any function g from D_b to D_a will induce a unique homomorphism g^+ from $D_{(a, t)}$ to $D_{(b, t)}$ with this scheme. Note that this is nothing other than the earlier Geach recipe of function composition for the transition

$$(b, a) \Rightarrow ((a, t), (b, t)) , \qquad \text{being} \qquad \lambda x_{(a, t)} \bullet \lambda y_b \bullet x(u_{(b, a)}(y)) .$$

As a special case, there will also be a correspondence between homomorphisms in type $((a, t), t)$ and arbitrary objects in type a, this time via the earlier Montagovian recipe of lifting.

The general virtue of the preceding result is that it provides a warning: much Boolean structure reported in logical semantics may be no more than an artefact of our categorial apparatus. But, there are also more concrete applications.

Example. Deflating Homomorphisms.
To begin with, the above observation explains the dual role of individuals in Keenan & Faltz 1985: once as ordinary citizens of D_e, and once as homomorphisms in the noun phrase domain $D_{((e, t), t)}$. And a similar observation holds for properties of individuals, in type $D_{(e, t)}$, which become precisely the homomorphisms in the type of intransitive verb denotations, being $(((e, t), t), t)$ (compare the relevant type change in Chapter 4).

Here is another kind of question. Frans Zwarts has asked why there seem to be no determiner expressions denoting homomorphisms. Here, the above analysis provides an explanation. Such homomorphisms would correspond to ordinary objects living in type $((e, t), e)$. But, no such 'choice functions' appear among the basic lexical items of natural language.

Moreover, this analysis can be made a little more pregnant. The above argument may be directly refined to show that, in fact,

There exists a canonical bijection between logical homomorphisms in type $((a, t), (b, t))$ and arbitrary logical objects in type (b, a).

Here, *logicality* stands for the permutation invariance of Chapters 2 and 10. Thus, homomorphic logical determiners would have to correspond to logical objects in the type $((e, t), e)$. But of the latter, there are none (unless one happens to live in a trivial hierarchy starting from just one individual object).

Finally, consider the earlier reflexivizer "self" (cf. Chapters 5, 10), whose semantic recipe was

$$\lambda R_{(e, (e, t))} \bullet \lambda x_e \bullet R(x)(x) .$$

As was stressed in Keenan & Faltz 1985, this is a homomorphism - whose type is $((e, (e, t)), (e, t))$, or equivalently, $((e \bullet e, t), (e, t))$ - which is obviously logical as well. And, these two semantic features determine reflexivization completely:

Proposition. Reflexivization is the *only* logical homomorphism occurring in type $((e, (e, t)), (e, t))$.

Proof. The reason is as before. Such logical homomorphisms correspond uniquely to logical items in type $(e, e \bullet e)$. And of the latter, there is only one, being the 'duplicator'

$$\lambda x_e \bullet \langle x, x \rangle .$$

11.3 Monotonicity

Boolean structure is tied up with the notion of implication, or inference, between linguistic expressions. And again, 'inference' occurs quite freely among linguistic expressions, not being confined to sentences:

"walk slowly" implies "walk" ,
"with a bread knife and without compunction" implies
"with a knife" as well as "without compunction" .

The appropriate general notion of inference here is that of an *inclusion* $\subseteq$ on all Boolean type domains:

General Boolean Inclusion

on D_t , $\subseteq$ is numerical $\leq$,
on D_e , $\subseteq$ is the relation of identity ,
on $D_{(a, b)}$, $f \subseteq g$ iff, for all $x \in D_a$, $f(x) \subseteq g(x)$ in D_b .

This notion reduces to what one would expect in concrete cases. For instance, on sets of individuals, in type (e, t) , it amounts to set-theoretic inclusion.

Now, a well-known semantic property expresses a certain sensitivity to this kind of inference. This may be illustrated by the case of quantifier expressions. For instance, a determiner like "some" has the following behaviour:

"some" AB , $A \subseteq A^+$, $B \subseteq B^+$ imply "some" A^+B^+ .

Likewise, "all" has a combination of 'upward' and 'downward' inference:

"all" AB , $A^- \subseteq A$, $B \subseteq B^+$ imply "all" A^-B^+ .

(See van Benthem 1986, 1987C for a more extensive logical theory of such 'monotone' determiners.) Technically, we define:

Monotonicity

f in type (a, b) is *upward monotone* if,

for all $x \subseteq y$ in D_a , $f(x) \subseteq f(y)$ in D_b .

f is *downward monotone* if the correspondence is inverse:

for all $x \subseteq y$ in D_a , $f(x) \supseteq f(y)$ in D_b .

And these notions can be extended in an obvious manner to deal with simultaneous monotonicity in a number of stacked arguments (as is the case already with the above determiners).

Remark. Triviality for Individual Types.

Note that monotonicity only has a bite for functions having arguments in Boolean types: since the notion of inclusion reduces to *identity* on non-Boolean types of the form

$(a_1, (..., (a_n, e)...))$.

Monotonicity arises very naturally in connection with polymorphism. Indeed, one reasonable constraint on possible type changes in our categorial apparatus might seem to be the following:

" Semantic recipes τ for taking objects of type a to new objects of type b should at least preserve inferential relationships. That is,

if $u \subseteq v$ in D_a , then we should have ' $\tau(u) \subseteq \tau(v)$ ' in D_b . "

In other words, what we are asking is that the transition function $\lambda u \cdot \tau(u)$ corresponding to the semantic recipe τ be upward *monotone*.

This form of preservation is not guaranteed, however, by the categorial mechanism of Chapters 5 and following. As may be computed easily, the recipe for a Geach transition

from (a, b) to ((c, a), (c, b)) ,

has it - whereas that for the Montague transition

from a to ((a, b), b)

lacks it in general. For a counter-example to the latter, consider the case of $a = (e, t)$ and $b = t$. Even when $U \subseteq V$, there is no reason why $\{\mathbb{A} \mid U \in \mathbb{A}\}$ should be a subfamily of $\{\mathbb{A} \mid V \in \mathbb{A}\}$: witness the singleton family $\{U\}$.

What is the reason for this difference? On closer inspection, it has to do with the *syntactic form* of the two lambda recipes involved. That for the Geach transition reads

$\lambda x_{(c, a)} \bullet \lambda y_c \bullet u_{(a, b)}(x(y))$,

where the parameter '$u_{(a, b)}$' occurs in what may be called 'positive' syntactic position, namely the function head. By contrast, the recipe for the Montague transition, being

$\lambda x_{(a, b)} \bullet x(u_a)$,

has its parameter occurring in an argument position.

Here is a more precise definition of the relevant notion of *positive occurrence* of a variable in a lambda term:

- x occurs positively in x itself, and in no other variable,
- if an occurrence of x is positive in τ, then it also positive in compounds $\tau(\sigma)$ and $\lambda y \bullet \tau$ (with the bound variable y distinct from x).

What this amounts to is that positive occurrences must lie at the syntactic head position already described in the proof of the Finiteness Theorem for Λ in Chapter 9.

Now, it is easy to show by induction that:

Proposition. If the variable u occurs positively in the lambda term τ , then the corresponding function $\lambda u \bullet \tau(u)$ is upward monotone.

But, the converse does not hold for the full lambda calculus. For a start, there are superficial counter-examples to this, such as

$(\lambda x_{(e, t)} \bullet x(v_e))\,(u_{(e, t)})$,

which is monotone in $u_{(e, t)}$. But here, the term becomes equivalent, after lambda conversion, to a term with $u_{(e, t)}$ indeed in positive position. More serious, however, is the following case.

Example. Monotonicity Without Positive Occurrence.
The following term τ is monotone in the parameter $x_{(t, t)}$, without having a definition in which this parameter occurs only positively:

$$\tau = x_{(t, t)}(x_{(t, t)}(y_t)) .$$

As to the first assertion, note that the domain $D_{(t, t)}$ is a four-element Boolean algebra with atoms $\lambda x_t \cdot x_t$ ('identity') and $\lambda x_t \cdot \neg x_t$ ('complement'). Now, it suffices to check the cases where, for some fixed y, $\tau(x)$ has value 1, replacing x by some $\subseteq$-larger function. And indeed, there, τ always retains the value 1.

The second assertion is by analysis of normal forms for the above expression, as employed already in Chapter 9. (Note that no conversion leading to normal form will affect positive occurrences.) In this case, the normal form can only involve the types t, (t, t). Moreover, it cannot begin with a lambda: whence it must be an application $\alpha(\beta)$. Here, α cannot be a lambda term, and neither can it be an application: since the latter would have type t. Therefore, it is a variable of type (t, t) : i.e., it must be $x_{(t, t)}$. Repeating this argument for the term β, and continuing, one finds that the normal form must look like this:

'some sequence of $x_{(t, t)}$, followed by one final y_t'.

Now, if $x_{(t, t)}$ were to occur only positively here, then just one shape would qualify, namely

$$x_{(t, t)}(y_t) .$$

And, the latter is clearly not equivalent to the above term τ. (For a counter-example, put in the complement function.)

This example illustrates the difficulties in obtaing a syntactic 'Preservation Theorem' for a semantic notion in higher-order logics.

Nevertheless, here is also where the earlier idea of a 'Semantic Hierarchy' of *fragments* of the full Lambda Calculus returns: certain natural fragments turn out to be better-behaved in this respect.

Theorem. A lambda term τ in the Lambek fragment Λ is semantically monotone with respect to some parameter u_a of a Boolean type a if and only if τ is logically equivalent to a term having u in positive head position.

Proof. The crux is to prove the assertion from left to right. So, consider any lambda normal form for τ. If the occurrence of u_a (which must be unique) is not positive there, then the following procedure will find a counter-example to semantic monotonicity for τ.

We define two variable assignments A_1, A_2, starting with

$$A_1(u_a) = 0_a , \qquad A_2(u_a) = 1_a ,$$

where $0_a, 1_a$ are the constant zero- and one-functions in the type $a = (a_1, ..., (a_n, t)...)$. (E.g., $0_a = \lambda x_{a1} \bullet ... \lambda x_{an} \bullet 0_t$.)

Consequently, $A_1(u_a) \subseteq A_2(u_a)$.

Now, here is the idea of the construction to follow. We extend A_1, A_2 progressively, making the same choices on further free variables, working upward in the construction tree of τ. As u_a does not occur positively, some first governing functor must be encountered, where we shall disturb the inclusion between the A_1- and A_2-values for the term processed so far. This non-inclusion is then preserved until the whole τ is reached: and we have the desired counter-example.

The following facts about normal forms may be recalled:

- each application $\alpha(\beta)$ is of the form
  'leading variable, followed by a string of suitable argument terms, of which β is the last',
- each subterm α occurs in a maximal *context* of the form
  $$\lambda y_1 \bullet ... \lambda y_k \bullet [\alpha(\alpha_1)...(\alpha_m)] ,$$
  which is either the whole term τ or the argument of some application.

Now, by the assignment made to u_a, its context γ will also denote the constant zero- or one-function in its type, say b - where A_1, A_2 may be extended arbitrarily on the relevant free variables. Next, a governing functor is encountered, with some leading variable v.

Case 1: the type of v ends in t.

Then, assign any function to v which yields value 1 for all tuples of arguments having 0_b at the position of γ, and value 0 for all others.

Case 2: the type of v ends in e.

Then assign any function which marks 0_b at the position of γ by some arbitrary value $e_1 \in D_e$, and 1_b by some e_2 distinct from e_1.

Other free variables within the scope of v may be given arbitrary values.

As a result, the A_1-value of the v-term is no longer included in its A_2-value. And the same will hold for the full context of v, which may have some further lambdas in front. (By the definition of inclusion for functions, $A_1(\lambda y \bullet \sigma) \subseteq A_2(\lambda y \bullet \sigma)$ would imply $A_1(\sigma) \subseteq A_2(\sigma)$.) Note that the domain of the assignment may be taken to decrease in this step.

Finally, if there are no further higher contexts, the non-inclusion can be preserved by making appropriate stipulations for leading variables, using values 0, 1, e_1, e_2: but now, without the earlier reversal between the Boolean values.

This procedure relies heavily on the *single bind* property of the Lambek fragment. For, the latter ensures that no clashes can occur between assignments made at some stage and those necessary in a wider context: as no free variable occurs twice. ■

The preceding result is reminiscent of Lyndon's well-known preservation theorem for first-order predicate logic, stating that a formula ϕ containing a predicate letter P is semantically monotone with respect to increases in the denotation of that predicate if and only if it is logically equivalent to a form in which P occurs only positively. But there is also a difference. In predicate logic, formulas may contain certain *constants*, namely Boolean connectives and quantifiers, whose meanings play a role in the definition of positive (and negative!) syntactic occurrence. Consequently, e.g., P is allowed inside argument positions of Boolean expressions, without thereby incurring non-monotonicity.

What this suggests for the present case is the following. We now allow constant parameters in our terms, and define 'positive' and 'negative' occurrences in the obvious way, counting argument positions of upward monotone parameters positive, and those of downward monotone parameters negative. We conjecture that the above preservation theorem will go through for Lambek terms with Boolean constants. (For a mathematical use of such a formalism, see Mundici 1989 on definability by monotone Boolean formulas without repeated variables.)

There are only few Lyndon-type results for higher formalisms beyond first-order predicate logic. Of course, the first-order analysis itself generalizes to even the full type theory L_ω, when the latter is taken in its many-sorted first-order guise. A less routine example is the preservation result in Makovsky & Tulipani 1977 for monotonic formulas in a first-order logic extended with generalized quantifiers. As for the full lambda calculus, the earlier 'functional completeness' result for monotone objects in a pure t-type hierarchy might be characteristic for necessary and sufficient shapes of definition.

This topic will acquire a more practical motivation in the next Section.

11.4 Inference and Type Change

It is one thing to propose a general theory of type change and categorial combination. But, there are also other general syntactic and semantic mechanisms in natural language. And what one ought to investigate as well is the *interaction* of these various phenomena. Here is one prominent example:

How does categorial combination interact,
or interfere, with the process of *inference* ?

We shall approach this issue by means of the earlier notion of monotonicity.

The above analysis provides us with the tools to actually create a *combined calculus* of syntactic parsing and logical inference, computing eventual inferential behaviour of the expression being analyzed. This calculus is best developed through a series of illustrations.

Example. Monotonicity Marking.

1 In the basic sentence pattern

"Bela dances",

the predicate "dances" occurs in upward monotone position. This may be seen by replacing it by a weaker predicate. For instance, a valid consequence is

"Bela moves" .

The explanation for this lies in the monotonicity behaviour of the basic pattern of categorial analysis:

- In function / argument combinations,
  the functor always occurs positively.

This may be marked in derivation trees:

$$\frac{e \qquad (e,t)^{+}}{t}\ .$$

The validity of this principle shows in the following contrast. In cases where "dances" is no longer the functor, monotonicity need not hold: witness the sentence

"No princess dances" ,

which does not imply that no princess moves.

2 What is the monotonicity behaviour of the other main categorial rule, being function abstraction? For instance, in a verb phrase

"find a unicorn" ,

the categorial derivation was like this:

$$\dfrac{\dfrac{\dfrac{e^1 \quad (e, (e, t))}{(e, t)} \quad ((e, t), t)}{t}\ \text{COND, withdrawing 1}}{(e, t)}\ .$$

Intuitively, the noun phrase is the functor here, which occurs in positive position. And this is borne out by the predictions: this verb phrase implies

"find an animal" .

Here, another general marking principle is at work:

- The body of a function abstraction
  is always positive in the abstract.

This may be displayed as follows:

$$\dfrac{\dfrac{\dfrac{e \quad (e, (e, t))^{+}}{(e, t)} \quad ((e, t), t)^{+}}{t^{+}}}{(e, t)}\ .$$

The reason why this makes the occurrence of the final noun phrase positive also involves another general principle - this time, concerning monotonicity marking along branches of derivation trees:

- A lexical item occurs in a marked position if it is connected
  to the conclusion at the root via an unbroken sequence of markings.

3 Still, the above analysis cannot be all there is to the computation of monotonicity effects. For instance, already the first example presented also shows upward monotonicity for the first argument:

"Bela dances" implies

"Bela or Anna dances" .

The reason here is as follows. If "Bela or Anna" is taken to be a new noun phrase, then the Boolean conjunction must have connected two lifted individuals in the Boolean type ((e, t), t) . Therefore, on this view, the subject of the sentence was the functor after all, via a Montague lifting:

$$\dfrac{\dfrac{\dfrac{e \qquad (e, t)^1}{t}\ \text{COND, withdrawing 1}}{((e, t), t)^{+}} \qquad (e, t)}{t\,.}$$

But then, how to explain the earlier monotone behaviour of the verb phrase? What this requires is a further rule of marking, this time *inside* functional types created by the rule of Conditionalization:

- If the type occurrence withdrawn occurred positively in the conclusion, then give the resulting conditional type a positive marking on its argument.

The point of this lies in the following further rule:

- In function applications where the functor has a positively marked argument, the argument supplied is counted positive.

In the above case, this works out as follows:

$$\dfrac{\dfrac{\dfrac{e \qquad (e, t)^{+}}{t}\ \text{COND}}{((e, t)_{+}, t)^{+}} \qquad (e, t)^{+}\ (!)}{t\,.}$$

4 But, this still leaves an earlier case of monotonicity unaccounted for. For, the verb "find" in "find a unicorn" occurs positively too, witness the valid inference to

"lose-or-find a unicorn" .

An explanation of this fact involves, not the general monotonicity effects of the categorial calculus as such, but a further phenomenon: namely , the occurrence of

- Special monotonicity effects for certain lexical items.

In the preceding case, the lexical source is the determiner "a" . (Observe that, by contrast, "find no unicorn" does not imply "lose-or-find no unicorn" .) Such monotonicity effects too may be encoded in the relevant lexical types. For instance, for the doubly upward monotone determiner "a" , this will look as follows:

$((e, t)_{+}, ((e, t)_{+}, t)$;

whereas, e.g., the doubly downward monotone "no" has

$((e, t)_{-}, ((e, t)_{-}, t)$.

Then, the full analysis tree becomes this:

$$\frac{\dfrac{e \qquad (e, (e, t))^{+}}{(e, t)^{+(!)}} \qquad \dfrac{((e, t)_{+}, ((e, t)_{+}, t)^{+} \qquad (e, t)^{+ (!)}}{((e, t)_{+}, t)^{+}}}{t^{+}}$$

(e, t) .

Of course, since lexical items can also introduce negative marking, the above rule of computation has to be refined, adding one final feature, namely

- The obvious algebraic rules of calculation along branches

  ++ = + +- = - -+ = - -- = + .

5 As a final example, consider the case of a complex subject noun phrase with an adjective, as in

"every blonde player" .

Here, the analysis will run as follows:

$$\frac{((e, t)_{-}, ((e, t)_{+}, t)^{+} \qquad \dfrac{((e, t), (e, t))^{+} \qquad (e, t)}{(e, t)^{-}}}{((e, t)^{+}, t)}$$

What this analysis produces is the following final marking in the noun phrase:

every + blonde - player .

This supports inferences to

"every or almost every blonde player" ,

"every very blonde player" .

What it does not support, however, is the perhaps plausible downward inference to

"every blonde master-player" .

For, there is no monotonicity marking on the common noun: its sequence of markings having been interrupted at the top.

This is in fact how things should be in general. For instance, when the adjective is a context-dependent one, such as "young" , there is no inference from "player" , either way: neither downward to "every young master-player" (since youthfulness for master players may have different standards than for players in general) , nor upward to "every young player-or-watcher" .

Why then the downward monotonicity in the former case? This is rather an effect of the absolute adjective "blonde" : its lexical marking should have been

$((e, t)_{+}, (e, t))$,

after which the previous method of computation would have given the common noun its intended negative marking. ∎

The preceding example has shown how a monotonicity calculus can be set up, on top of a parsing mechanism. Of course, many further lexical items will come with special markings, notably the Booleans themselves.

Another use of such a system is the inferential comparison of different readings for expressions, as analyzed in Chapter 9.

Example. Inferences from Different Scope Readings.
The above calculus will predict the following outcome for the small scope direct object reading of the sentence

"No priest knows a way out":

No + priest - knows - a - (way out) - ,

but another one, also correct, for the wide scope reading:

No + priest - knows - a + (way out) + .

●

One interesting practical question is which monotonicity markings are common to all possible readings for an expression. Indeed, Sanchez 1990 makes an attempt at formulating a calculus of monotonicity inference that can already operate without having to resolve all scope ambiguities beforehand, i.e., without always supplying full categorial derivations. This seems realistic, in that human linguistic information processing can take place at various levels of partial resolution of ambiguity. (Fenstad et al. 1987 provide another logical implementation for this idea.)

Actually, the linguistic side of things is still more subtle here. It is well-known that monotonicity is involved in the distribution of so-called 'polarity items' in natural language. For instance, roughly speaking, a negative polarity item like "any" wants to occur in negative syntactic contexts, whereas a positive polarity item like "some" needs positive occurrence. Thus, the above calculus of monotone occurrence is also relevant to the distribution of such expressions. Now, Frans Zwarts has pointed out that the cognitive function of polarity expressions may be taken to consist in a *reduction of ambiguity*. If one says that

"No priest knows any way out" ,

the negatively polar subexpression "any way out" should occur negatively, and hence only the above small scope reading qualifies. Conversely, with a positively polar direct object and a negative subject, as in "No priest believes some miracles" ,
the wide scope reading is obligatory for the direct object.

Besides monotonicity, there are other important inferential mechanisms in natural language too. Van Benthem 1987B in fact presents a view of natural language as a conglomerate of different inferential systems, having varying complexity, and being located at different levels of linguistic analysis. It would be of interest to study their interaction with categorial combination too.

A case in point is the earlier notion of Conservativity for determiners, and its polymorphic spreading. For instance, in the work of Charles Saunders Peirce, Monotonicity and Conservativity were key principles driving what may be called a "natural logic" of mental representational forms (cf. Sanchez Valencia 1989). In a sense, of course, the analysis of general Domain Restriction given in Chapter 10 is an answer to this particular question: although it remains to provide a simple efficient algorithm for calculating what restricts what.

There is more to the design of natural logic than what has been observed so far. For instance, the above account would have to be supplemented with an account of anaphoric principles identifying certain denotations of constituents. But the present Chapter may at least have conveyed some idea of its Boolean aspects.

V VARIATIONS AND EXTENSIONS

The framework of the previous Parts has been that of a standard extensional Type Theory, employing only two truth value types and a ground type of individual objects. Evidently, many, if not most, of the results obtained so far do not depend essentially on this assumption: and further basic types could be introduced without changing the earlier theory. One important example is that of *intensional* types, arising upon addition of a new base domain of what may be called "possible worlds", "situations" or "states" . The resulting structure will be investigated in Chapter 12.

Note that, in fact, it is also possible to *re-interpret* the old individuals in various ways: for instance, as groups or chunks of some underlying stuff - which makes for additional flexibility in the application of our earlier ideas. (See van Benthem 1989G for further elaboration of this line of thought.) Moreover, we can change the number of inhabitants, even of the old truth value domain, shifting to a three- or four-valued logic (cf. Langholm 1987).

Another source of variation lies in the models themselves. As was noticed in Chapter 2, there also exist 'general models' for our language, rather than just standard function hierarchies. Up till now, this direction has not really been pursued here, for lack of significant applications. But, passing on to such general models might be inevitable, once certain new empirical phenomena are taken into consideration. For instance, the linguistic process of *nominalization* turns verbs into objects:

"mov-*ing*" has the following structure in categorial morphology

$(e, t) \quad ((e, t), e) \;\Rightarrow\; e$.

Now, many authors believe that the nominalization process is *one-to-one*. But what that means is that the function [[-ing]] should be an injection from domains of type (e, t) to those of type e . And that is impossible, because the cardinality of power sets exceeds that of their ground domain. Thus, such a view can only be implemented, if at all, on general models allowing the predicate domain to be equinumerous with that of the individuals.

Further variations rather concern the mechanism of categorial combination itself. For instance, instead of having functional types, there may be occasions where the use of *relational types* has advantages. These will be formed from one ground type e using formation of finite sequences $(a_1, ..., a_n)$ denoting n-ary relations over the Cartesian product of type domains $D_{a1} \times ... \times D_{an}$. (See Doets & van Benthem 1983, Muskens 1989 for technical details and illustrations.)

Perhaps most conspicuous among the latter kind of variations is the introduction of polymorphism, not just into the process of derivation, but also into the basic type assignments themselves, through the device of introducing *type variables*, for which further substitutions can be made in the process of parsing-cum-interpretation. The resulting logical systems will be studied in Chapter 13.

12 Intensionality

With many linguistic expressions, the picture of individuals in one situation is not sufficient for an adequate semantics. Notably, temporal constructions presuppose histories of changing situations over Time, while modal expressions require a space of logically possible situations, in addition to the actual one. And similar 'ranges' of situations occur in recent more 'dynamic' semantics for computer programs (which change states of some computing device) or epistemic instructions (which modify knowledge states) - as will be amply demonstrated in Part VI of these Lectures.

The general picture here is that of a base domain D_s of intensional 'indices', usually provided with some relational pattern, such as 'precedence' in the case of Time, 'alternativeness' with modalities, or 'reachability' for programs. In this chapter, we shall survey a few logical themes that arise in this setting.

12.1 The Mechanism of Intensional Interpretation

As a general strategy of intensionalization, one can re-interpret the former type t as standing for, not truth values, but *propositions*. By the familiar identification of the latter with sets of indices, then, these become objects of type (s, t) [with 't' now taken again in its old truth value sense] . Thus, types formerly assigned to expressions undergo a uniform intensionalization *, via the replacement

$$t^* = (s, t) .$$

This process is studied in van Benthem 1987D - which shows, amongst others, that this rule produces essentially the same outcomes as Montague's well-known principle of intensionalization for initial syntactic types. Moreover, no new syntactic structures are created in this way:

For all extensional {e, t}-types X, a ,

$X \Rightarrow a$ is L-derivable if and only if $(X)^* \Rightarrow a^*$ is L-derivable.

A more sophisticated calculus of type change-cum-intensionalization, designed to alleviate the combinatorial extravagances of Montague Grammar, has been proposed in Hendriks 1989.

In a sense, the operation * is itself a form of type change. But, there does not seem to be any uniform meaning shift associated with it. Some extensional items become indeed 'intensional' by mere derivational polymorphism. An example is Boolean negation. For, $(t, t)^*$ equals $((s, t), (s, t))$, which can be derived via the Geach rule, with a meaning

$$\lambda x_{(s, t)} \bullet \lambda y_s \bullet \neg_{(t, t)}(x(y)) .$$

Formally, of course, this is identical to the earlier shift from sentential negation to predicate negation.

In general, however, there may be several options for the meaning of items in intensional contexts. For instance, the determiner "every" has its former extensional type $((e, t), ((e, t), t)))$ intensionalized to

$$((e, (s, t)), ((e, (s, t)), (s, t))) .$$

One conservative choice here is to employ a standard categorial derivation from the former to the latter, which can be done in the calculus LPC with Contraction. Its associated meaning will be the following natural lifting:

$$\lambda x_{(e, (s, t))} \bullet \lambda y_{(e, (s, t))} \bullet \lambda z_s \bullet \mathrm{EVERY}_{((e, t), ((e, t), t)))}(x(z))(y(z)) .$$

But, there is also another, truly intensional, 'law-like' universal quantifier, stating that the relevant inclusion has to hold in all situations, not just the current one. And finally, of course, certain expressions, such as modal operators, only make sense at all in a truly intensional type, since they refer essentially to global properties of the whole pattern D_s of states.

There is some further fine-structure to this picture. First, as to an adequate amount of polymorphism for intensional types, it seems that the latter behave a bit more freely than the extensional types encountered so far. In particular, the use of Contraction seems reasonable in many cases. Therefore, Prijatelj 1989 studies an extended calculus LPC_S, being the calculus LP with Contraction added for the primitive type s only. Its proof theory is like that of LPC (cf. Chapter 8), but simpler in some interesting ways, due to the restriction on Contraction. Thus, in the Categorial Hierarchy, the choice of structural rules need not be all-or-nothing matter: they may also be added 'locally', for specific kinds of type only.

An interesting alternative may be found in Morrill 1989, who uses a *modal operator* $\Box a$ to encode intensional types (s, a) , and then exploits such analogies as that between a Geach transition $(a, b) \Rightarrow ((s, a), (s, b))$ and the well-known inference rule of 'Modal Distribution':

from $\alpha \rightarrow \beta$ to $\Box\alpha \rightarrow \Box\beta$.

Then, the above intensionalization * will correspond to the modal rule of Necessitation:

if α is a theorem, then so is $\Box\alpha$.

Thus, one obtains what may be called a 'modal logic' of intensionalization, which can be regarded as an occurrence-based variant of standard modal logic. The system also demonstrates an earlier theme (cf. Chapter 6), namely the freedom to introduce further logical constants into the categorial machinery.

The semantics for such systems involves only a certain *fragment* of the full lambda calculus over three initial types. This is reminiscent of a theme already raised in Gallin 1975, in connection with Montague Grammar. The intensional types needed for the purposes of Montague Grammar, formed only a fragment of the full intensional type theory - and one may study its semantic peculiarities.

Our next topic illustrates that certain surprises may lie in store in an intensional framework. For instance, here is a perhaps unexpected question.

12.2 What is Extensionality?

How can we recognize the old extensional objects in a wider intensional setting? This may seem asking for the obvious - but in fact, it is not so clear what a general answer should be. To be sure, there are some standard examples in the literature:

1. Extensional *predicates*, in type (s, (e, t)) , are those functions on the possible worlds domain D_s whose value does not vary over different worlds. Thus, they denote essentially one rigid predicate of individuals in the domain $D_{(e, t)}$.
2. Extensional *propositional operators*, in type ((s, t), (s, t)) , are those functions which are directly derived from some truth function by a simple lambda recipe, as with the above negation.
3. Extensional *adjectives*, in type ((s, (e, t)), (s, (e, t))) , are those functions which compute their values on some intensional predicate locally in each world by using only the extension of the predicate at that world.

(Contrast the extensional adjective "blue-eyed" with the genuinely intensional cases of "future" or "imaginary".)

But now, what is the *general* underlying notion?

Here, the considerations of earlier Chapters may be used. For a start, a reasonable 'linguistic' proposal for defining extensionality would be this:

I An object in the $\{e, t, s\}$ - hierarchy is *extensional*$_1$ if it can be defined by some lambda term employing only parameters from the pure $\{e, t\}$ - subhierarchy.

This proposal fits the above examples:

1 $\lambda x_s \bullet P_{(e, t)}$

2 $\lambda x_{(s, t)} \bullet \lambda y_s \bullet \neg_{(t, t)}(x(y))$

3 $\lambda x_{(s, (e, t))} \bullet \lambda y_s \bullet ADJ_{((e, t), (e, t))}(x(y))$.

Moreover, it allows us to *classify* all extensional items of a certain kind, using the methods of Chapter 9. For instance, by inspection of lambda normal forms, it may be shown that

All extensional adjectives must be of the above form 3 .

But, a more structural answer is possible too. It seems reasonable to characterize the extensional objects mathematically as those items in the $\{e, t, s\}$ - hierarchy which do not care about the underlying s-structure (cf. Chapters 2, 10):

II An object is *extensional*$_2$ if it is invariant for all permutations of the base domain D_s , lifted canonically to all other type domains.

Or, even stronger than this, one might argue for invariance with respect to binary *relations* over indices in D_s , as defined in Chapter 2 : which gives us, say, *extensionality*$*$. Here is an illustration of the latter notion.

Example. Extensional$*$ Adjectives.

Any relation invariant adjective is 'local' , in the sense that,

If P, Q are intensional predicates of type $(s, (e, t))$,
and x, y are indices such that $P(x) = Q(y)$,
then also $A(P)(x) = A(Q)(y)$.

To see this, define a binary relation R on D_s as follows:

$$R = \{ (x, y) \} .$$

Then, $P R Q$ iff $P(x) = Q(y)$ (by the canonical lifting of R to this type). Hence, as A is relation invariant, $A(P) R A(Q)$: i.e., $A(P)(x) = A(Q)(y)$.

Finally, from local adjectives A , it is easy to extract an object $A^{\#}$ in type ((e, t), (e, t)) such that

$$A(P) = \lambda x_s \bullet A^{\#}(P(x)) .$$

◆

By the earlier results of Chapter 2, extensionality$_1$ implies extensionality$_2$, but not vice versa.

Finally, here are some logical questions about these notions. First,

- Is it *decidable* whether a given object is extensional?

Even in a hierarchy starting from finite base domains, the answer to this is unknown.

Next, there is a more modest issue:

- In which types will there be extensional items at all?

This may be decided via the following recursive characterization.

Proposition. The types in which extensional items occur are precisely those of the following forms

$$a = (a_1, ..., (a_n, x)...)) ,$$

with a final type x satisfying one of the following clauses:

$x = t$

$x = e$

$x = s$, and at least one a_i does not contain extensional items.

Proof. This assertion may be proved by induction on types, like the analogous characterization of all extensional types containing logical items which is given in van Benthem 1988B.

First, any type ending in t contains extensional items: for instance, the constant function assigning value 1 . Likewise, types ending in e contain the extensional constant function assigning some fixed individual in all cases.

As for the third case, suppose that all argument types a_i contained extensional items, say, $u_1, ..., u_n$. Then, no object f in type a can be extensional. For, consider its value on the n-tuple of arguments $(u_1, ..., u_n)$: say, x . Now, let π be any permutation of D_s shifting x to some other index y . Then,

$\pi(f)(u_1, ..., u_n) = \pi(f)(\pi(u_1), ..., \pi(u_n))$ (by the extensionality of these arguments) $= \pi(f(u_1, ..., u_n))$ (by the definition of $\pi(f)$) $=$

$\pi(x) = y \neq x = f(u_1, ..., u_n)$.

Therefore, $\pi(f) \neq f$, and hence, f is not permutation invariant in the required sense for extensionality.

Now, conversely, suppose that some argument type a_i contains no extensional items. Then, by the inductive hypothesis, a_i must have a final type s , wile each of its argument types (if any) contains some extensional item. For the sake of illustration, assume that there is just one argument type, containing some extensional object u . Then, the following scheme defines an extensional object in the original type a :

$\lambda x_{a1} \bullet ... \lambda x_{an} \bullet x_{ai}(u)$.

◆

Next, we turn to a consideration of some of the earlier basic semantic properties in an intensional setting.

12.3 Denotational Constraints

We shall consider two basic semantic properties introduced in earlier Chapters, namely logicality and monotonicity.

The notion of *permutation invariance*, originally introduced as a feature of logical objects, was mentioned already in connection with extensionality. But, as a general constraint on intensional operators, it has little to recommend itself. For instance, van Benthem (1986A, Chapter 3) shows that

> The permutation invariant operators on sets of points
> are precisely the set-theoretic Boolean ones.

Nevertheless, in certain richer settings, permutation invariance does provide a significant condition on 'logicality' of intensional operators: in particular, in a 'dynamic' logic of programming and control.

Example. Logical Operators on Programs.

Some basic operations on programs (cf. Chapters 16, 17) are *Composition* (;) , Boolean *Choice* (IF THEN ELSE) or *Iteration* (WHILE DO). With programs viewed as denoting transition functions on computer states, in type (s, s) , the latter have types in a pure {s, t} type theory:

Composition	((s, s), ((s, s), (s, s)))
Choice	((s, t), ((s, s), ((s, s), (s, s))))
Iteration	((s, t), ((s, s), (s, s))) .

All of these operations are again permutation invariant, with respect to permutations of the state domain D_s . And conversely, one can also enumerate possible logical programming constructs via this notion, using earlier techniques, weeding out implausible cases via suitable additional denotational constraints.

Moreover, there is also polymorphism and type change here. For instance, conditional Choice has a denotational recipe which is really just an inflated version of a simpler object in the lower type (t, (s, (s, s))) :

$$\lambda x_{(s,\,t)}\bullet\, \lambda y_{(s,\,s)}\bullet\, \lambda z_{(s,\,s)}\bullet\, \lambda u_s\bullet\ \iota v\bullet\ (\ (x(u) \wedge v=y(u)) \vee (\neg x(u) \wedge v=z(u))\ )\ .$$

[Here, the ι stands for the definite description operator.]

Another example would be the step from binary composition of unary functions to more general forms of composition encountered in practice, such as

'Compose binary f with two functions of the same arity' .

The meaning of such more general compositions can be derived via a straightforward derivation in the Lambek Calculus with Contraction.

Finally, in order to account for the phenomenon of *indeterminism*, the type of programs may be changed to that of relations between states, that is (s, (s, t)) . The above programming constructs then also adapt their types. And the systematic study of important programming operations will become like that of Relational Algebra (cf. Chapter 16). For instance, many examples found in practice, such as Composition or Indeterministic Choice are *continuous* in the sense of Chapter 10. By the enumeration given there, then, we may classify all logical continuous operations on programs.

There are further general analogies between the logic of programs and the themes studied in the preceding Chapters, which we must forego here. For instance, the Categorial Hierarchy also seems a suitable medium for studying the syntax, and polymorphism, of programming languages. Some further semantic analogies will be found in Chapters 16, 17. ●

As was observed before, basic intensional domains will usually come provided with some additional relational structure. Therefore, for instance, in the case of Time, basic temporal operators in type ((s, t), (s, t)) will not satisfy permutation invariance: as permutations of points in time may destroy the fundamental temporal order. They will retain another form of invariance, however:

Temporal operators are invariant
for *automorphisms* of the temporal order.

Van Benthem (1986, Chapter 5) contains a classification of such operations on the structure of the real numbers, using additional denotational constraints such as Continuity and various forms of Monotonicity.

A similar additional structure may be found with intensional domains of 'information states' ordered by 'inclusion', as will be considered in Chapter 15. Basic operators in this setting, such as the universal modality

$$\Box(X) \quad = \quad \{\, x \in D_s \mid \text{for all } y \in D_s \text{, if } x \subseteq y \text{, then } y \in X \,\}\,,$$

will only satisfy invariance for inclusion automorphisms of the intensional ground domain.

The latter notion can be defined in a completely general manner, just as was done with permutation invariance in Chapter 2. That is, starting from any inclusion automorphism on individual states, and the identity on truth values, the same canonical lifting to arbitrary type domains works as before. And then, 'invariance' again means being a fixed point under the action of all such lifted automorphisms in the relevant domain.

All the usual modal notions, first-order or higher-order, are invariant in this sense, because of the following observation about intensional type hierarchies.

Proposition. Each object defined by a formula of the finite type theory $T_\omega(\subseteq)$ having one parameter $\subseteq$ denoting inclusion among states is invariant for inclusion automorphisms.

There is also a partial converse, which may be proved by a simple modification of an analogous argument concerning permutation invariance in Chapter 10.

Proposition. In the intensional hierarchy over a *finite* base domain D_s, all automorphism invariant objects have an explicit definition in $T_\omega(\subseteq)$.

As a final example of a different semantic phenomenon, we point out a basic distinction concerning the intensional behaviour of the other important denotational constraint of *monotonicity*. As this notion was used before, it referred to Boolean inclusion within one single model, or world. For instance, the determiner "every" was upward monotone in its right-hand argument:

"every" AB , $B \subseteq B^+$ imply "every" AB^+ .

What this does not mean, however, is that "every" is *persistent* in the sense that universal statements true in one situation should also be true in larger situations. For, new individuals may actually refute the earlier regularity. Thus, the additional inclusion structure on the domain D_s creates new semantic distinctions, such as that between monotonicity and persistence.

In particular, in the richer intensional perspective, propositions themselves may come in different classes of closure behaviour with respect to inclusion. Some are upward preserved along $\subseteq$, while others need not be. This topic will be considered in more detail in Chapters 16, 17 below.

12.4 Comparing Different Intensional Domains

Finally, we consider the issue of invariance from yet another angle, namely in terms of comparisons between different models, rather than re-shufflings of one single structure.

It is one thing to introduce a notion of a structured intensional domain, temporal, epistemic, or otherwise. But, one should also provide some criterion of identity, telling us when two such domains can be considered 'equivalent'. For instance, in computer science, one does not want to distinguish between two machine models, or information structures, which can 'simulate' each other. Now, one candidate for such an equivalence relation has a strong backing in contemporary computer science (but also, e.g., in set theory), namely *bisimulation* . As it happens, this notion existed already in Modal Logic (cf. Chapter 15), whence we shall follow the presentation from that field:

Modal Bisimulation and Invariance

Let $\mathbb{M}_1$, $\mathbb{M}_2$ be two modal structures, each consisting of an intensional domain with an inclusion relation $\subseteq$ and a 'valuation' from proposition letters to subsets of those domains (these are abstract patterns of possible information growth, with the actual facts 'spread' over them). Now, a relation C between $\mathbb{M}_1$, $\mathbb{M}_2$ is called a *bisimulation* if it satisfies the following conditions:

1 if w_1Cw_2 then w_1 , w_2 carry the same valuation on proposition letters

2a if w_1Cw_2 , $w_1 \subseteq v_1$, then there exists v_2 such that $w_2 \subseteq v_2$, v_1Cv_2

2b analogously, in the other direction .

For instance, the identity relation is a bisimulation between any model and its so-called 'generated submodels'.

Now, modal formulas φ are *invariant for bisimulation*, in the sense that

$$\text{if } w_1Cw_2, \quad \text{then } \mathbb{M}_1 \models \varphi[w_1] \text{ iff } \mathbb{M}_2 \models \varphi[w_2] .$$

And this invariance property is characteristic for the modal formalism, as will be shown in Chapter 15 (cf. van Benthem 1985).

The presentation so far concerned the language of Modal Logic: that is, a small first-order fragment of the full intensional type theory over $\{s, t\}$. But, as before, a type-theoretical framework is well-suited for generalizing the essential semantic phenomena at this modal ground level. We would like to express, for instance, that certain higher-order 'modal' operations over our base domains $(D_s , \subseteq)$ are bisimulation invariant too.

Here are some illustrations.

Example. Bisimulation in Higher Types.

Let C be a bisimulation between two models $\mathbb{M}_1$, $\mathbb{M}_2$ on domains $D^1{}_s$, $D^2{}_s$.
Intuitively, two propositions φ_1 , φ_2 'correspond' under C if always

$$w_1Cw_2 \text{ implies that } \varphi_1(w_1) \text{ iff } \varphi_2(w_2) .$$

And then, an operator on propositions may be considered 'bisimulation invariant' if it takes C-corresponding propositions to C-corresponding propositions. Note that the so-called 'p-Morphism Lemma' in Modal Logic states essentially just this fact for propositional operators which are definable in the standard modal formalism.

To take another type of modal operator, we have bisimulation invariance for the following 'modal projection' of dynamic propositions:

$$\lambda R_{(s, (s, t))} \bullet \lambda x_s \bullet \exists y_s \supseteq x_s \bullet R(x, y) .$$

What this means is that, for each pair R_1 , R_2 of 'C-corresponding' relations [i.e., whenever w_1Cw_2 , v_1Cv_2 , then $R_1(w_1, v_1)$ iff $R_2(w_2, v_2)$], their modal projections are C-corresponding propositions as above. ■

A proper generalization of this notion goes as follows, taking a cue from the 'general relations' of Chapter 2. Starting from some relation C between $D^1{}_s$ and $D^2{}_s$, we define a family of relations $\{C_a\}_{a \in TYPE}$ between objects in corresponding type domains in the hierarchies built over these two intensional ground domains :

type s $\quad w_1 C_s w_2$ iff $w_1 C w_2$

type t $\quad x C_t y$ iff $x=y$

type (a, b) $\quad f C_{(a, b)} g$ iff for all x, y such that $x C_a y$:

$$f(x) C_b g(y) .$$

For instance, for the types (s, t) and (s, (s, t)), this coincides with the above notion of 'C-correspondence'.

Now, an expression E in any type a is *bisimulation invariant* if its denotation, viewed as a function from models to a-type objects in their type hierarchy, has the following property:

For all bisimulations C between models $\mathbb{M}_1, \mathbb{M}_2$,

$[[E]]^{\mathbb{M}1} \; C_a \; [[E]]^{\mathbb{M}2}$.

Again, this fits the above examples of modalities.

And in fact, there is also a generalization of the usual p-Morphism Theorem, lifting modal bisimulation invariance at the ground level to arbitrary types:

Proposition. Any closed term in a typed lambda calculus containing Boolean parameters as well as restricted modal quantification of the form $\exists y_s \supseteq x_s$ defines a bisimulation invariant expression.

Proof. By induction on the construction of such lambda terms, it is easily proved that the following stronger assertion holds:

If a term τ_a in this format has the free variables $x_1, ..., x_n$,
and A_1, A_2 are assignments in $\mathbb{M}_1, \mathbb{M}_2$, respectively,
such that $A_1(x_i) \; C_{a_i} \; A_2(x_i)$ (where a_i is the type of x_i ; $1 \leq i \leq n$),
then $[[\tau]]^{\mathbb{M}1}[A_1] \; C_a \; [[\tau]]^{\mathbb{M}2}[A_2]$.

🍎

In line with the analysis of Chapter 15, it would be of interest to determine the precise fragment of this applied Lambda Calculus having a parameter $\subseteq$ for inclusion that corresponds to bisimulation invariant expressions.

Of course, not every important modal construction passes this test (see the more complex modal constructions of Chapter 15). For instance, the 'update' mode to be introduced in Chapter 17, operating in type ((s, t), (s, (s, t))) :

$$\mathrm{ADD}(P) \;=\; \{\, (x, y) \mid x \subseteq y \ \text{and} \ P(y) \,\}$$

is not bisimulation-invariant. The reason is that bisimulations will not preserve inclusion in a suitably strict fashion. Another counter-example is a basic relational operator like the earlier Composition. Of course, one can introduce stronger notions of invariance to deal with such cases, but these will not be pursued here.

The point of this final excursion has been merely to show how some of the central notions of Modal Logic can be lifted to become part of the general type-theoretic setting of this Book.

13 Variable Polymorphism and Higher Type Theories

In recent research on type theories, one conspicuous feature is the use of *variable types*, without a fixed place in the earlier hierarchies. These give rise to 'second-order' calculi, in which abstraction and quantification is also possible over descriptions of types themselves. This generality has again to do with polymorphism. Certain functions seem quite general, not bound to any specific type at all. For instance, intuitively, there is only one identity function behind the family

$$\{ \lambda x_a \bullet x_a \mid a \in \text{TYPE} \} ,$$

namely, some variable procedure

$$\lambda x \bullet x .$$

And the latter may be viewed as standing for a function assigning each type domain its appropriate identity function:

$$\Lambda y \bullet \lambda x \bullet x_y .$$

Categorial Grammar has also had several encounters with variable types. This happened for various reasons. For a start, categorial parsing often uses Logic Programming as a congenial methodology (cf. Moortgat 1988), and the latter brings with it the use of type variables in searching the Gentzen space of possible derivations. But, there are more intrinsic connections too. For instance, when *learning* a natural language with an underlying categorial grammar, one may start with some global hypotheses about the relevant types, represented by schemata with variables, which then become more detailed as new information comes in. This topic of what may be called 'discovery procedures' for categorial grammars will be considered below. But in fact, there are also full-fledged extensions of the categorial paradigm itself, based on the interplay of two mechanisms: namely, the earlier 'derivational polymorphism' as well as 'variable polymorphism'. The resulting Categorial Unification Grammar will also be considered here as to its logical properties.

In the course of the presentation to follow, we also point at connections with the theory of polymorphism and type assignment for programming languages. All this requires extension of the earlier categorial apparatus. Thus, a number of logical calculi with variable types and appropriate rules of inference will be studied, ascending up to full second-order type theories.

13.1 Categorial Equations

In understanding the categorial structure of a language, there is actually a phase of learning the basic assignment. But, how does one find that in the first place? This issue is studied in van Benthem 1987*, Buszkowski 1987, as well as Buszkowski & Penn 1989. Here are some key features of the process.

In practice, type assignment may start from various kinds of data. We have some rough empirical idea concerning categories of expressions and their possible combination, which may be stated in the form of 'categorial equations'.

Example. Basic Sentence Patterns.

Let verb phrases be given the variable type x, and noun phrases y. Corresponding equations might be

$$x + e \Rightarrow t$$

$$y + x \Rightarrow t,$$

where e, t are already known types. Here, a simplest solution might be as follows:

$$x = (e, t), \qquad y = e.$$

But, the proper solution (cf. Chapter 3) will be enforced, once we also have a sentence operator z behaving as follows:

$$z + t \Rightarrow t$$

$$y + z \Rightarrow y.$$

Then, we must have

$$x = (e, t), \qquad z = (t, t), \qquad y = ((e, t), t).$$

As it stands, however, this example is not precise. What counts as a 'solution' depends on some prior statement of the admissible range of types for the variables, as well as the specification of some categorial engine for driving the reduction relation $\Rightarrow$. For instance, the equation

$$x + x \Rightarrow x$$

is not solvable in the Ajdukiewicz system, and trivially solvable in the Intuitionistic conditional logic (all types x qualify). For the Lambek calculus, however, the space of solutions is more interesting, typically involving such zero-count items as

$$x = (t, t).$$

Now, a proof-theoretic analysis of these matters yields the following outcome.

Theorem. Solvability of categorial equations is decidable for the Ajdukiewicz system, as well as for the Intuitionistic conditional logic.

Proof. First, we may observe that, if a given system of equations has any solution at all, then it has a solution involving only the primitive types occurring in the system itself.

Now, the second assertion of the Theorem depends on the logical finiteness of an intuitionistic conditional logic over finitely many proposition letters ('Diego's Theorem'; cf. Chapter 8). Therefore, all possible solutions can be enumerated, and then checked one by one, using the known decidability of Intuitionistic propositional logic.

As for the Ajdukiewicz system, this is not a logically finite calculus, and hence a different argument is needed.

Consider an equation of the form

$a_1 + ... + a_n \Rightarrow b$,

involving possibly complex types constructed out of primitive types and additional variables. The following procedure finds the possible solutions, involving the least constraints on those variables:

First, there is only a finite number of possible proof shapes, which arise by giving a complete specification of the function / argument hierarchy among the terms $a_1, ..., a_n$. It suffices to consider the latter, one by one. Then, to see if the variables can support such a proof shape, one inspects all steps in the relevant tree, noting down successive constraints as they arise. The crucial step is as follows:

If some intermediate result A is to combine with B to $A(B)$, where A, B have been treated already, then

1 A has to be a type (A_1, A_2) where A_1 can be unified with B, and the *most general unifier* σ is to be performed everywhere: one then continues with $\sigma(A_2)$ instead of $A(B)$,

or

2 A is a variable, which is set equal to (A_1, A_2) for new variables A_1, A_2 everywhere: one continues with A_2 instead of $A(B)$.

At the end, the outcome is compared with the desired b, performing a final (most general) unification, if possible.

Correctness and termination of this algorithm are evident. ◆

For the intermediate Lambek Calculus L , however, the situation is not so clear. On the one hand, this system is not logically finite, whence the earlier knock-down argument does not apply. But on the other hand, it has an infinity of possible proof shapes, because of the Conditionalization rule: which may make for an infinite regress in searching for solutions. Therefore, here is an open

Question: Is solvability of categorial equations in the Lambek Calculus *decidable*?

There is also another natural version of these issues, however, which leads to a somewhat different technical perspective.

13.2 Type Assignment

Our initial linguistic data may even comprise information about desired *constituent structures*. This is quite plausible in actual language learning: we are becoming acquainted with structured objects, not flat strings. (Compare the suggestion made in Chapter 3 about the possibility of having a phonologically motivated 'initial constituent structure' for discourse, upon which the grammar will act subsequently.)

Technically, then, one is given lambda terms with variable types attached, where the latter are still to be determined, so as to form well-formed expressions of the typed lambda calculus. In this case, our problem simplifies considerably. One illustration will in fact convey the essence of the method.

Example. Terms to be Adorned.

1 Here is a first example:

$\lambda z \bullet x(z)(y(z))$.

A most general solution may be obtained here through the following type identities:

$a_x = (a_z, u)$, $a_y = (a_z, v)$, $u = (v, w)$.

This system has a most general solution, obtainable through the well-known *Unification Algorithm*:

a_z can be arbitrary

$a_y = (a_z, v)$ for arbitrary v

$a_x = (a_z, (v, w))$ for arbitrary w .

2 Here is a second example:

$\lambda z \bullet x(z)(x)$.

No solution exists here. And indeed, the Unification Algorithm will report failure.

The solution problem for variable categories in this second form is much simpler than the earlier version:

Proposition. The type assignment problem is *decidable* for all categorial calculi.

Moreover, the standard Unification algorithm will produce *most general* solutions, of which every particular solution must be a special case.

Remark. Stratifying Untyped Terms.
The preceding problem is equivalent to one from Lambda Calculus and Combinatory Logic (cf. Hindley & Seldin 1986). Already in the sixties, Haskell Curry proposed a unification algorithm for determining precisely the set of 'stratifiable' type-free lambda terms, admitting of adornment to a well-formed term of the typed lambda calculus. ◆

Type assignment is a well-known technique in the semantics of programming languages. For instance, Damas 1985 presents type assignment algorithms for the programming language ML in close analogy to Curry's procedure. These will take a programming construct such as

" let *compare* f g = $\lambda x \bullet f(g(x))$ "

and then compute the following type for the *compare* function:

((a, b), ((c, a), (c, b))) .

Some interesting comparisons may be drawn here between the linguistic and the computational perspective. In programming, type assignment acts as a kind of initial 'filter' on a pre-structured program text, whose desired functional dependencies are already indicated, which determines if processing is possible at all. The linguistic equivalent would be to have a test for 'interpretability' of expressions, without actually carrying out the full semantic computation. Moreover, type assignment procedures themselves involve phenomena of wider interest. Two important examples are 'coercion', forcing different expressions to assume the same type, or the detection of 'over-loading', where different occurrences of the same expression must be assigned different types, if processing is to succeed. (Damas in fact has one algorithm where different occurrences of the same variable in a lambda term may receive different types. Then, for instance, even a 'self-applied' term like x(x) becomes typable.) This whole perspective of interlacing type assignment and semantic computation seems linguistically significant too.

13.3 Categorial Unification Grammar

One basic motivation behind the flexible Categorial Grammar of Chapters 3, 4 has been the polymorphism of the Boolean particles. What the various systems in the Categorial Hierarchy provide for this purpose is what may be called *derivational polymorphism*. One basic type is assigned, and then all others are derived, with a canonical transfer of meaning. But, it might also be natural to get all these polymorphic occurrences in one fell swoop, by employing a suitable form of *variable polymorphism*, using schemata with type variables. An example might be

$$(x, (x, x))$$

for conjunctive coordination. (Note that this covers both Boolean and non-Boolean cases.) And, when desired, certain restrictions may be encoded by specifying terms a litttle further. For instance, another proposal for categorizing negation has been the variable type

$$((x, t), (x, t)) ,$$

reflecting its attachment to Boolean types only, ending in a final t.

How are such variable types to be used in linguistic analysis? Here is where *Unification* comes in again. The idea is that variable types adapt to their environment to the minimal extent that is needed for categorial combination. For instance, in the following derivation, the type of negation gets adapted so as to allow a function application:

$$\dfrac{((x, t), (x, t)) \qquad \dfrac{e \qquad (e, (e, t))}{(e, t)}}{(e, t)}\ x = e$$

The particular adaptation may still depend on various options encountered later on, witness the possible continuations of the following phrase:

" Cedric ignored a comment $and_{(x, (x, x))}$ ... "

"insult" :	x = (e, t)
"two grins" :	x = ((e, t), t)
"rebutted another one" :	x = (e, t) .

Another example is the treatment of quantifier phrases in Emms 1989, who assigns them the variable type ((e, x), x) ,

which then functions as follows in the problem of direct objects NPs discussed in Chapters 3, 4:

"a present	pleases	every child"	
((e, x), x)	(e, (e, t))	((e, x), x)	x = (e, t)
		(e, t)	x = t
t			

These are the ideas behind the 'categorial unification grammars' developed in Calder, Klein & Zeevat 1987 and Uszkoreit 1986, whose inspiration goes back to polymorphic programming languages such as the above-mentioned ML .

13.4 Categorial Logics with a Substitution Rule

Now, a closer analysis of these systems shows that the logical analogy set forth in Chapter 4 still holds good. For instance, in the simplest variant of categorial unification grammar, the basic rule is essentially this:

a and (b, c) imply $mgu_{[a, b]}(c)$,
where $mgu_{[a, b]}$ is a *most general unifier* of the types a and b .

But this is precisely the inference rule of *Resolution* as employed in contemporary logic programming and theorem proving. (Indeed, the whole paradigm should rather be called 'Resolution Grammar'.) Moreover, a more liberal variant of the system also allows resolution variants of further categorial rules from the calculus L . For instance, the Geach rule becomes

(a, b) and (c, d) imply $(mgu_{[b, c]}(a), mgu_{[b, c]}(d))$.

From a more general logical point of view, what is going on here is essentially this. We are still using the old categorial calculi, but now provided with an additional

Rule of *Substitution*:

"Replace any variable type by any one of its substitution instances" .

Thus, there are two basic categorial processes now:

$\Rightarrow$	'derivable transition'
$\triangleright$	'substitutional specification' .

Remark. Resolution Explained.
In this perspective, the rule of Resolution may be viewed as providing a *normal form* for categorial derivation admitting a Substitution rule (cf. Robinson 1965). A free use of Modus Ponens steps and Substitutions may be replaced by a sequence of Resolution steps (possibly followed by one final substitution).

Several logical questions arise now concerning the old categorial calculi with an added rule of Substitution. In particular, which increase in complexity is incurred in what, after all, is supposed to be an efficient computational mechanism? Here, outcomes are comparable to those already encountered with solvability of categorial equations. For reasons of scientific geography, let us call the Ajdukiewicz system with Substitution 'Edinburgh Grammar', and the Lambek variant 'Stuttgart Grammar' . The term 'Amsterdam Grammar' might then go to the Intuitionistic conditional logic with substitution.

Theorem. Amsterdam Grammar and Edinburgh Grammar are both decidable.

Proof. The first assertion is proved just like the corresponding one for categorial equations. Due to logical finiteness, only finitely many substitution instances of the premises can be relevant, and all these fully specified cases can be checked separately.

As to the second assertion, given a question if b follows from $a_1, ..., a_n$, one first considers the finite number of more general types b' such that $b' \triangleright b$. Then, in each of these cases, the issue is whether the premises $a_1, ..., a_n$ derive b' via Resolution. As there is only a finite number of potential proof shapes for this (depending on which binary combinations are made in which order), all possibilities can be checked effectively.

But again, there remains an open problem, which has proved intractable so far:
Question: Is Stuttgart Grammar decidable?

By itself, the calculus L is decidable (cf. Chapter 9) - but what we would need here is some effective bound on the space of substitutions.

Remark. Dangers of the Substitution Rule.
Adding a rule of Substitution to the full intuitionistic propositional logic (including further connectives, such as disjunction) makes the latter system *undecidable*. This follows from the construction of an undecidable finitely axiomatized 'intermediate propositional logic' in Shehtman 1978.

In fact, Substitution may even make the Ajdukiewicz base system undecidable: once it is combined with another interesting principle (cf. Chapter 9), namely that of *Iteration* of premises. For, the general problem whether

$$\{a_1^*, ..., a_n^*\} \Rightarrow b$$

is derivable via the Substitution rule plus Modus Ponens is none other than the central question of *Hilbert-style axiomatics*:

" Does a certain theorem follow from a set of axiom schemata
using Modus Ponens as the sole rule of inference? "

And for general conditional logic, the latter problem is undecidable (Singletary 1964). ●

Finally, as for other proof-theoretic properties (cf. Chapter 8), the *recognizing power* of even Edinburgh Grammar is still terra incognita. (Henk Zeevat and Michael Moortgat have proposed counter-examples to context-freeness.)

13.5 Variable and Derivational Polymorphism

What we have now are two kinds of mechanism for describing polymorphism, namely derivation and substitution. A natural question is then to which extent these two processes subsume each other. Here is what happens in the Lambek Calculus.

In one direction, there is an obvious non-inclusion.

Proposition. S-polymorphism does not reduce to D-polymorphism.

Proof. Here is a counter-example. The variable type x for any expression generates the substitution set of *all* types. But, no finite set of types can generate the latter via derivations in L , as these would only allow a corresponding finite set of *count* patterns (cf. Chapter 6), whereas the totality of types exhibits an infinite variety of these. ●

In the converse direction, there is no inclusion either. For instance, in a purely functional type formalism, no single scheme with variables can capture the polymorphism of negation, since the latter involves arbitrary depth of nestings

$((a_1, (a_2, ..., (a_n, t)...)), (a_1, (a_2, ..., (a_n, t)...)))$.

Still, with *product* types, all these may be rewritten to the flattened form

$((a_1 \bullet ... \bullet a_n, t), (a_1 \bullet ... \bullet a_n, t))$,

for which the polymorphic type $((x, t), (x, t))$ indeed suffices.

In fact, it might seem that S-polymorphism does subsume D-polymorphism. After all, is not this the moral of the earlier result about variable type schemata forming 'most general solutions' to systems of categorial equations? But, there is a confusion of levels here. 'Principal type schemata' in Curry's sense refer to families of *readings* for some given expression, whereas we are after the description of families of types derivable from a given one. And the truth of the matter is as follows.

Proposition. D-polymorphism does not reduce to S-polymorphism.

Proof. Consider the set of types derivable from the single type t in the Lambek Calculus. This includes at least the following infinite sequence

$((t, t), t)$, $((t, t), ((t, t), t))$, etcetera.

Now, suppose that some finite set of variable type schemata described exactly all L-consequences of t . Then, by the Pigeon Hole principle, at least one of these schemata τ must have infinitely many instances among the above types ' $((t, t)^i, t)$ ' :
say, starting with a smallest index i . Using this observation, we deduce the following facts about τ - whose general shape already looks as follows:

$(\tau_1, ..., (\tau_k, \phi)...))$,

with k final brackets following some 'tail' ϕ which is either a primitive type or some type variable:

1 $k \leq i$
2 ϕ must be a type variable
3 the arguments τ_m can only be type variables or terms of the form (u, v) with u, v type variables or primitive types
4 ϕ must occur in at least one term τ_m .

Here, assertions 1, 2, 3 are obvious. As for 4 , suppose that ϕ does not so occur. Then, the schema τ would have substitution instances which cannot be L-derivable from the type t . These may be produced by substituting some new primitive type s for ϕ in the schema τ , and then substituting arbitrarily for all τ_m using only pure t-types. For, such a type can never be derivable from t in L , because of the disparity in s-count.

Using these facts, a contradiction is obtained as follows. Let τ also have some type $((t, t)^{i+j}, t)$ for an instantiation, with $j \geq 2$. Then, it follows that, to obtain the latter type, a term with at least two right-most brackets is to be substituted for ϕ : whose corresponding substitution in some initial τ_m will yield a term whose bracketing pattern is unlike that of $((t, t)^{i+j}, t)$. ◆

Actually, the issue of polymorphic ranges for expressions in natural language is even more subtle than may have appeared so far. There are many plausible *restrictions* on full substitution, both for empirical and for logical reasons, which are still to be spelt out.

For a discussion of some curious technical effects of allowing full substitution on apparently innocuous variable types, see van Benthem 1989*. For instance, two more constrained types may combine to form a less constrained one, as in

$$((x, t), y) + ((x, t), y) \Rightarrow y .$$

But certainly, one intuitive constraint on the whole combinatorial process should be that types produced in the course of a derivation can only become progressively more 'specific', in some suitable sense.

As for a more empirically oriented illustration, consider, for instance, the following plausible variable type for quantifier phrases:

$$((x, t), t) .$$

An argument for this variable polymorphic assignment in the lexicon may be this. Depending on the nature of the verb phrase to follow, the quantifier can remain at the individual level:

x becomes e ,

or move to the level of groups and collections, where

x becomes (e, t) .

(Recall the discussion of 'Collectivization' at the end of Chapter 5.)

Now, for a start, the latter option should not be taken to mean that a noun phrase could just take any ((e, t), t) type argument. Otherwise, ill-formed 'sentences' would have to be accepted like

" Every fool no fool "

$\underline{((x, t), t)}$ $\underline{((e, t), t)}$ $x = (e, t)$

t .

But also, the range of admissible types for x in the process of Collectivization seems highly constrained, namely, to something like the following sequence of levels:

$x_1 = e$, $x_2 = (e, t)$, $x_3 = ((e, t), t)$,

and in general, $x_{n+1} = (x_n, t)$.

But then, we have the following fact, showing that we have gone beyond the resources of either variable of derivational polymorphism:

Fact. The family $X = \{x_1, x_2, ...\}$ is neither an S-set nor a D-set.

Proof. That no finite set of variable type schemata can produce exactly this set may be shown by a syntactic argument similar to that given above. (This time, one should focus on the growing sequence of left-most brackets.)

Next, as for derivational polymorphism, there is no obvious difficulty with counts here: as the family X displays only two alternating count patterns. But, there is another obstacle. If X were of the form $\{ b \mid a \Rightarrow b$ is L-derivable $\}$ for some type a , then some element x_n would already L-imply all others. In particular, this x_n implies $x_{n+1} = (x_n, t)$. But then, x_{n+1} would be provable by itself in the Intuitionistic conditional logic, since $(u \rightarrow (u \rightarrow v))$ implies $(u \rightarrow v)$ in that system. Now, the latter statement is refutable. For, no x_n is even classically provable, witness the following sequence of counter-examples by truth table valuations:

$V(e) = 1$, $V(t) = 0$ refutes all x_n with n even

$V'(e) = V'(t) = 0$ refutes all x_n with n odd .

13.6 Semantic Interpretation of Variable Types

One issue omitted so far was that of the *meaning* of the whole apparatus of variable types. At least, this needs investigation if the latter is to be something more than a convenient computational device.

What is needed now is a *second-order* language with variable types and two lambda abstractors:

λ as usual , and Λ over type variables .

For instance, one expects the following account of polymorphic negation in the variable type $((x, t), (x, t))$:

$$\Lambda x\bullet\ \lambda y_{(x,\,t)}\bullet\ \lambda z_x\bullet\ \neg_{(t,\,t)}(y(z))\ .$$

This illustrates the general schema for derivations in the earlier categorial calculi with a substitution rule. With any derivation of a transition

from $a_1, .., a_n$ to b ,

there will be a term associated of the form

$\Lambda x_1\bullet\ ...\ \Lambda x_m\bullet$ " L-free λ-term of type b " ,

where $x_1, ..., x_m$ are the free type variables in b .

And the semantic interpretation of such terms presents no difficulty.

This is still an innocuous form of second-order polymorphism, arising out of 'universal generalization' over ordinary categorial validities. But, even basic lexical items themselves can already have genuinely polymorphic meanings, witness the earlier logical quantifiers in type $((x, t), t)$. For instance, the existential quantifier might be described as

$$\Lambda x\bullet\ \lambda y_{(x,\,t)}\bullet\ \exists z_x\bullet\ y(z)\ .$$

Such examples suggest that there is something more to variable polymorphism than purely procedural convenience.

For a more detailed semantic account of interpreting derivations with variable types, see Hindley & Seldin (1986, Chapter 16), Barendregt & Hemerik 1990, who present, amongst others, the general machinery of a *second-order polymorphic lambda calculus* $\lambda 2$. Its 'type schemes' include all type constants and type variables, as well as everything that can be formed out of these by means of function types (a, b) and 'universal generalizations' $\Delta x\bullet a$. Accordingly, in addition to the standard rules of Function Application and Function Abstraction, basic principles of calculation will now also include rules for manipulating these additional resources:

- *Instantiation*

  If a term τ has type scheme $\Delta x\bullet a$, and b is an arbitrary type scheme, then $\tau(b)$ has type $[b/x]a$
- *Generalization*

  If the conclusion that a term τ has type scheme a has been reached without using assumptions in which the type variable x occurs free, then the term $\Lambda x\bullet\tau$ has type scheme $\Delta x\bullet a$.

Whether any genuinely higher-order parts of such systems are involved in understanding the workings of natural language, is an interesting open question. Presumably, in line with the earlier methodology of Chapters 4, 5, the art would be to locate suitable *fragments* of such powerful type theories.

We conclude with an illustration of a rather mild form of variable polymorphism encountered in practice, showing how working with such a fragment of a higher type theory affects the earlier semantic analysis of Chapter 9.

Example. Boolean Variable Polymorphism.
Suppose that Booleans have been added to the language of the Lambek Calculus in variable types (x, x) and (x, (x, x)) . What will happen then to the earlier Finiteness Theorem, concerning readings for L-derivable transitions? Here is a typical answer:

Proposition. If b is L-derivable from standard types $a_1, ..., a_n$ as well as variable Boolean type occurrences (x, x) , (x, (x, x)) , then there are still only finitely many interpretations of this transition, modulo logical equivalence.

Proof. This may be shown by a judicious combination of the earlier inductive argument for the Finiteness Theorem and the known logical finiteness of Boolean operations on any truth value type domain.

The only relevant new cases in the earlier argument are as follows (demonstrated for the variable type (x, (x, x))) :

- $u_{(c, (c, c))}$, where b = (c, (c, c))
- $\tau^1{}_{(c, b)}(\tau^2{}_c)$,

  with a leading variable in τ^1 corresponding to some occurrence of (x, (x, x)) . That is, this term has the form $u_{(c, (c, c))}(\sigma^1{}_c)(\tau^2{}_c)$.

Now, in general, the two argument terms displayed here need not have reduced complexity, and they might start again with Boolean operators. Still, the general case will be that of a 'Boolean skeleton' of eventual components which are all of the three kinds listed in the original proof, to which the inductive hypothesis of the Finiteness Theorem does apply. And of such Boolean skeletons, there will be only finitely many too, up to logical equivalence. ♠

To conclude, here is a more daring perspective.

Digression. Constructive Type Theories and Linguistic Meaning.

Already, the linguistic enlistment of even more complex higher-order type theories than $\lambda 2$ has been proposed for the purpose of understanding natural language. For instance, Pereira 1990 discusses parsing with such formalisms. In particular, Ranta 1990 advocates the use of a constructive 'Martin-Löf Type Theory' (cf. Martin-Löf 1984), which mixes the formation of terms and types to a greater extent . The latter has an additional mechanism for accumulating 'contexts' of type assignments $\tau : A$. These are involved in key principles of higher-order typing, such as the introduction rule for abstracts living in Cartesian products of families of types:

$$\frac{(x : A)\ b_x : B_x}{\lambda x \bullet b_x :\ \Pi_{x \in A} \bullet B_x}$$

Accumulation of contexts may be a good model for various dynamic phenomena in natural language, such as anaphoric connection. Moreover, this mechanism may also provide a deeper *explanation* for the occurrence of denotational constraints, as encountered in Chapter 10. For instance, Conservativity might emerge as follows. Interpreting a determiner statement D AB involves creating a context for A first, which can then be used subsequently in creating one for the second argument predicate B .

Another important feature of a type theory like this is its constructive character. The earlier notion of 'formulas-as-types' motivates a ubiquitous dual reading of type assignments $\tau : A$

- object τ has semantic type A
- derivation τ is a proof, or an algorithm, realizing assertion A .

Such constructive, proof-oriented views of natural language understanding, already defended in Dummett 1976, Hintikka & Kulas 1983, Kracht 1988, are coming to the fore nowadays as a potential challenger to the reigning model-theoretic paradigm. ◆

VI TOWARD A LOGIC OF INFORMATION

This final part of our study has a somewhat broader topic than the theory of Categorial Grammar and natural language semantics developed so far. In a sense, it may be seen as a small 'book within a book', whose guiding ideas are set forth below.

Logic Meets Information Processing

A noticeable tendency in current logical research is the move away from still reflection of abstract truth to a concern with the structure of *information* and the mechanism of its *processing*. Thus, two aspects come to the fore which used to be thought largely irrelevant for logical analysis, namely the actual linguistic detail of presentation of premises and the actual procedures for setting up arguments. Accordingly, logical analyses will now have to operate at a level where, e.g., the *syntax* of occurrences of propositions matters - and likewise it will be occupied, not just with declarative structure, but also with matters of argumentive *control*.

In the form of a slogan, many people nowadays believe that

Natural Language is a Programming Language

for effecting cognitive transitions between information states of its users.

What we can observe in the literature is a number of independent attempts at creatimg conceptual frameworks allowing us to capture significant features at this level, while still retaining a workable logical theory. One thing which many of these newer approaches have in common is the failure of certain so-called *structural rules* found in standard logic, such as

Monotonicity $$\frac{X \Rightarrow A}{X, Y \Rightarrow A}$$

or even

Contraction $$\frac{X, B, B \Rightarrow A}{X, B \Rightarrow A}$$

In standard calculi, these seem harmless, and evident, book-keeping rules: but now, their failure becomes a very general *symptom* (though by no means the essence) of operating at a finer-grained level of logical analysis. Thus, we become explorers of a landscape, so to speak, of logic underneath the usual classical or intuitionistic base systems: a landscape which has many analogies with the Categorial Hierarchy studied so far.

The sources of these newer systems are diverse. Some motivations are *proof-theoretic*, with prime examples in Relevance Logic (cf. Dunn 1985), which drops Monotonicity, or Linear Logic (cf. Girard 1987), which also drops Contraction of premises. Even more radically, from a syntactic *linguistic* perspective, one must also drop the structural rule licensing Permutation of premises (cf. Lambek 1958). Put differently, Relevant Logic is still concerned with *sets* of premises, Linear Logic with *bags* (or 'multi-sets'), and our earlier linguistic paradigm of Categorial Grammar in general with ordered *sequences* of premises. (Of course, bags are again central in a more semantic type-combining system like the Lambek Calculus with Permutation.) And the latter 'sequential' level of detail also arises with motivations of a more *computational* nature, in what may be called the Dynamic Logic of inference and interpretation (cf. Harel 1984, van Benthem 1988B). More information on these various approaches will be found in the course of this Part.

Now, the purpose of the following Chapters is not to start with any sacrosanct calculus of 'sub-standard inference', trying to understand its secrets, but rather to explore a number of *models* for the structure of information and its processing, reflecting various intuitions that we have about the latter. These will come in the following forms. First we consider *language models* ('L-models') focussing on syntax and occurrence (Chapter 14). Then, in Chapter 15, we move on to more abstract *information models* ('I-models'), arising from L-models through a certain measure of collapse of syntactic detail. Alternatively, this Chapter may be seen as broadening the scope of Intuitionistic Logic, being the traditional 'guardian' of information and verification in the setting of the foundations of mathematics. Next, we move from information structures to procedures operating on them. Processing of information will be central in more 'dynamic' *relational models* ('R-models' , Chapter 16) reflecting the control structures of programming transitions between information states. Chapter 17 is then devoted to a perspective from *Dynamic Logic* integrating these two strands, and relating them to standard logic, in a suitably general type-theoretic setting.

There is still a natural extension of our earlier concerns in all this, in that the Categorial Hierarchy itself is a prime example of a logical landscape without obligatory structural rules. Moreover, it turns out that, in particular, the Lambek Calculus itself admits of procedural re-interpretation, and thus, categorial calculi may turn out to describe cognitive procedures just as much as the syntactic or semantic structures which provided their original motivation.

In all these cases, some common logical issues arise. Most conspicuously, there is a proliferation of *logical constants*, creating richer languages beyond the classical core, and also a variety of natural kinds of *valid consequence*, with corresponding choices in designing logical calculi of inference. (In fact, the two kinds of enrichment are interrelated, as we shall see.) We shall propose some systematic perspectives on the choices involved - whilst also investigating the potential of existing research programs (Categorial Grammar, Modal Logic, Relation Algebra) for adapting to this wider purpose.

The various kinds of model will be compared and integrated at the end of this Part, in an attempt to create one basic picture, or framework for a logical theory of information. The resulting system is a 'dynamic logic' of information processing, inspired by similar calculi in the semantics of programs - an analogy which seems only natural, given the earlier conception of natural language as a vehicle for cognitive programming. In particular, this system allows for coexistence of earlier 'static' views of propositions and the newer 'dynamic' ones. But again, the proposal made here does not stand or fall with the adoption of some unique preferred 'base calculus' of information-processing oriented logic.

To return to our opening sentence, there are many observable 'tendencies' in any given science at any given time: and most of them prove ephemeral fashions. Nevertheless, there is some reason to believe that we are better off here. Basing logic on the processing of information merely continues a historical development already begun in constructive logics, be it in a more radical manner. Moreover, it is a good sign that the new perspective, once grasped, allows one to make sense of various scattered precursors in the literature, including such diverse topics as Quantum Logic (cf. Dalla Chiara 1985), where testing one occurrence of a premise may not yield the same result as testing another - or the study of the Paradoxes (cf. Fitch 1952), where problematic arguments rest essentially on such classical structural rules as Contraction on Liar sentences at different stages of the paradoxical reasoning.

So, to summarize the purpose of this Part:
We want to *signal* an emerging shift of emphasis in current logical research toward the phenomenon of information processing, then extract various *analogies* among different strands involved here, and finally propose a suitable *general perspective.*

14 Language Families

Although information is certainly something more abstract than concrete linguistic form, it nevertheless proves useful to start with a study of Syntax, i.e., informational code, in order to get below the surface of standard logic. And therefore, we start from the perspective of this Book so far, trying to broaden it beyond too specific syntactic necessities of natural languages.

What we have been working with is the paradigm of *Categorial Grammar*, in which category assignments match an underlying semantic type-theoretic structure, whose grammatical analyses turned out to correspond closely to derivations in weak calculi of implication. Thus, categorial parsing was indeed identical with logical deduction. Nevertheless, this still left the question *what kind of* deduction corresponds to grammatical parsing, and here various options emerged. Indeed, from a linguistic perspective, there need not be one single 'best' categorial calculus at all. Lambek's own calculus certainly turned out to be a natural basis, but, as has been demonstrated in previous Chapters, various syntactic and semantic phenomena may involve strengthenings or weakenings. Therefore, a better picture is that of Chapter 4: being a *Categorial Hierarchy* of calculi underneath, but ascending up to the standard systems of logic. This variation is precisely an asset in making intra-linguistic, or cross-linguistic, comparisons of complexity between phenomena in natural language: and presumably, the same holds for informational structure generally.

Viewed as systems of logic, these categorial calculi are rather poor, employing only a few basic logical constants, namely, implication and conjunction. For instance, disjunctions or negations are missing. Nevertheless, a noticeable tendency in the recent linguistic literature has been to introduce further operations on types: either as 'compiled' versions of complex functional types (like the 'exponential' $\uparrow$ at the end of Chapter 6), or for more convenient storage of information (as with 'disjunctive' category structures in computational linguistics), or even for high-lighting new phenomena (compare the 'intensionalizer' $\Box$ in Chapter 12, or Moortgat 1988 on gapping operators).

To get a more systematic view here, one should ask what categorial systems in the syntactic mode are supposed to describe, and what would be the natural operations on categories then. As it happens, a notion of model having the required generality may be found by returning to an analogy with more standard concerns in Mathematical Linguistics (cf. Chapter 8). Subsequently, we can also 'step back' a bit, asking what would be the best way of describing such models, and then evaluating how the design of our earlier calculi fits into this enterprise.

14.1 Language Models

That a richer logic of types should lie behind Categorial Grammar becomes clear once we realize that the basic structures in formal linguistics are *families of languages*

$\{L_a \mid a \in A\}$

over some finite alphabet of symbols, which are closed under certain natural operations (cf. Hopcroft & Ullman 1979). The latter may be systematized roughly as follows:

Boolean operations

$-, \cap, \cup \qquad \bot, T$

Order operations

$\bullet, \backslash, / \qquad 1 \qquad *$

As for the latter, we have (with juxtaposition indicating concatenation)

$L_a \bullet L_b$	$=$	$\{ xy \mid x \in L_a, y \in L_b \}$	(product)
$L_a \backslash L_b$	$=$	$\{ x \mid \forall y \in L_a : yx \in L_b \}$	(left inverse)
L_b / L_a	$=$	$\{ x \mid \forall y \in L_a : xy \in L_b \}$	(right inverse)
L_1	$=$	$\{ <> \}$	(empty sequence)

and the Kleene star * denotes finite *iteration* as usual.

These operations are natural, e.g., in the sense that the family of all *regular* languages is closed under them. Now, we say that

An L-*model* is any family of languages over some finite alphabet which is closed under the above operations.

Next, we need a suitable notion of validity.

A sequent of types $X \Rightarrow a$ is *valid* in L-models if, for every interpretation $[\![\ ]\!]$ sending primitive types to arbitrary languages and complex types to the obvious compounds, it holds that $[\![\bullet X]\!] \subseteq [\![a]\!]$.
Here, '$\bullet X$' is the concatenation product of X (with the stipulation that $\bullet\emptyset = 1$).

Valid principles on this account will include all Boolean laws, as well as typical principles of the Lambek Calculus such as

$a \bullet (a \backslash b) \Rightarrow b$, or $a \Rightarrow (b / a) \backslash b$.

In fact, here is a straightforward observation:

Proposition. The Lambek Calculus is sound for interpretation in L-models.

The converse is still an open question: as we shall see later on.

Of course, the L-interpretation also produces further validities for other logical constants. For instance, it is of interest to compare the behaviour of our two 'conjunctions': with $\bullet$ satisfying the Gentzen laws of the Lambek Calculus, and Boolean $\wedge$ rather the following two:

$$\frac{X, A \Rightarrow B}{X, A \wedge C \Rightarrow B} \qquad \frac{X \Rightarrow A \qquad X \Rightarrow B}{X \Rightarrow A \wedge B}$$

Note also that the Kleene star behaves somewhat like an S4 *modality*, in that we have the validity of

$a \Rightarrow a^*$, $a^*, a^* \Rightarrow a^*$.

In fact, this operator licences a structural rule which is not valid as such in the Lambek Calculus which has played an important role in earlier Chapters, viz. *Contraction*:

$$\frac{X, a^*, a^* \Rightarrow b}{X, a^* \Rightarrow b} .$$

The analogy with modality is not perfect, however, in that iteration respects neither conjunction nor disjunction:

not $(a \cup b)^* \Rightarrow a^* \cup b^*$, *not* $a^* \cap b^* \Rightarrow (a \cap b)^*$.

Even so, iteration is obviously an interesting logical operation in its own right.

Thus, whatever its precise family resemblances, the logic of L-models has an independent interest as an object of investigation. For, the ordering operations seen central to syntax, and even iteration becomes quite natural once we move from single sentences to *texts*. (E.g., texts themselves have the iterate t^* of the sentence type t.)

Remark. In this connection, logical proof-theoretical structures, such as the above sequents are already *textual* objects, with the *comma* as a separator. And the latter operator itself needs interpretation before it makes sense to discuss validity or non-validity of such principles as the usual structural rules. Thus, in a sense, the popular observation about loss of these rules is too simple-minded. E.g., Contraction fails if we interpret the comma as a concatenation product - but it would remain valid if we had treated the comma via Boolean conjunction. We shall return to this choice point in Chapter 16. ♠

Finally, it is also quite possible to introduce further operations on L-models. For instance, two useful notions are

$\pi(L)$ = all *permutations* of sequences in L ,

$\iota(L)$ = all *mirror images* of sequences in L .

These will again exemplify logical laws, such as

$\pi(a \cup b) \Leftrightarrow \pi(a) \cup \pi(b)$, $\pi\pi(a) \Leftrightarrow \pi(a)$,

$\iota(-a) \Leftrightarrow -\iota(a)$, $\iota\,\iota(\alpha) \Leftrightarrow a$.

Next, we consider what happens with undirected categorial calculi, which have been at the focus of attention in this book.

14.2 Numerical Models

When full Permutation is allowed, the only information left about a string of symbols is the number of occurrences of each basic symbol in it. (Compare the use of numerical occurrence vectors in Chapter 8.) Thus undirected categorial calculi invite consideration of *numerical models*, defined as follows:

There is a family of sets of vectors in N^k (where k is the size of the alphabet) which is closed under the Boolean operations as well as vector addition and its converses providing the obvious interpretation for the order operations.

For instance, now

$$L_{a \bullet b} = \{ x+y \mid x \in L_a ,\ y \in L_b \}$$

$$L_{a \to b} = \{ x \mid \text{for all } y \in L_a ,\ x+y \in L_b \} .$$

If we want to have a smooth notion of $L_{a \to b}$, however, we shall need unlimited *subtraction*: which would require having the integers $\mathbb{Z}$ rather than the natural numbers N (even though this move would lose us a straightforward linguistic interpretation).

Valid principles over these models are of independent mathematical interest. For instance, how does the logic change across successive 'dimensions' for our numerical occurrence vectors?

Example. Numerical Satisfiability.
To get a feel for systems like this, it is useful to show, e.g., that the following set of formulas is satisfiable in N^1:

$$\{ p, -(p\bullet p), \ p\bullet p\bullet -p \} .$$

◆

More systematically, again, the earlier soundness result extends to interpretation in N- or $\mathbb{Z}$-models, for the Lambek Calculus LP admitting a Permutation rule.

The latter observation generalizes the use of so-called primitive type *counts* (cf. Chapter 6) as a check on derivability. Thus, after all, the above, seemingly uninterpretable mathematical generalization toward negative numbers makes sense: as type counts may be negative.

Example. Counts Via Numerical Models.
Consider two primitive types e, t. Assign the following two singleton sets of vectors:

$$L_e = \{\langle 1, 0\rangle\}, \qquad L_t = \{\langle 0, 1\rangle\} .$$

Then, for each complex type a, the inductively computed integer value of L_a becomes a singleton set $\{\langle x, y\rangle\}$ with

x is the e-count of a, y is the t-count of a.

E.g., ((e, (e, t)), t) goes to $\{\langle +2, 0\rangle\}$, ((e, t), (e, e)) goes to $\{\langle +1, -1\rangle\}$. ◆

But also, we can quickly check further non-derivabilities which did not show up in the pure count system.

Example. Logics of Special Numerical Structures.
The sequent ((e, t), t) ⇒ e has equal counts on both sides: which is the necessary condition for Lambek derivability induced by count. Nevertheless, it is not derivable - as may be established by proof-theoretic analysis. But now, we can also provide a counterexample, with

$$L_t = \mathsf{N}, \qquad L_e = \varnothing .$$

(Observe that $L_{(e,\,t)} = \mathsf{N}$, $L_{((e,\,t),\,t)} = \mathsf{N}$.) Thus, the implicational logic on numerical models has at least 'truth-value counterexamples' - and hence it must be contained in the classical conditional logic. We conjecture that it is in fact equal to the conditional logic axiomatized in the non-directed Lambek Calculus.

It would be of interest to see if this numerical interpretation yields further algorithms for reducing the Gentzen search spaces encountered in categorial parsing. (See Moortgat 1988 on the use of the original counts for the latter purpose.)

Now, the above models suggest a number of systematic logical questions.

14.3 Description Languages and Calculi of Inference

So far, we have used type sequents as a format for stating properties of language families. But why not reconsider this legacy from earlier Parts of this Book? After all, there are many other possibilities.

Perhaps the most obvious formalism for describing L-models or N-models is that of a standard first-order logic over the appropriate similarity type. For instance, here are two relevant observations:

Proposition. The first-order theory of concatenation on expressions from a *one*-symbol alphabet is equivalent to Additive Arithmetic.

The first-order theory on *two* symbols, however, becomes equivalent to the True Arithmetic of addition and multiplication.

Thus, the one-symbol case is decidable, whereas two symbols introduce highly non-effective complexity. (Current practice in mathematical linguistics, of concentrating on one-symbol 'pilot cases', may therefore be misleading.)

Proof. For the first assertion, it suffices to equate sequences with their length, and observe that concatenation becomes addition then.

For the second assertion, it suffices to provide a first-order definable encoding from $(\mathsf{N}, +, \bullet)$ into the two symbol syntactic structure. (The converse embedding from syntax into numbers is provided by the usual techniques of arithmetization.) For the universe, take the subdomain of all sequences consisting only of occurrences of the first symbol, say a . Then, addition has an obvious definition via concatenation. As for multiplication, the following trick employs only first-order concatenation-definable notions:

For two a-sequences x , y , construct the sequence z as follows

— b y bb — b y y bb — ··· b y ··· ybb ,

where the parts — stand for successive non-empty initial segments of x , with the interleaved parts between successive boundaries b and bb receiving an additional copy of y in each step.

The product value may then be read off at the end. ●

Remark. As Kees Doets has pointed out, this result was also obtained by Quine 1946. ●

By contrast, removal of ordering information in sequences decreases complexity: the first-order theory of N-models is embeddable into Additive Arithmetic in an obvious way, and hence it is decidable by Presburger's Theorem.

Next, one can go up to more complex higher formalisms, such as the *monadic second-order* logic over L-models or N-models, allowing quantification over sets of expressions or vectors as well. This is in fact what lies behind the earlier validity of propositional sequent principles. For instance, the validity of

$$a \bullet (a \backslash b) \Rightarrow b$$

corresponds to that of the following second-order principle

$$\forall A \forall B \forall x: \quad \exists y \exists z: (x = yz \wedge Ay \wedge \forall^{A} u: Buz) \rightarrow Bx .$$

But note that this is only a small Horn-type fragment of the full second-order formalism, which need not be subject to general negative complexity results about the latter. For instance, the question as to effective axiomatizability of the *universal* ($\Pi^1{}_1$)-*fragment* of monadic second-order logic for $\backslash, /, \bullet$ over L-models appears to be open.

Thus, in the present general perspective, our earlier categorial calculi formalized a certain fragment of the second-order logic of L- or N-models. And one obvious question is how far such axiomatizations can be complete.

Against this background, it is instructive to consider some completeness results obtained in the tradition of Categorial Grammar. This continues an approach to categorial semantics first touched upon in Chapter 5, but postponed there in favour of a direct lambda semantics for derivations.

Categorial calculi like Lambek's may be modelled 'cheaply' via some suitable notion of *algebra* (obtainable via a Lindenbaum construction). In particular, one may use so-called 'residuate semigroups'. An improvement was obtained in Dosen 1985, using 'residuate semigroups spread over partially ordered semigroups', i.e., structures

$$M = (|M| , \leq , \bullet , \backslash , /)$$

defined over some partially ordered semigroup $(|G|, \leqq, \bullet)$ as follows:

$$|M| = \{ A \subseteq |G| \mid \forall b \in A, a \leqq b: a \in A \}$$

$$A \bullet B = \{ c \in |G| \mid \exists a \in A, b \in B: c \leqq a \bullet b \}$$

$$A \backslash B = \{ c \in |G| \mid \forall a \in A: a \bullet c \in B \}$$

$$B/A = \{ c \in |G| \mid \forall a \in A: c \bullet a \in B \}$$

Next, Buszkowski 1986 did even better, by proving the equivalence between Lambek-derivability and validity over 'residuate semigroups spread over semigroups', where the partial order $\leqq$ is just *identity*. The latter kind of structure comes already quite close to our L-models, which are spread over *free* semigroups. As Buszkowski has shown, the $\backslash, /$ Lambek calculus is complete with respect to the latter L-models too; but the question is open when we add the product $\bullet$. (An up-to-date survey of matters so far may be found in Dosen 1990.) Moreover, no results seem to be known for the case where we add the Boolean operators.

14.4 The Proper Logical Constants

What we can no longer assume in the new context is that the old set of logical operators from standard logic will be sufficient. And in fact, we saw several variants for the old conjunction, as well as various new kinds of operator. Now, can we find some *systematic perspective* on this, which will allow us to formulate issues of *expressiveness* and *functional completeness*?

One possible approach here is proof-theoretic. One can try to generalize the analysis of general 'formats' of introduction rules for operators (as Zucker & Tragesser 1978), showing how some distinguished set of operators defines all possibilities. This suggestion has been taken up in Wansing 1989.

Another approach is model-theoretic, referring to the earlier L- or N-models. The set of all a priori possibilities may then be seen as embodied in the appropriate first-order language of concatenation, which served as a medium for defining the earlier operations. For instance, the Booleans are definable via, e.g.,

$$A \cap B \quad : \quad \lambda x \bullet Ax \wedge Bx ,$$

and the ordering operations via, e.g.,

$$A \bullet B \quad : \quad \lambda x \bullet \exists y \exists z (Ay \wedge Bz \wedge x = yz) .$$

In general, this formalism supplies *infinitely* many non-equivalent possibilities for defining operations on languages. Nevertheless, there are some natural special classes to be considered. For instance, we can look at special syntactic *fragments* with a reasonable motivation.

In particular, a useful perspective may be borrowed here from Modal Logic (cf. Gabbay, Pnueli, Shelah & Stavi 1980, Immerman & Kozen 1987): we may view our types or propositions as variable-free notations for languages described by first-order concatenation formulas employing only some fixed *finite* number of variables. In particular, with the linguistic operations considered so far, truth definitions involved at most 3 variables over sequences. And conversely, such finite variable fragments, at least for pure *predicate*-based first-order languages, always admit of redescription in terms of some finite functionally complete operator notation. In the present case, the result is an ascending *hierarchy of logical constants*, arising from the introduction of a ternary concatenation predicate, and determination of successive complete operator sets for linguistic operations

$$\lambda x \bullet \varphi (A, B, C, \ldots ; x)$$

definable by a schema φ employing 1, 2, 3, ..., n, ... variables.

(At the ground level, the Boolean operations are already complete for the one-variable case where φ uses only x itself.)

A third approach would be to locate some special semantic characteristics of admissible logical constants, thereby cutting down on the number of a priori schemas for linguistic operations.

We shall return to these options for approaching logicality in more detail in the following Chapters. For the moment it will suffice to have established, contra some doubts voiced in the literature, at least the possibility of thinking about these issues in a systematic manner.

14.5 Meanings of Derivations

One basic tool in our preceding study of categorial calculi, and indeed their application to natural language semantics, is the correspondence, developed in Chapter 5, between categorial derivations and *terms* from a *lambda calculus* allowing function application and abstraction (as well as pairing and projection).

From this viewpoint, we are not only interested in valid inferences, but also in the different ways that may be available for deriving the inference (its 'readings', to extend a linguistic concept). As it stands, however, this correspondence only works for functional and product types. Can it also be extended tot the other new logical operators encountered above?

In fact, this can be done relatively easily, by extending the Lambda Calculus with suitable operations matching additional operators such as Boolean conjunction or disjunction. Thus, we can code up more general logical derivations, and bring out their intuitive differences: say, that between the following two routes to the same valid inference:

$$-A\cap(A\cup B) \qquad -A \qquad -(A\cap -B)\,,$$
$$-A\cap(A\cup B) \qquad B \qquad -(A\cap -B)\,.$$

Example. Encoding Another Conjunction.

In Chapter 5, it was shown how sequential categorial conjunction corresponds to the formation of 'products' of types. But it is also possible to encode a second conjunction by means of a new kind of pairing plus projection:

$$\frac{X \Rightarrow a \qquad X \Rightarrow b}{X \Rightarrow a\cap b} \qquad \frac{\tau^1{}_a\,[u_X] \qquad \tau^2{}_b\,[u_X]}{[\,\tau^1{}_a[u_X] \mid \tau^2{}_b[u_X]\,]}$$

$$\frac{X,\, a,\, Y \Rightarrow b}{X,\, a\cap c,\, Y \Rightarrow b} \qquad \frac{\tau_b\,[u_X, u_a, u_Y]}{\tau_b\,[u_X, \pi_L(u_{a\cap c}), u_Y]}$$

$$\frac{X, a, Y \Rightarrow b}{X, c \cap a, Y \Rightarrow b}$$ likewise,

where the formation of the new products presupposes equality of parameter sequences in both factors.

◆

Nevertheless, the key issue is whether such enriched lambda calculi provide a natural *independent* intuition concerning more general logical derivations.

Remark. Recognizing Power With Additional Logical Constants.
Other more linguistically inspired topics, studied earlier in this Book, may have some meaning in this wider setting too. For instance, what would be the more general logical import of the notion of *recognizing power* which formed the topic of Chapter 8? In particular, does the addition of Boolean operators change the recognizing power of our earlier categorial paradigm? ◆

There is more to the comparison between the dominant model theory of our earlier Parts and the present semantic approach to categorial calculi. For instance, would not the functional domains D_a of Chapters 2, 5 *themselves* provide another direct semantics for type sequents, inducing a notion of validity for sequents

$a_1, ..., a_n \Rightarrow b$

along the lines of that for L- or N-models? There is a subtle difference here. The earlier notion of validity rested on an explication of the following form:

There exists some constructive (lambda-definable) *procedure* for converting, in any functional model, an arbitrary sequence of objects in the domains $D_{a1}, ..., D_{an}$ into an object in the domain D_b.

With the notions of validity in this Chapter, however, one such procedure is fixed beforehand, namely 'concatenation' of successive strings or 'addition' of vectors (or partial 'composition' of successive relational arrows in Chapter 16), whose result delivers a combined object in the conclusion type, without any further transformation.

This distinction has repercussions on the lambda term correspondence sketched above, whose motivation was clear in set-theoretic or categorial models in the D-style. Within the present L- and N-models (as well as the I- and R-models of subsequent

Chapters), different lambda recipes for computing denotations will tend to collapse: they merely represent equivalent ways of seeing that the object constructed via our fixed convention lives in the target type domain. This shows in the following observations:

- the two 'readings' of the string $(t/t)\ t\ (t\backslash t)$ with different scopes for the left and right operators will merely indicate two ways of grouping in concatenation,
- the lambda abstraction in the derivation of the sequent $t \Rightarrow (t/(t\backslash t)))$ does not transform the object in the t-domain: the latter itself is precisely the unique object having the desired functional behaviour.

Thus, the new notions of validity rest more on the special choice of models involved, in particular, on their operational structure.

14.6 <u>Appendix</u> **Linear Logic**

An emphasis on computational processing and the proper level of syntactic detail involved therein is also characteristic of the current research line of so-called 'Linear Logic'. This area shows a number of striking resemblances with Categorial Grammar, especially in the extended sense developed here. Indeed, the latter may be viewed as a linguistic paradigm taking a somewhat more liberal view of syntax than ordinary linguistic theories (in being willing to countenance permutations) - while the former is a logical paradigm taking syntax rather more seriously than is done in ordinary logical approaches: two converging deviations which explain their rapprochement. The present appendix will merely point at a number of analogies.

First, we need some concrete system for the purpose of comparison. There is already some choice here, since linear logic has, at least, intuitionistic and classical variants. We shall not go into the relative merits of 'Linear A' or 'Linear B' here.

Instead of providing a full motivation, we refer to Girard 1987, Lafont 1988 for the proof-theoretical and computational background of the following basic 'intuitionistic' linear calculus, which allows Permutation of premises: [the notation for connectives has been modified for the sake of typographical consistency]

Axioms

$A \Rightarrow A$ $\qquad$ $\Rightarrow 1$ $\qquad$ $X, \bot \Rightarrow A$ $\qquad$ $X \Rightarrow T$

Rules

First, there are two conjunctions:

$$\frac{X \Rightarrow A \qquad Y \Rightarrow B}{X, Y \Rightarrow A \bullet B} \qquad \frac{X, A, B \Rightarrow C}{X, A \bullet B \Rightarrow C}$$

$$\frac{X \Rightarrow A \qquad X \Rightarrow B}{X \Rightarrow A \cap B} \qquad \frac{X, A \Rightarrow B}{X, A \cap C \Rightarrow B} \qquad \frac{X, A \Rightarrow B}{X, C \cap A \Rightarrow B}$$

Note that these two notions would collapse in the presence of the usual structural rules of standard logic. Next, there is disjunction and implication:

$$\frac{X, A \Rightarrow B \qquad X, C \Rightarrow B}{X, A \cup C \Rightarrow B} \qquad \frac{X \Rightarrow A}{X \Rightarrow A \cup B} \qquad \frac{X \Rightarrow A}{X \Rightarrow B \cup A}$$

$$\frac{X \Rightarrow A \qquad Y, B \Rightarrow C}{X, Y, A \rightarrow B \Rightarrow C} \qquad \frac{X, A \Rightarrow B}{X \Rightarrow A \rightarrow B}$$

Next comes Girard's 'exponential' or 'modality', absorbing some behaviour of the earlier structural rules into its logical rules (and thereby allowing for a translation from standard logics into linear systems):

$$\frac{X \Rightarrow A}{X, !B \Rightarrow A} \qquad \frac{X, !B, !B \Rightarrow A}{X, !B \Rightarrow A}$$

$$\frac{X, A \Rightarrow B}{X, !A \Rightarrow B} \qquad \frac{!A \Rightarrow B}{!A \Rightarrow !B}$$

The intuitive motivation is that $!A$ stands for arbitrary finite iterations of A-type propositions. (We skip the rules here for the dual operator $?A$.)

Finally, we add

$$\frac{X \Rightarrow A}{X, 1 \Rightarrow A} .$$

This calculus as well as several variants has been under intensive investigation recently (cf. Abrusci 1988A, Sambin 1988). For present purposes, it will suffice to note the obvious resemblance, and indeed identity, of its $\bullet, \rightarrow$ fragment with that of the undirected Lambek Calculus LP . This analogy can be extended, as it turns out, to a useful comparison between the existing proof theory for categorial grammar and that for linear logic. (For a first survey, see Ono 1988.)

A case in point are the earlier-mentioned completeness theorems for categorial calculi in L-models. These are very reminiscent of recent work on the semantics of linear logic. Roughly speaking, the results in Sambin 1988, Abrusci 1988A seen comparable to Dosen's kind of completeness theorem, be it for a richer kind of language. In particular, they are forced to employ a notion of 'closure' in the definition of a product, which involves something like Dosen's ordering. It seems that one cannot do better than this in general, restricting to more special simple model classes (as Buszkowski was able to do for the pure Lambek Calculus). The reason is an example in Abrusci 1988A of a distributivity principle that is non-derivable in the basic linear logic, even though it is valid in every 'simple' intuitionistic topophase structure (with the closure operation reducing to identity). One single perspective unifying these various approaches has been presentred recently in Ono 1990.

The exact extent of the analogies between the metatheory of categorial grammar and linear logics remains to be established.

To conclude , there is a simple connection here with our earlier L-models.

Proposition. The following correspondence provides a sound interpretation for the corresponding part of linear logic:

(Boolean) $\perp : \perp$, $T : T$, $\cap : \cap$, $\cup : \cup$,

(order) $\bullet : \bullet$, $1 : 1$.

There is no direct categorial analogue of linear $\rightarrow$: but \ and / are just right for the directed, non-permuted calculus of linear logic proposed in Abrusci 1988B. Moreover, there is no analogue for linear ! as it stands, although Kleene iteration seems to have a similar infinitary flavour, while validating the structural rule of Contraction - as was already observed before. Nevertheless, * failed to validate some of the necessary laws for modalities. But probably, some suitably defined iterative operator on formal languages will do the job.

It remains to comment on the connection between Linear Logic and our categorial logics. The basic notion of an occurrence-based calculus of inference without structural rules is common to both. But then, this is an old idea in any case, which has been discovered independently several times over the past decades (see the survey Ono 1988 on the pioneering work of Grishin or Weyrauch). What Linear Logic has added here is the conscious pursuit of new varieties of logical constants in such a framework, an idea which had been only slowly emerging in Categorial Grammar (although it was consciously pursued earlier on in Relevance Logic, which already knew most, if not all, linear connectives: cf. Dunn 1985, Avron 1988). These additional resources are now being explored systematically for possible linguistic applications: see Morrill 1990 for a first round of examples. Finally, what seems definitely new in Linear Logic is the insight that logical constants may be used to encode various additional structural inferences after all, on the basis of a system lacking them as such. This trade-off is still a mysterious phenomenon, in that the 'exponentials' which seem typically needed for encoding 'standard' structural behaviour have turned out to make propositional linear logic *undecidable* (cf. Lincoln et al. 1990), unlike the standard propositional logics themselves.

The latter interplay is significant in various fields, not just as a technical trick, but also as a substantial issue. Already in linguistics, one finds a natural distinction between grammatical mechanisms that are supposed to 'run freely' in the syntax of natural languages, such as function application or perhaps also function composition, and operations that need explicit lexical encoding, such as conversion of argument positions when passivizing sentences. And there are some intriguing border-line cases, such as identification of argument positions, which seems sometimes permitted freely, by judicious uses of the Contraction rule, but which needs mostly explicit encoding, via reflexivizers such as "self". (The variable-free formulation of first-order predicate logic in Quine 1971 is relevant too, which lexicalizes permutation, identification and expansion of argument positions by means of explicit operators, even inside logic.)

But similar issues also arise in contemporary computer science. 'Loss of classical structural rules', such as Monotonicity, has been a focus in the current literature on non-standard varieties of inference in Artificial Intelligence (such as default logics or circumscription). What has been neglected, however, is the probability that such varieties of inference also engender varieties of logical constants beyond the classical core. And once the latter are taken into account, there is a genuine issue of encoding non-standard modes of inference: whether in new notions of consequence or in the special behaviour of certain operators.

15 Modal Logic of Information Patterns

Now, let us abstract away from syntactic occurrences, and move to a level of analysis which is traditionally considered more appropriate for locating information structures. Here, we can take a lead from the tradition - as there are already established systems of logic based on information structures, noteworthy examples being Intuitionistic Logic, Relevant Logic and in fact already Modal Logic.

15.1 Modal Logic of Information Patterns

The simplest kind of information structure is a pattern of states or stages, ordered by inclusion ('possible growth'):

$$(I, \subseteq)$$

Arguably, informational inclusion should at least be a *partial order*. A richer perspective in the literature also has a notion of addition ('merging') of states or pieces of information, embodied in *semi-lattices*

$$(I, \subseteq, +),$$

allowing one to define inclusion in the usual manner:

$$x \subseteq y \quad \text{iff} \quad x+y = y .$$

We shall stick mainly with partial orders here, for the following reason. The latter structures allow for a natural operation of addition as well, through the notion of a *supremum* in the inclusion ordering:

$$x+y = z \qquad \text{iff} \qquad z = \sup(x, y) .$$

In general, the latter will only be partially defined, of course: but this seems just right for addition of information pieces - where, for instance, contradictory evidence need not admit of meaningful merging.

Both approaches agree on the following crucial deviation from the earlier Chapter, however: addition of information pieces is an undirected, *symmetric* operation. Abstract information is not sensitive to any ordering. Some people would even take this property to be the fundamental divide between 'linguistics' and 'logic'.

In this Chapter, we shall use the term I-*models* mainly for the above partial orders, possibly carrying 'valuations' (see below), but sometimes also for semi-lattices over them.

Now, a number of questions arises similar to those encountered before. Here is a short 'tour d'horizon', motivating the more extensive discussion in later Sections.

- What kind of logical operators are suitable here?

One simple natural formalism over information models is that of a propositional *Modal Logic*, that may be evaluated on I-models

$$\mathbb{M} = (\mathrm{I}, \subseteq, \mathrm{V})$$

having a valuation V for the proposition letters spread over their state pattern. The key clauses in its truth definition are

$$\mathbb{M} \models \Box\varphi[w] \quad \text{iff} \quad \mathbb{M} \models \varphi[v] \ \textit{for all} \ v \supseteq w$$
$$\mathbb{M} \models \Diamond\varphi[w] \quad \text{iff} \quad \mathbb{M} \models \varphi[v] \ \textit{for some} \ v \supseteq w\,.$$

The minimal logic valid here is S4 , or S4 plus Grzegorczyk's 'induction axiom' Grz (cf. Bull & Segerberg 1984) if one restricts attention to finite partial orders only.

This formalism only looks 'forward' along the inclusion order. But cognitive operations like retraction or revision will also involve looking back at previous, poorer information states. A simple extension doing that would become a *Tense Logic* having a future operator $\mathbb{F}$ corresponding to the earlier modality, but also a past dual $\mathbb{P}$:

$$\mathbb{M} \models \mathbb{P}\varphi[w] \quad \text{iff} \quad \mathbb{M} \models \varphi[v] \ \textit{for some} \ v \subseteq w\,.$$

Finally, in addition to standard Boolean conjunction, Relevant Logic has an 'intensional conjunction' which may be understood as a new binary modal connective, whose informational reading is as follows:

$\mathbb{M} \models \phi + \psi\,[w]$ iff

there exist u, v *such that* $w = \sup(v, u)$, $\mathbb{M} \models \phi\,[u]$, $\mathbb{M} \models \psi\,[v]$.

Thus, there is a rich structure of logical constants, even on information models collapsing addition of identical states.

These seem to create a hierarchy of ever more complex statements about information patterns. In the limit, a natural general format of semantic truth conditions then becomes a *first-order language* referring to the binary order $\subseteq$ as well as having unary predicates over stages.

Example. Unary and Binary Modalities.

The usual unary modalities are expressible as follows:

$\Diamond p$, $\mathbb{F}p$ $\lambda x \bullet \exists y\, (x \subseteq y \wedge Py)$

i.e., ' p is still an epistemic possibility'

$\mathbb{P}p$ $\lambda x \bullet \exists y\, (y \subseteq x \wedge Py)$

i.e., ' p has been accepted once' .

And here is the above binary modality, involving a 'join' of information:

$p+q$ $\lambda x \bullet \exists y\, \exists z\, (Py \wedge Qz \wedge$ ' x is the supremum of $\{y, z\}$ ') .

The first-order format even suggests natural generalizations here, such as a dual binary operator involving a 'meet':

$p*q$ $\lambda x \bullet \exists y\, \exists z\, (Py \wedge Qz \wedge$ ' x is the infimum of $\{y, z\}$ ') .

Also, e.g., using suprema reflecting 'addition' of information stages, a notion of implication may be defined as in Chapter 5:

$p \rightarrow q$ $\lambda x \bullet \forall y\, (\, Py \rightarrow Q\, \sup(x, y)\,)$.

🍎

This general first-order language allows us to step back, as it were, and re-examine the *design* of existing logics, asking whether their founding fathers managed to bring out enough of the informational phenomena that are of interest to us. One more specific question to be addressed later on then is how this full first-order language can be layered into a meaningful hierarchy of more or less complex logical constants.

- How can we describe more special informational phenomena?

One important phenomenon in information processing is that of informational *persistence*:

Which statements, once obtained, retain their truth when our information grows?

Here is precisely where another major existing formalism comes into its own, namely *Intuitionistic Logic* (see the survey Troelstra & van Dalen 1988.) All its statements are persistent along the inclusion order. And indeed, we shall prove that it may be viewed as being precisely the 'persistent fragment' of our basic modal language. Thus, it pays to specialize as well as generalize. (Of course, this also means that intuitionistic logic has no favoured status as a logic of information, as it treats only part of the relevant operators. One reason why it does not see these additional possibilities is that a 'logic of pure progress' will collapse distinctions, such as that between $p+q$ and $p \wedge q$.)

Another interesting phenomenon of information flow is exhibited by 'additive' propositions, satisfying the condition of *cumulation*:

if $\mathbb{M} \models \varphi\,[u]$ *and* $\mathbb{M} \models \varphi\,[v]$ *and* $w = \sup(u, v)$,
then $\mathbb{M} \models \varphi\,[w]$.

For instance, which syntactic forms guarantee this behaviour, starting from atomic propositions already having it? One example are the persistent formulas: but are there others involving $+$?

Remark. 'Information' as a Mass Term.
This type of question has strong analogies with calculi of mass terms, plurals or temporal aspect in the semantics of natural language, where persistence and additivity are fundamental notions (cf. Krifka 1989). In the metaphysics of ordinary language, "information" is a kind of stuff, that can flow or leak, be lost or found.

- What calculi of inference are appropriate over information models?

Whatever choice is made, the general analysis given in Chapter 14 still applies: validity in the above calculi will involve fragments of the universal monadic second-order closure of the above first-order language. Still, there is a difference with the case of L-models too. The above I-models form an elementary class, as the defining axioms for partial orders or semi-lattices are *first-order*. Therefore, we have this

Fact. On the universe of I-models, logical validity in any modal formalism within the first-order language is effectively axiomatizable.

Proof. The first-order transcription of any modal principle is valid iff it holds in all I-models. And since the latter form an elementary class, this is equivalent to a problem of consequence in ordinary first-order logic, where effective axiomatizability holds by the Completeness Theorem.

The art remains, of course, to find enlightening purely modal axiomatizations.

Perhaps more interesting is the issue whether this informational perspective suggests *new* notions of valid inference, different from the standard ones. For instance, already Chapter 5.1 contained a procedure of successive addition of pieces of information supporting the premises of an inference, and then checking whether the conclusion holds in their *sum*.

The deviant behaviour of such varieties of consequence will be studied later on in this Chapter, as well as in the next.

15.2 A Hierarchy of Logical Constants

We shall examine two systematic perspectives on a possible hierarchy of logical constants for information models.

Bounds on Variables and Numbers of Semantic Registers

Our first perspective takes its point of departure in a well-known fact about Modal Logic. The basic modal language can be translated (as was already suggested in the above) into a standard *first-order* one involving a binary relation $\subseteq$ as well as unary predicates P corresponding to the proposition letters p . For instance, the modal formula

$$\Box\Diamond\Box p$$

will then correspond to the first-order formula

$$\forall y\,(x\subseteq y \rightarrow \exists z\,(y\subseteq z \wedge \forall u\,(z\subseteq u \rightarrow Pu)))\,.$$

(Van Benthem 1985 develops the full mathematical theory of this correspondence.) Next, the point is that the first-order formulas needed for this are special in several ways. For instance, all their quantifiers occur 'restricted' to R . But more significantly, they can make do with a very small number of distinct variables:

two variables suffice.

To see this, translate proposition letters as Px or Py as the case may be, respect Boolean operations as usual, and treat the modality as follows:

$$\tau(\Box\varphi) \;=\; \forall y\,(x\subseteq y \rightarrow \tau(\varphi))$$

if y is the free variable in $\tau(\varphi)$

(otherwise, use $\forall x\,(y\subseteq x \rightarrow \tau(\varphi))$).

Example. Economizing On Variables.

The translation of $\Box\Diamond\Box p$ can be written more economically, but equivalently, as

$$\forall y\,(x\subseteq y \rightarrow \exists x\,(y\subseteq x \wedge \forall y\,(x\subseteq y \rightarrow Py)))\,.$$

Thus, ordinary modal logic is part of a 2-variable fragment of a first-order language over information structures. The semantic import of this restriction may be stated as follows:

> Computing semantic truth conditions for modal statements involves only two 'registers', or put differently: it only involves inspecting patterns of at most two states at any one time.

The basic modal language does not exhaust this 2-variable fragment of the full first-order formalism. For instance, translations of its tense-logical extension with $\mathbb{F}$ and $\mathbb{P}$ will also fall within the latter. Therefore, it must have further semantic characteristics, that will be brought out below.

Richer fragments with 3 variables will become necessary, for instance, with extended tense-logical formalisms, having binary operators "Since" and "Until". And writing out the earlier clauses for an addition modality in terms of suprema will turn out to involve essentially 4 variables, i.e., four states in their interaction.

Now, Gabbay 1981 has shown that, quite generally,

Theorem. There exists an effective one-to-one correspondence between finite intensional operator formalisms and fixed k-variable fragments of the above first-order logic over I-models.

Here is a complete description of the first two steps in the resulting hierarchy, working inside some fixed model $\mathbb{M}$.

Proposition. All operations of the form $\lambda x \bullet \varphi(x, A_1, \ldots, A_n)$ with φ first-order, employing only *one* variable are definable by Boolean combination of the A_i.

With *two* variables, the following set of modal operators is functionally complete:

$$\overrightarrow{\Diamond}^{+} p \quad : \quad \lambda x \bullet \exists y\, (x \subsetneq y \wedge Py)$$

$$\overleftarrow{\Diamond}^{+} p \quad : \quad \lambda x \bullet \exists y\, (y \subsetneq x \wedge Py)$$

$$I\, p \quad : \quad \lambda x \bullet \exists y\, (x \not\subseteq y \wedge y \not\subseteq x \wedge Py) .$$

Proof. With one variable (x).
Subformulas of the form $\exists x\alpha$ are closed and hence have a fixed truth value. Thus, only a Boolean combination remains.

With two variables (x, y).
Consider innermost subformulas starting with, say, a quantifier $\exists y$. Such formulas of the form $\exists y\ \alpha(y, x)$ may be rewritten as Boolean combinations of cases

$$\exists y \bigwedge\{(\neg)\ x\subseteq y\ ,\ (\neg)y\subseteq x\ ,\ (\neg)Py\}$$

using general logic. But then, by the partial order axioms, the above three possibilities clearly suffice. And this procedure may be repeated outward. ■

In general, however, no general functional completeness result holds over the present class of models, allowing arbitrary finite widths of branching.

Theorem. For the full first-order language over information models, no finite functionally complete set of operators exists.

Proof. This may be shown using the method of *Ehrenfeucht Comparison Games* with Pebbling presented in Immerman & Kozen 1987.

As is stated in Part VII of this Book, two models are indistinguishable by first-order sentences up to quantifier complexity n iff the second player in an Ehrenfeucht comparison game over these models has a winning strategy in any play over n rounds. And this analysis may be refined as follows:

> Indistinguishability by sentences up to depth n employing only some fixed set of k variables amounts to the existence of a winning strategy for the second player in a modified Ehrenfeucht game over n rounds, where each player receives k *pebbles* at the start, and can only select objects, in the two models being compared, by putting one of these pebbles on them.

Now, any finite operator formalism has a first-order transcription involving only some fixed finite number k of variables over states. Therefore, if any such formalism were functionally complete, this would mean that the full first-order language over our universe of I-models would actually be logically equivalent to one of its k-variable fragments. But the latter kind of reduction is impossible:

Let the first-order sentence ϕ state the existence of a top node having at least k+1 distinct immediate successors. Consider any two models consisting of a top node with k and k+1 immediate successors, respectively, and the same valuation at all states. It is evident that the second player has a winning strategy in the Ehrenfeucht comparison game between such models with k pebbles, over an arbitrary finite number of rounds. Hence, no k-variable sentence distinguishes between these two I-models: whereas ϕ can.

Thus, one part of the art in Modal Logic is to locate suitably expressive fragments which still admit of an enlightening variable-free operator analysis.

Varieties of Bisimulation

Another general perspective on the Modal Hierarchy of logical constants and their expressive power starts from special semantic characteristics exhibited by modal formulas.

A first example of this has to do with 'search'. Intuitively, deciding the truth or falsity of some modal statement φ at an information state w involves only surveying states 'accessible' from w via successive steps in the inclusion pattern $\subseteq$. The special case with only steps 'into the epistemic future' induces a constraint which is embodied in the well-known "Generation Theorem" of Modal Logic, stating that modal formulas are invariant, at any state w, between evaluation in a full model $\mathbb{M}$ and evaluation in the smallest $\subseteq$-closed submodel that is 'generated by' w. (In general information processing, of course, 'epistemic retreats' and general 'wavering' are admissible too.)

Behind this constraint, there lies a more general semantic issue of invariance which was already discussed in Chapter 12. There, the notion of *bisimulation* was introduced as a plausibble behavioristic identification between informationally 'equivalent' models. And the well-known "p-Morphism Theorem" of Modal Logic says that modal formulas are invariant for bisimulation. We repeat the relevant

Definition. Bisimulation for Basic Modal Logic.

A relation C between two models $\mathbb{M}_1, \mathbb{M}_2$ is a *bisimulation* if it satisfies the following conditions:

1 if w_1Cw_2, then w_1, w_2 carry the same valuation on proposition letters

2a if w_1Cw_2, $w_1 \subseteq v_1$, then there exists v_2 such that $w_2 \subseteq v_2$, v_1Cv_2

2b analogously, in the other direction.

Now, modal formulas φ are *invariant* for bisimulation:

Proposition. For any bisimulation C between two I-models, if $w_1 C w_2$, then $\mathbb{M}_1 \models \varphi[w_1]$ iff $\mathbb{M}_2 \models \varphi[w_2]$.

And this property is completely characteristic for the basic modal formalism (cf. van Benthem 1985):

Theorem. A first-order formula $\varphi = \varphi(x)$ is (equivalent to) a translation of a modal formula if and only if it is invariant for bisimulation.

Proof. (Sketch) 'Only if'. By a straightforward induction on modal formulas, where the back-and-forth clauses take care of the modality.

'If'. Let $\mathbb{m}(\varphi)$ be the set of modal consequences of φ. We prove that, conversely,

$$\mathbb{m}(\varphi) \models \varphi,$$

from which the desired definability follows by Compactness.

So, let $\mathbb{M} \models \mathbb{m}(\varphi)[w]$. By a standard model-theoretic argument, there must be $\mathbb{N}$, v satisfying

- $\mathbb{N} \models \varphi[v]$,
- $(\mathbb{M}, w)$, $(\mathbb{N}, v)$ verify the same modal formulas.

Now, take any two *countably saturated* elementary extensions of $\mathbb{M}$, $\mathbb{N}$: say, $\mathbb{M}^*$ and $\mathbb{N}^*$. In such saturated models, the following stipulation defines a bisimulation:

" $(\mathbb{M}^*, x)$ verifies the same modal formulas as $(\mathbb{N}^*, y)$ ".

For instance, if x' is any $\subseteq$-successor of x in $\mathbb{M}^*$, then each *finite* subset Δ of its modal theory is satisfiable in some $\subseteq$-successor of y in $\mathbb{N}^*$: since the modal formula $\Diamond \bigwedge \Delta$ holds at x and therefore also at y. But then, its *full* modal theory must be satisfiable in some $\subseteq$-successor y' of y, by Saturation: and such a state y' is the required match for x' in the zigzag clause of bisimulation.

But then we have, successively:

$\mathbb{N} \models \varphi[v]$ $\quad$ $\mathbb{N}^* \models \varphi[v]$ (elementary extension)

$\mathbb{M}^* \models \varphi[w]$ (bisimulation invariance) $\quad$ $\mathbb{M} \models \varphi[w]$ (elementary descent).

Example. Finite Models.
Since finite models are always saturated, the above is illustrated more concretely in their midst. For finite $\mathbb{M}$, $\mathbb{N}$, the following two assertions are equivalent:

1. $(\mathbb{M}, w)$, $(\mathbb{N}, v)$ verify the same modal formulas ,
2. some bisimulation between $\mathbb{M}$ and $\mathbb{N}$ connects w to v .

The same definition of C as above works here, by a simple direct argument. ♦

These were the characteristic semantic properties of the basic modal formalism, that will have to be modified in richer logics of information structure. For instance, tense-logical formulas will not be invariant for the above bisimulations. Their parallel evaluation in two models requires back-and-forth clauses, not just with respect to successors in the inclusion ordering, but also with respect to predecessors. Then, a next level of bisimulation might involve proper respect for 3-state configurations of 'betweenness' in information patterns, and so on.

Thus, a new hierarchy arises, of various kinds of bisimulation between models, respecting ever more details of their information patterns, whose corresponding invariances become ever less constraining on first-order statements. The syntactic effects of these invariances may be determined inside the full first-order language, along the lines of the earlier characterization of the basic modal fragment.

In principle, this second classification is independent from the earlier finite-variable approach. Nevertheless, in Chapter 17, the latter too will be analyzed via invariance for a matching model-theoretic notion of semantic simulation, being ' k-*partial isomorphism*'. The present forms of bisimulation then turn out to be specialized varieties of the latter.

15.3 Some Steps of the Ladder

Let us illustrate the above considerations by climbing a few steps of the k-variable hierarchy (using bisimulation as a minor theme).

- First, consider the level with k=2 . In studying special informational phenomena, one can specialize even further, restricting attention to very special modal formulas. A prime example here is the earlier notion of *Persistence*:

  Which formulas φ have the property that always
  if $\mathbb{M} \models \varphi[w]$ and $w \subseteq v$, then $\mathbb{M} \models \varphi[v]$?

Persistence is a well-known feature of formulas in *intuitionistic* logic: and indeed, we have the following

Observation. The persistent modal formulas are exactly those definable using the intuitionistic connectives $\wedge, \vee, \rightarrow, \bot$ (with each proposition letter occurring in the scope of some $\rightarrow$).

Proof. In one direction, intuitionistic formulas $\varphi \rightarrow \psi$ (i.e., $\Box(\varphi \rightarrow \psi)$ in the modal reading) are persistent, and so are their compounds with $\wedge, \vee$ and $\bot$.

Conversely, let φ be persistent: and hence equivalent to $\Box\varphi$.
Now, rewrite φ using some well-known modal equivalences to a form as described. The key observation here is that any formula $\Box\alpha$ is equivalent to some form

$$\Box \bigwedge\bigvee\{(\neg)p,\ (\neg)\Box\beta\},$$

and hence to

$$\bigwedge\Box\ \bigvee\{(\neg)p,\ (\neg)\Box\beta\},$$

i.e., to a conjunction of forms

$$\Box(\bigwedge\{p,\ \Box\beta\} \rightarrow \bigvee\{p, \Box\beta\}$$

(with the falsum $\bot$ used for empty disjunctions). $\blacksquare$

- The next level k=3 introduces several natural kinds of statement about information patterns which go beyond the resources of the standard modal language. For instance, when 'updating' an information state with the proposition φ, the new relevant states would be those where φ has become true for the first time: and this requires a comparison with a 3-configuration of the initial state, the first φ-state, and all states lying in between. Thus, in addition to succession, *betweenness* along the information ordering becomes important.

A suitable modal language for this level introduces a binary 'temporal' operator "Until" as follows:

$\mathbb{M} \models \mathbb{U}\varphi\psi[w]$ iff *there exists some* $v \supseteq w$ *with* $\mathbb{M} \models \varphi[v]$
and for all u *with* $w \subseteq u \subsetneq v$: $\mathbb{M} \models \psi[u]$.

Example. Updating.
The formula

$$\mathbb{U}(\varphi \wedge \psi)\neg\varphi$$

expresses that ψ holds at some updated φ-state. $\blacksquare$

This language is appropriate for describing the behaviour of programs over time (cf. Goldblatt 1987), involving properties like 'safety', 'liveness' or 'absence of unsolicited response' . Thus, it also seems useful as a description language for information processing in a more general sense. Here is one further illustration.

Example. Necessary and Sufficient Conditions.
To say that φ is a *sufficient* condition for ψ may be rendered, at least to a reasonable approximation, within the basic modal logic:

$$\Box(\varphi \to \Diamond\psi) .$$

This is in fact what computer scientists call a 'liveness' property (namely, that φ 'enables' ψ to occur).
To say that φ is a *necessary* condition for ψ , however, involves inspection of the past of ψ states. A first attempt might read

$$\Box(\psi \to \mathbb{P}\varphi) ,$$

where $\mathbb{P}$ was the past analogue of $\Diamond$. But, we want only past occurrences within the future of the initial point of evaluation, and hence, the appropriate formalization must involve "Until" :

$$\neg \mathbb{U}\psi\neg\varphi .$$

●

Calculi of logical deduction in the $\{ \Diamond , \mathbb{P} , \mathbb{U} \}$ formalism may be found in Goldblatt 1987 or Burgess 1982. (But note that some of their axioms depend on *linearity* of the underlying temporal ordering: something which is of course not assumed here for information patterns.)

The underlying first-order language of this 'temporal' formalism indeed employs 3 variables, witness a typical translation clause like that for "Until" :

$$\tau(\mathbb{U}pq) \quad = \quad \exists y(x \subseteq y \wedge Py \wedge \forall z(x \subseteq z \subsetneq y \to Qz)) .$$

What about plausible semantic invariances via bisimulation for such formulas at the 3-variable level?

One enriched notion of *strengthened bisimulation* satisfies all earlier clauses, as well as a certain respect of betweenness:

3a if w_1Cw_2 , v_1Cv_2 and $w_1 \subseteq u_1 \subseteq v_1$,
then there exists some u_2 with u_1Cu_2 , $w_2 \subseteq u_2 \subseteq v_2$,

3b analogously in the opposite direction .

This results in additional power of discrimination:

Example. Finer Distinctions.
Two frames may be bisimulation equivalent, without admitting strengthened bisimulation:

■

Even so, some 3-variable statements do not transfer under this mode of identification between models.

Example. Common Successors.
Statements involving connections like $x \subseteq z \wedge y \subseteq z$ are vulnerable to the above bisimulations. Here is a graphical illustration:

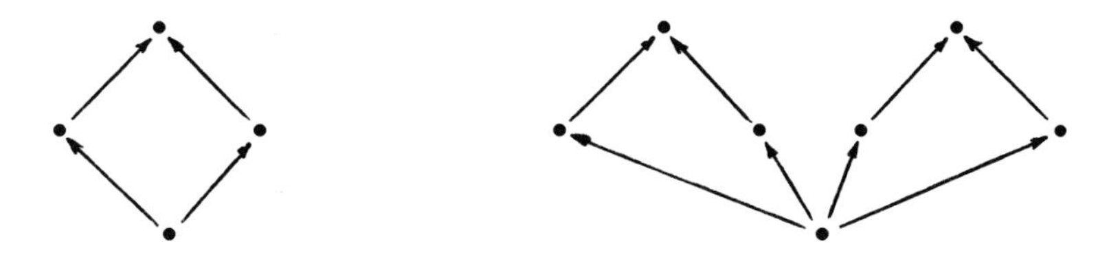

There is an obvious strengthened bisimulation between these frames: and yet only the one on the left validates 'confluence':

$$\forall x \forall y \forall z \, ((x \subseteq y \wedge x \subseteq z) \rightarrow \exists x \, (y \subseteq x \wedge z \subseteq x)) .$$

■

All statements in the "Until" formalism are indeed invariant for strengthened bisimulations. Even so, this semantic notion does not quite 'fit' - as becomes clear with technical difficulties in adapting the characterization theorem for the original notion of bisimulation. And so far, more sophisticated candidates have turned out to fare no better.

The main difficulty appears to be this. One is trying to characterize a genuine 3-variable fragment by focussing on its formulas with one free variable only. This seems inappropriate, and it would be much easier to work with a formalism allowing formulas both with one free variable (*properties* of states) and with two free variables, expressing binary *relations* over states. This change would fit in well with existing 'multi-dimensional modal logics', having several indices of evaluation simultaneously. As we shall see in Chapters 16,17, there is also a good deal of independent motivation for making such a move, which imports explicit dynamic aspects of information processing into our modal formalisms, and further relevant results will be presented there.

- Next, we consider the level $k=4$. This is the natural habitat for our earlier modal notion of 'addition' of information:

$$\mathbb{M} \models \varphi+\psi\,[w] \quad \text{iff} \quad \textit{there exist } u, v \textit{ such that } w = \sup(u, v),\ \mathbb{M} \models \varphi[u],\ \mathbb{M} \models \psi[v] .$$

When spelt out, this becomes

$$\exists y \exists z\, (y \subseteq x \wedge z \subseteq x \wedge \varphi(y) \wedge \psi(z) \wedge \neg \exists u\, (y \subseteq u \wedge z \subseteq u \wedge u \subsetneq x)\,) .$$

This fragment, too, may be analyzed in terms of semantic invariances. For instance, a *strong bisimulation* has all the earlier features of strengthened bisimulations, while it also 'respects suprema' in the following sense:

4a if $w_1 C w_2$ and $w_1 = \sup(u_1, v_1)$, then there exist u_2 , v_2 such that $w_2 = \sup(u_2, v_2)$ and $u_1 C u_2$, $v_1 C v_2$,

4b and vice versa .

Here is a small illustration of its effects:

Proposition. Two finite models $(\mathbb{M}, w)$, $(\mathbb{N}, v)$ are strongly bisimulation equivalent if and only if they verify the same extended modal formulas (of the $\Diamond$, + language) in their roots w, v .

Proof. The 'only if' direction is a simple induction.

As for 'if': here is the crucial observation. Suppose that x, y verify the same modal formulas, and $x = \sup(u, z)$. Assume that there are no states s, t with $y = \sup(s, t)$ such that both u, s and z, t verify the same modal formulas. For each of these finitely many cases, then, pick some formula α with $u \models \alpha$, $s \not\models \alpha$ or a formula β with $z \models \beta$, $z \not\models \beta$. Then, x, y will disagree on the modal formula $\mathbb{A}\alpha + \mathbb{A}\beta$. ◆

- The temporal analogy also highlights the 'backward' direction of information processing. In a general theory of information, both forward and backward directions of search through information patterns will be essential. Many important cognitive activities involve surveying the 'epistemic past', or even back-and-forth movement. The 'idealized mathematician' of intuitionistic logic may move steadily ahead, but ordinary mortals will be found plodding on zigzagging epistemic trajectories.

Example. Conditionals.
A popular folklore account of possibly counterfactual conditionals is the so-called 'Ramsey Test', which runs like this:

> "Assume the antecedent. Or, if this is not consistently possible, go back to the first previous stage where it was still possible. Then see if the consequent always follows from the antecedent".

Its formulation involves an obvious backward past version "Since" ($\mathbb{S}$) of the earlier operator "Until" :

$$\neg\mathbb{S}\ (\Diamond\varphi \wedge \neg\Box(\varphi\rightarrow\psi)\ ,\ \neg\Diamond\varphi)\ .$$

🍎

Another illustration is the recent logical work on epistemic operations changing knowledge states (cf. Gärdenfors 1987). Here, ***revision*** of information is just as important as addition. And indeed, this may be rendered in our enriched modal formalism just as well.

Example. Addition and Subtraction.
A knowledge state updated by φ validates just those ψ which satisfy

$$\neg\mathbb{U}\ (\varphi \wedge \neg\psi,\ \neg\varphi)\ .$$

The reverse process is sometimes called 'epistemic contraction'. A knowledge state 'downdated' by φ validates just those ψ which satisfy

$$\neg\mathbb{S}\ (\neg\varphi \wedge \neg\psi,\ \varphi)\ .$$

🍎

Indeed, our modal formalism *by itself* already forms a systematic theory of updating, downdating and revision, with their interactions. For instance,

> Does an update followed by a downdate
> with respect to the same proposition cancel out?

Such an issue may be formulated in the above modal language (cf. van Benthem 1989C):

$$p \rightarrow \neg \mathbb{U} ((q \wedge \neg \mathbb{S} (\neg q \wedge \neg p, q), \neg q) .$$

Thus, its validity becomes a matter of standard universal validity over information models. (A related modal reduction, to a system of 'propositional dynamic logic', has been elaborated in Fuhrmann 1990.)

Of course, special purpose axiomatizations for these cognitive phenomena (like those provided by Gärdenfors himself) may still be of independent interest.

As a last case of looking behind, the earlier binary notion of 'addition' of information admits of an obvious downward relative too. In addition to suprema, one may just as well consider *infima* in the information ordering, giving rise to a dual modality $\varphi * \psi$ expressing a kind of generalized disjunction.

- Finally, we mention the existence of enrichments of the basic modal formalism to be found along a somewhat different route.

Digression. D-Logic.

There exists a relatively simple addition to the basic modal language which removes some of its most striking failures of expressive power. The idea (due independently to Gargov et al. 1987, Koymans 1989) is to introduce a *difference operator* D , whose semantic truth condition reads as follows

$$\mathbb{M} \models D\varphi[w] \quad \text{iff} \quad \mathbb{M} \models \varphi[v] \ \textit{for some}\ v \neq w .$$

This is again a modality, satisfying the usual Distribution axiom

$$D(\varphi \vee \psi) \leftrightarrow D\varphi \vee D\psi ,$$

as well as principles of symmetry and 'pseudo-transitivity'.

Adding D to the basic modal language will already allow us to define all *universal* first-order frame properties of the inclusion ordering $\subseteq$. In particular, its basic property of *anti-symmetry*, which was beyond the resources of the basic modal formalism, is now expressed as follows:

$$\forall x \forall y ((x \subseteq y \wedge y \subseteq x) \rightarrow y = x) :$$

$$(p \wedge \neg Dp \wedge \Diamond (q \wedge \Diamond p)) \rightarrow q .$$

15.4 Further Information Structures

• The mere partial orders of the preceding Sections may be extended, so as to become upper *semi-lattices*, having an operation of addition:

$(W, \subseteq, +)$.

This perspective is familiar from Relevance Logic (cf. Dunn 1985).

If one wants to take advantage of this richer structure, suitable modalities must be introduced exploiting it. Here is one candidate, resembling an earlier one that was living at the 4-level of information over partial orders:

$\mathbb{M} \models \varphi + \psi\,[w]$ iff *there exist* u, v *with* $w = u+v$ *such that* $\mathbb{M} \models \varphi[u]$, $\mathbb{M} \models \psi[v]$.

One argument for having an independent operation of addition derives from its importance in setting up natural alternative information-based notions of *inference*.

Standard logical consequence says that the commas in a sequence of premises signal an ordinary *conjunction* :

- $\varphi_1, \ldots, \varphi_n \models \psi$
  if, in all information models $\mathbb{M}$, at each state w ,
  $\mathbb{M} \models \varphi_1[w]$ *and* ... *and* $\mathbb{M} \models \varphi_n[w]$ *implies* $\mathbb{M} \models \psi[w]$.

But, in many recent proposals for logics oriented toward information processing (as reviewed in van Benthem 1989A), one rather thinks of successive *addition* of the information supplied by the premises, so that their *sum* enforces the conclusion:

- $\varphi_1, \ldots, \varphi_n \models^* \psi$
  if, in all information models $\mathbb{M}$, at all states $w_1, \ldots, w_n$,
  $\mathbb{M} \models \varphi_1[w_1]$ *and* ... *and* $\mathbb{M} \models \varphi_n[w_n]$ *implies* $\mathbb{M} \models \psi[w_1 + \ldots + w_n]$.

The former notion of consequence has all the familiar classical properties, encoded in the usual structural rules. By contrast, the second notion loses some. In particular, the property of Monotonicity fails:

$\Diamond\neg p \models^* \Diamond\neg p$, but not $\Diamond\neg p , p \models^* \Diamond\neg p$.

We shall return to such options and their structural behaviour in Chapter 16.

Remark. Purely technically, there is no compelling reason for axiomatizing two different notions of consequence here. For, the second notion may be *defined* in terms of the first, if we let the commas induce explicit summation rather than Boolean conjunction:

$$\varphi_1 + \ldots + \varphi_n \Vdash \psi .$$

●

• But, one can go further, and *change* the present modal perspective, so as to make addition the only central notion, rather than inclusion among information states. Basic structures will then be 'frames'

$$(W, R)$$

with a ternary relation of 'partial addition' of information states. On these structures, the modality $+$ evaluates as follows:

$\mathbb{M} \models \phi+\psi\, [w]$ iff *there exist* u, v *with* Ruvw *such that* $\mathbb{M} \models \phi\, [u]$, $\mathbb{M} \models \psi\, [v]$.

Moreover, the inclusion order among states may then be derived as usual:

$w \subseteq v$ iff Rwvv .

This modal logic too has a perfectly ordinary development, as may be seen in van Benthem 1989B . For instance, $+$ is distributive, so that it has a familiar 'minimal logic', valid on the most general class of frames. Then, a well-known virtue of the modal perspective emerges: against such a minimal background, additional axioms may now be seen systematically as reflecting special features of information structure and processing.

Here are some illustrations of expressive power in this formalism.

Example. Expressive Strength of the Additive Modality.

- The formula

  $$(\varphi+\psi)+\chi \rightarrow \varphi+(\psi+\chi)$$

  corresponds on R-frames to a requirement of Associativity in the following form:

  $$\forall x \forall y \forall z \forall u \forall v \quad (Ryzx \wedge Ruvy) \rightarrow \exists s\, (Rusx \wedge Rvzs) .$$

- Likewise,

  $$\varphi+\varphi \rightarrow \varphi$$

  expresses Idempotence:

  $$\forall x \forall y \quad Ryyx \rightarrow x=y .$$

- Having the ordinary modalities referring to an inclusion $\subseteq$ after all, one can also enforce connections such as the following:

  $$\varphi + \psi \rightarrow \mathbb{P}\varphi \wedge \mathbb{P}\psi$$

  says exactly that

  $$\forall x \forall y \forall z \quad Ryzx \rightarrow (y \subseteq x \ \wedge \ z \subseteq x) .$$

●

Behind these examples lies a general definability result:

Theorem. All modal principles of the following form define first-order conditions on $R, \subseteq$ (which are effectively obtainable from them):

$$\varphi \rightarrow \psi ,$$

with φ a compound of proposition letters, $\vee$, $\wedge$, $+$, $\Diamond$, $\mathbb{P}$ and ψ an arbitrary formula in which each proposition letter occurs only positively.

Its proof is a straightforward extension of that for the Sahlqvist Theorem in ordinary Modal Logic, as the formulas listed retain semantically 'continuous' antecedents and 'monotone' consequents (cf. van Benthem 1989B).

Remark. Procedural Re-Interpretation.
This formalism is useful under other interpretations too. For instance, it can also serve as a formalism for a procedural logic, with the objects in W standing for *transition arrows* and R for their *partial composition*. In the latter guise, it will return in Chapter 16, as an alternative to Relational Algebra. ●

15.5 Appendix **What are Modalities?**

Viewing matters from the perspective of a full first-order language over our information models raises some interesting general questions. For instance,

What *is* a 'modality'?

There are more or less restrictive answers here, but one candidate would be this, involving a central denotational constraint from Chapters 11, 12:

> A *modality* is any function on sets defined by some first-order schema $\lambda x \bullet \varphi(x, A_1, ..., A_n)$ which is *continuous* in the sense of commuting with arbitrary unions of its arguments A_i $(1 \leq i \leq n)$.

Continuity expresses a requirement of 'local computability' : as may also be seen from the following syntactic characterization (cf. Appendix 1 to Chapter 2).

> A first-order formula φ is *distributive* if it has been constructed from conjunctions of atoms $A_i y$ in which no predicate A_i occurs more than once, and arbitrary formulas in which no A_j occurs $(1 \leq j \leq n)$ using only $\vee, \exists$.

A typical example, of course, is the modality $\Diamond$ itself, being $\lambda x \bullet \exists y\, (x \subseteq y \wedge Ay)$.

Proposition. A first-order formula defines a modality if and only if it is equivalent to a distributive form.

Proof. (This argument, due to Peter Aczel, simplifies a more complex ancestor.)

First, all distributive forms define continuous functions. In one direction, this is clear from the positive occurrence of all A_i, and the resulting semantic *monotonicity*. In the other, from a union to at least one of its members, the statement is clear for formulas

$\exists \overline{x} : \bigwedge\{ A_i x$, wholly A_j-free formulas $\}$ without iterated occurrences of the A_i, and disjunctions of these. But, all distributive formulas can be brought into such a form.

Conversely, suppose that φ defines a continuous operation. Then, there is a distributive equivalent, as shown in the following special case. Let $\varphi = \varphi(A_1, A_2)$. Then, the following semantic consequence holds, distinguishing cases as to (non-) emptiness of arguments:

$$
\begin{aligned}
\varphi \models\ & \exists x_1 \exists x_2\, (A_1 x_1 \wedge A_2 x_2 \wedge [\lambda y \bullet y = x_1 / A_1, \lambda y \bullet y = x_2 / A_2]\varphi) \\
\vee\ & \exists x_1\, (A_1 x_1 \wedge [\lambda y \bullet y = x_1 / A_1, \bot / A_2]\varphi) \\
\vee\ & \exists x_2\, (A_2 x_2 \wedge [\bot / A_1, \lambda y \bullet y = x_2 / A_2]\varphi) \\
\vee\ & [\bot / A_1, \bot / A_2]\varphi .
\end{aligned}
$$

This consequence uses the downward half of continuity (observing that each denotation $[[A_1]]$, $[[A_2]]$ is the union of its singleton subsets): that it is actually an equivalence, follows from the upward half (being monotonicity). ■

Remark. Finite Continuity.
The preceding argument fails if one merely requires the usual Distribution Axiom, which expresses commutation with *finite* unions of arguments. For instance, on linear orders, the following principle is finitely, but not fully continuous in A :

$$\forall x \exists y\ (x<y \wedge Ay)\ .$$

What would be an appropriate syntactic characterization in this case? 

Other questions arising at this level of generality have to do with the distinction between *monadic* and *polyadic* modalities. For instance, which modalities are genuinely binary, resisting decomposition into a Boolean compound of unary modalities? Here is one answer (cf. van Benthem 1989D for a proof):

Proposition. The binary operator "Until" is not definable in terms of unary modalities.

Other examples of genuine binary modalities occur, e.g., with the earlier 'minimal updating'. Saying that all future φ are ψ can be done with a unary modality:

$$\Box(\varphi \rightarrow \psi)\ .$$

But, saying that all *first* φ in the future are ψ amounts to an essentially binary connection of the form

$$'\mu(\varphi,\ \psi)'\ .$$

Remark. Infinitary Modal Operators on Information Models?
On the view just stated, binary modalities become more like *generalized quantifiers* over sets of worlds or states - and might be profitably studied as such. In general, however, this move would take us outside of the first-order representation language on information structures. But, that only raises one more interesting issue:

Does the modal logic of information also need
higher-order truth conditions eventually?

At least, the procedural logics of the next Chapter suggest the use of higher-order infinitary operations too, such as forming the transitive closure of transition relations.

16 Relational Algebra of Control

So far, we have concentrated on the logic of information structures. Now, we want to turn to the dynamical structure of information processing itself. Thus, the central topic becomes that of cognitive procedures and the control structures steering them. Again, logical phenomena will turn up similar to those encountered before. For a start, here are some pertinent illustrations from the recent literature.

16.1 Cognitive Transitions

In many recent publications, one can find attempts at formulating logics reflecting more dynamic procedural aspects of interpretation and inference - thus mixing 'declarative' and 'imperative' aspects within a single system of interpretation and inference. Examples are Gärdenfors 1988, Groenendijk & Stokhof 1988, Spohn 1988, Veltman 1989 (see also the more extensive surveys in van Benthem 1989E, 1990A which list many further historical ancestors).

The underlying pattern here may be described as follows. In standard logic, propositions stand for sets of possible worlds, or more generally, situations verifying their content. Thus, they are *properties* of information states. But now, we look at the effects of processing a proposition at a transient information state. Thus, dynamically, a proposition acts as a transformer on states, and its denotation will now rather be the binary *relation* consisting of its 'succesful transitions'.

A first manifestation of these ideas occurs in what may be called the

Dynamics of Interpretation

Example 1. Semantics of Programs

In the semantics of imperative programming languages, a program π denotes a binary relation between computer states, recording transitions consisting in shifts of register contents. This can be pictured in standard semantics, by equating data structures with models, and register states with variable assignments. A program then denotes a binary relation between Tarskian assignments. Traditional logical formalisms still serve here as a means of making static assertions about such states. An example of the resulting interplay are the usual 'correctness assertions' in the theory of program behaviour, which state that any succesful transition for the program whose starting assignment verifies some 'precondition' has a resulting assignment verifying a corresponding 'postcondition'.

But actually, dynamic aspects emerge just as well in the semantics of declarative programming languages. For instance, the usual successive approximation of predicate denotations in minimal Herbrand models for logic programs is itself another example of a dynamic sequential process.

Example 2. Anaphora: Changing Assignments.
Following Barwise 1987 or Groenendijk & Stokhof 1988, one can also interpret these traditional formalisms dynamically, via transition relations between assignments which may change in the course of evaluation of expressions. And a similar shift in viewpoint is possible when interpreting natural languages. This seems in fact closer to the actual anaphoric mechanism of the latter, with its changing relations between pronouns and objects temporarily denoted by them as a text proceeds.

One concrete system of dynamic interpretation for predicate logic in a model $\mathbb{M}$ with assignments a, b works as follows:

$[[Pt]]$	=	$\{ (a, a) \mid \mathbb{M} \models Pt\,[a] \}$	atomic test
$[[\exists x]]$	=	$\{ (a, b) \mid a =_x b \}$	random assignment to x
$[[\phi \wedge \psi]]$	=	$[[\phi]] \bullet [[\psi]]$	relational composition
$[[\neg\phi]]$	=	$\{ (a, a) \mid$ *for no* $b, (a, b) \in [[\phi]] \}$	strong failure

There are several deviations from standard dogma here. In particular, on this view, quantifications $\exists x\phi$ will obtain their dynamic effect through a composition of random assignment to x and subsequent succesful transitions for their matrix ϕ. Thus, quantifiers correspond to basic actions on states, not to procedural operations.

What tends to occur in this setting, as in our previous Chapters, is experimentation with both richer sets of logical constants and with several varieties of inference.

For instance, there are two ways now of understanding 'conjunction' of propositions. One is the above sequential *composition* of relations - the other is Boolean *intersection*, which has a more parallel flavour ('both procedures at the same time'):

$$[[\phi \cap \psi]] \quad = \quad [[\phi]] \cap [[\psi]] .$$

Moreover, entirely new operators may arise, such as a forward-looking *modality*:

$$[[\, F\, \varphi \,]] \quad = \quad \{\, (a, a) \mid \textit{for some assignment } b : (a, b) \in [[\, \varphi \,]] \,\}$$

or, in a more general version:

$$[[\, \mathbb{F}\, \varphi \,]] \quad = \quad \{\, (a, b) \mid \textit{for some assignment } c : (b, c) \in [[\, \varphi \,]] \,\} .$$

Next, there are various attractive versions of *valid dynamic consequence* too. For instance, one possibility would be to process the premises consecutively, and then see if the resulting transition is succesful for the conclusion. That is, the composed transition relation for the premises should be part of that for the conclusion:

$$\varphi_1, \ldots, \varphi_n \models^1 \psi \quad : \quad [[\, \varphi_1 \bullet \ldots \bullet \varphi_n \,]] \subseteq [[\psi]] .$$

Another option is to process the premises, and then see if the conclusion still has a chance of being executed:

$$\varphi_1, \ldots, \varphi_n \models^2 \psi \quad : \quad \textit{if } (a, b) \in [[\, \varphi_1 \bullet \ldots \bullet \varphi_n \,]] ,$$

$$\textit{then } (b, c) \in [[\psi]] \textit{ for at least one assignment } c .$$

Note the similarities with information-based consequence, as introduced in Chapter 15. In particular, again, the new notions of inference will tend to lose classical structural rules. In particular, neither $\models^1$ nor $\models^2$ satisfies Permutation or Monotonicity with respect to their premise sequences. This does not mean that all logical regularity is lost, however. For instance, the latter notion still permits addition of premises in front.

The second main logical manifestation of these procedural ideas occurs in the

Dynamics of Information Flow

Example 3. Changing Information States.

Assignments are not the only possible 'states'. In the recurrent folklore idea of propositions as transformations, states might be sets of models or possible worlds, representing ranges of uncertainty concerning the actual world. This has in fact been one of the guiding intuitions between possible worlds semantics. Then, dynamically, each successive proposition picks out some further subset of the current one. This idea was taken up formally in Heim 1982, as well as Spohn 1988, Veltman 1989. Again, we find a proliferation of useful logical constants as well as notions of inference, each with different formal properties. ●

Example 4. Dynamic Epistemic Systems.
Even though the 'ideology' of information-based modal logics (including intuitionistic logic or relevant logic) is couched in dynamic terms like 'growth' or 'search', with agents moving through information patterns, this dynamics remains a didactic fable: their usual implementations are static. By contrast, here is a dynamic version of a modal logic of information, of the kind introduced in Chapter 15, bringing out the aspect of information processing explicitly.

Consider possible worlds models $(S, \subseteq, V)$ of information states, ordered by inclusion. Now, one can read the modal operator $\diamond$ *itself* as a name for the inclusion relation ('random growth') . Moreover, atomic propositions q can be interpreted as standing for either for atomic tests, or for atomic 'updates', either random or 'minimal'. Compound formulas then correspond to complex instructions for evaluation, moving us along the inclusion ordering. Here are some clauses, analogous to the earlier ones for changing assignments:

$[[q?]] = \{ (s, s) \mid s \in V(q) \}$
$[[upd(q)]] = \{ (s, s') \mid s \subseteq s'$ *and* $s' \in V(q) \}$
$[[\mu\text{-}upd(q)]] = \{ (s, s') \mid s \subseteq s'$ *and* $s' \in V(q)$ *and no* u *strictly in between* s *and* s' *is in* $V(q) \}$
$[[\diamond]] = \subseteq$
$[[\phi \wedge \psi]] = [[\phi]] \bullet [[\psi]]$
$[[\neg\phi]] = \{ (s, s) \mid$ *for no* u , $(s, u) \in [[\phi]] \}$.

There is really a family of systems here, differing in their atomic repertoire and their procedural apparatus. For instance, other atomic operators might be useful too, such as backward-looking ones for *revision* or 'downdating':

$[[downd(q)]] = \{ (s, s') \mid s' \subseteq s$ *and* $s' \notin V(q) \}$
$[[\mu\text{-}downd(q)]] = \{ (s, s') \mid s' \subseteq s$ *and* $s' \notin V(q)$ *and each* u *strictly in between* s *and* s' *is in* $V(q) \}$.

Moreover, the updating or downdating modes might be extended to general procedural operations on arbitrary propositions; while other procedural operations might be added, such as 'choice'.

The logical theory of this dynamic modal family is still largely unknown. ◆

There are still further manifestations of procedural logic, occurring at the higher level of text and discourse. Indeed, in logical *pragmatics*, dynamic views have a much longer history, witness a classic like Stalnaker 1971 or the logical game theories of Lorenzen and Hintikka, with an inspiration going back eventually to Wittgenstein's procedural conception of language games. (Van Benthem 1991 gives a more general discussion of such alternative dynamic paradigms.)

Upon reflection, the variety of systems encountered in this procedural setting, with their attendant logical constants and notions of inference, may be understood as the effect of having both more purely descriptive notions and purely procedural operations of *control*. We shall now propose a convenient general setting for reflecting upon the latter phenomenon by itself.

16.2 Relational Algebra

One abstract pattern behind all the above systems involves structures that might be called R-*models*

$$(S , \{ R_p \mid p \in P \}) ,$$

i.e., families of transition relations on some carrier set of states, closed under suitable operations. These states can still be quite diverse: cognitive or physical, or mixtures of both - and likewise, the actions in P may range from value assignment in registers to digging for gold in one's garden. For the moment, we are after their common procedural 'superstructure'.

But then, we can use the existing work on *Relational Algebra* to get a better grasp of the relevant logical structure here (cf. Jónsson 1984, Németi 1990). In the theory of binary relations, one tries to generalize the Boolean Algebra of unary propositions by creating a suitable richer similarity type. Notably one has

Boolean operations

$$- , \cap , \cup \qquad \bot , \top$$

Order operations

composition $\bullet$ as well as *converse* $^{\smile}$

and one special relation, viz.

identity id .

Moreover, one can introduce also other operations, such as analogues of our earlier slashes

$$R\backslash S \quad = \quad \{(x,y) \mid \forall z: (z, x) \in R \Rightarrow (z, y) \in S\}$$

$$S/R \quad = \quad \{(x,y) \mid \forall z: (y, z) \in R \Rightarrow (x, z) \in S\}.$$

Such further operations are often available by definition. For instance, here are two reductions reminiscent of propositional logic:

$$R\backslash S \quad = \quad -(R^{\cup}\bullet S) \qquad\qquad R/S \quad = \quad -(R\bullet S^{\cup})\,.$$

Reasoning about these operations is a well-understood matter. Here are the basic laws of Relational Algebra:

i	all Boolean identities				
ii	$(R\cup S)^{\cup}$	$=$	$R^{\cup}\cup S^{\cup}$		
	$(-R)^{\cup}$	$=$	$-R^{\cup}$		
	$R^{\cup\cup}$	$=$	R		
	$id^{\cup}$	$=$	id		
iii	$(R\cup S)\bullet T$	$=$	$R\bullet T \cup S\bullet T$		
	$R\bullet(S\cup T)$	$=$	$R\bullet S \cup R\bullet T$		
	$R\bullet(S\bullet T)$	$=$	$(R\bullet S)\bullet T$		
	$R\bullet id$	$=$	$id\bullet R$	$=$	R
iv	$(R\bullet S)^{\cup}$	$=$	$S^{\cup}\bullet R^{\cup}$		
	$R^{\cup}\bullet-(R\bullet S)$	$\subseteq$	$-S$		

All relational operations so far have been first-order. But there are also natural *infinitary* operators on binary relations. These correspond to 'unlimited structures' in programming or cognition, such as endless repetition. The most prominent example is the ordering operation of *transitive closure*:

$$R^* \quad = \quad \{ (x, y) \mid \text{there exists some finite sequence of successive } R \text{ transitions linking } x \text{ to } y \}\,.$$

The latter will satisfy various algebraic principles too, witness the following sample

$$R\cup R^* \quad = \quad R^*$$

$$R^{**} \quad = \quad R^*$$

$$(R\cup S)^* \quad = \quad (R^*\bullet S^*)^*\,.$$

As another illustration of this notion, Pratt 1990 proposes a basic 'action logic' extending the first-order relational operations $\{+, \bullet, \backslash, /\}$ with $*$, which has typical principles such as the following 'Induction Axiom':

$$(R \backslash R)^* \quad = \quad (R \backslash R) .$$

The framework of Relational Algebra may now be used to analyze the earlier systems of dynamic logic. For a start, proposed 'logical constants' often turn out to be algebraic operators per se.

Example. Algebraic Logical Constants.
The earlier 'modality' $\mathbb{F}$ is definable as follows:

$$T \bullet R^{\cup} \quad = \quad \{(x, y) \mid \exists z: (y, z) \in R\} .$$

Likewise, 'strong negation' becomes definable:

$$[[\neg\phi]] \quad = \quad id \cap - \mathbb{F}[[\phi]] .$$

Moreover, it was shown already how the two earlier categorial implications may be defined algebraically. ●

And the same holds for proposed notions of inference. For instance, what was probably the most natural notion of validity for sequents may be explained as follows using relational composition:

$X \Rightarrow R$ is *valid* if, for all relational interpretations $[[\ ]]$, $[[\bullet X]] \subseteq [[R]]$.

This definition may still be reduced to the above equational format, via the equality

$$[[\bullet X]] \cup [[R]] = [[R]] .$$

Remark. Empty Premises.
In the absence of premises, the above notion is meant to reduce to the inclusion

$$id \subseteq [[R]] .$$

That is, the empty sequence of premises is interpreted as the identity relation. In order to justify this stipulation, one may note that the following assertions are all equivalent in Relational Algebra:

$$R \subseteq S \qquad id \subseteq R \backslash S \qquad id \subseteq S / R .$$

●

We noticed that, with this notion, failures occurred for structural rules of standard logic, such as Monotonicity. This may now be understood as a matter of Relational Algebra: the corresponding algebraic principle is just not generally valid. And indeed, we can study such issues more generally in this framework, noting similar failures for most other classical structural rules from earlier Chapters, such as Permutation or Contraction. (For instance, a contraction $R, R \Rightarrow R$ is only valid for *transitive* relations R.)

Alternative proposed notions of consequence may often be analyzed here by means of suitable *definitions*. For instance, the alternative dynamic notion of inference $\models^2$ introduced earlier on can be reduced to

$$\bullet X \Rightarrow \mathbb{F} R .$$

And the latter definition explains its various structural properties, which may be calculated by the above principles of Relational Algebra.

Now, the proposed use of this framework vis-à-vis concrete systems of dynamic interpretation and inference is as follows. It provides a procedural superstructure that is common to all approaches, enabling us to theorize about their general features. In addition, one can then determine which further properties of some specific system are due to its special features (e.g., the use of special types of transition relation only, or the selection of some special set of states, or idiosyncratic format of inference). Moreover, the general algebraic perspective provides a fresh look at the design of existing systems of logic, suggesting various natural extensions.

Example. Looking Back.

The operation of relational converse $^\cup$ does not seem to have any immediate analogue at the level of types or propositions, even though it occurs naturally in general action. Nevertheless, it does emerge at the level of texts, and conscious operations on information states: witness the 'revisions' and 'contractions' of Gärdenfors & Makinson 1988. Moreover, even the earlier model class semantics for information states invites consideration of further operations than just 'updating', rather requiring 'stepping back' undoing the effects of some earlier transformation. (For instance, an epistemic disclaimer like " unless ϕ " tells us to increase the range of possibilities again.) And eventually, even in the above dynamic predicate or modal logic, one might just as well study the full relational algebra of random assignment and atomic tests, including converse and even all Booleans. ◆

16.3 Procedural Logic

As in earlier Chapters, some broad logical topics now emerge.

- *Logical Constants*

Relational Algebra provides a completely general model for the procedural structure of activities. For instance, $\cup$ models choice, $\bullet$ sequential execution, $-$ 'avoiding' and $^{\smile}$ 'reversal' or 'undoing'. These notions occur across many human activities: not only cognitive ones, but also concrete actions like digging or playing. What makes these activities different is rather the state space on which they operate, and the appropriate repertoire of basic actions over the latter.

This generality may be made precise using a notion of logicality from Chapters 2, 10, 12. Algebraic operations are independent from any specific structure of states:

Proposition. All operations of Relational Algebra are *invariant under permutations* π of the state set S. In particular, with binary operations O:

$$\pi[O(R, S)] \quad = \quad O(\pi[R], \pi[S]) .$$

Next, we want to investigate the general structure of procedural logical constants. As in earlier Chapters, it is convenient to introduce a first-order description language here, having identity and a generous supply of binary predicate symbols. This is the natural vehicle for defining the earlier basic algebraic notions, such as

$R \bullet S \qquad \lambda xy \bullet \exists z\, (Rxz \wedge Szy)$

$id \qquad \lambda xy \bullet\ x=y .$

One immediate issue arising here is the matter of *functional completeness*:

Can we find some natural 'complete' set of relational operations?

There is no general satisfactory answer to this issue (cf. Németi 1990). But at least, the similarity type for relational algebra is well-chosen in the following sense (Maddux 1983):

Proposition. Each first-order definable operation $\lambda xy \bullet \varphi(x, y, R)$ on binary relations employing only *three* variables $\{x, y, z\}$ can be written using only

$$- ,\ \cap ,\ id ,\ \bullet ,\ ^{\smile} .$$

The general perspective here is again that of 'finite variable fragments' of the above semantic first-order language, as encountered already in Chapter 15. And technical outcomes are similar here to those observed before (see van Benthem 1990A for details):

Proposition. There is a genuine ascending hierarchy of k variable fragments, without any finite functionally complete set covering all.

The use of this Procedural Hierarchy lies again in its fine-structure. In particular, one can measure proposed relational operators as to their definitional complexity.

Example. Dynamic Predicate Logic.
Of the two operations in the earlier dynamic version of predicate logic, composition $R \bullet S$ involves essentially three variables in its definition:

$\lambda xy \bullet \exists z\, (Rxz \wedge Szy)$.

But strong failure $\neg R$ needs only two. Its prima facie definition

$\lambda xy \bullet x=y \wedge \neg\exists z\, Ryz$ is equivalent to $\lambda xy \bullet x=y \wedge \neg\exists x\, Ryx$.

🍎

More generally, the two-variable level has the following explicit description:

Proposition. The following set of relational operations is complete for the *two-variable* fragment of the above first-order language:

all Booleans Conversion Identity

as well as the earlier existential Modality $\mathbb{F}$ ('domain') .

The proof is by straightforward inspection of possible shapes of first-order definitions in $\{ \neg , \wedge , \exists \}$ for relational operations employing only two variables, starting from innermost existential quantifiers and working outward.

Remark. Limitations of Fragments.
There are certain disadvantages to imposing very tight k variable restrictions. For, some of the well-known properties of first-order logic do not hold for finite variable fragments. For instance, Ildikó Sain has recently shown that Beth's Definability Theorem fails for Relational Algebra: the promised explicit definition for some notion defined implicitly at the 3 variable level may lie outside of the fragment. 🍎

• *Notions of Inference*

First, we consider the logical behaviour of the basic notion of dynamic inference $\models^1$. Here, again, the categorial calculi of earlier Parts turn out to be relevant. In particular, we have an observation similar to those made in Chapter 14:

Proposition. Under the relational Boolean / order interpretation, both the Lambek Calculus and (the relevant part of) Linear Logic are sound.

This amounts to an embedding of these calculi into Relational Algebra, by the analysis of the preceding Section. It is still an open question if this interpretation is *complete*, even for the Lambek Calculus with implications and product only. (Dunn 1990 presents a general theory of representation that would at least model the latter system completely in terms of *ternary* relations.)

Remark. Two Senses of Dynamics.
In earlier Parts of this Book, the 'dynamic' aspect of categorial calculi resided in their derivations, which were interpreted as lambda-definable procedures over objects in type domains D_a. But now, the type domains R_a are themselves procedures, with valid sequents expressing certain relations between these. By contrast, the role of different derivations for the same sequent has diminished (compare a similar remark in Chapter 14): two different lambda terms, or two derivations merely display two ways in which sequential composition of arrows can establish the desired inclusion. ♠

The more urgent general issue is to obtain a systematic view of natural varieties of dynamic inference in a relational setting. In the literature so far, at least the following explications have been put forward for the notion '$\phi_1, \ldots, \phi_n \models \psi$':

I Process all premises successively, then see if the resulting transition is succesful for the conclusion.

II Process all premises successively, then see if a transition can still be made for the conclusion.

III See if the conclusion 'holds' at all states where all premises 'hold', (where a proposition *holds* at a state if that state is among its fixed points).

IV See if the conclusion holds at all states arrived at by processing the premises starting from some state of vacuous information.

These various notions of inference have quite different formal properties, as may be measured by their behaviour with respect to the usual structural rules. But are they really independent contenders? Recall the 'linear logic strategy' (cf. the discussion at the end of Chapter 14): various notions of inference, with different structural behaviour, may often be encoded on top of one austere basic notion, by introducing suitable new connectives. And indeed, all of the above can be reduced to type I via a translation into Relational Algebra using some obviously definable operators:

II the range of the composition of the $[[\phi_i]]$ is contained in the domain of $[[\psi]]$

III the diagonal of the composition of the $[[\phi_i]]$ is contained in $[[\psi]]$

IV the range of the composition of the $[[\phi_i]]$ applied to some empty information state is contained in the diagonal of $[[\psi]]$.

Accordingly, many structural properties of such notions can be studied within Relational Algebra itself.

Remark. Describing Properties of Inference.

Structural features of inference can be purely combinatorial, such as Contraction or Cut, but they may also involve Boolean notions. For instance, the above inference I satisfies 'Conjunction of Conclusions'

$$\frac{\phi_1, \ldots, \phi_n \models \psi_1 \qquad \phi_1, \ldots, \phi_n \models \psi_2}{\phi_1, \ldots, \phi_n \models \psi_1 \wedge \psi_2}$$

This is even valid in the following strong 'local' sense:

If the two inclusions corresponding to the premises hold in a relational model, then so does that of the conclusion.

Now, validities of this local kind may still be described via equations within Relational Algebra itself. The reason is as follows (cf. Németi 1990 or van Benthem 1990A):

On relational set algebras, arbitrary universal statements over Boolean compounds of relational equalities may be reduced to (conjunctions of) such equalities.

●

In our view, there are only two intuitively independent varieties of inference in the present setting, from premises $P_1, ..., P_n$ to a conclusion C. These are the above dynamic variant, involving *sequential* processing of premises:

$$P_1 \bullet ... \bullet P_n \subseteq C,$$

together with a classical variant, involving *parallel* processing of these premises:

$$P_1 \cap ... \cap P_n \subseteq C.$$

Then, it becomes of interest to see exactly which clusters of structural rules are exemplified by these two notions. As will be proved in Appendix 2 to this Chapter, the answer is as follows:

Proposition. The structural rules for classical inference are determined precisely by
Reflexivity Cut Monotonicity Contraction.

Other classical principles, such as Permutation, are derivable from these four.

Proposition. The structural rules for dynamic inference are determined precisely by Reflexivity and Cut only.

This emphasis on structural rules has a reason. In the course of this Book, these principles have changed their role completely. From simple domestic properties of standard logics, they became important choice points in the Categorial Hierarchy. And now, we may view them as important characteristics of kinds of reasoning, and look for their 'natural clusterings'.

- *Classifying Natural Kinds of Action*

Finally, the relational perspective brings out various kinds of dynamic behaviour for propositions or actions, and allows us to classify them. For instance,

- an action R is *stable* if $\forall x \forall y\, (Rxy \rightarrow Ryy)$
  i.e., $\mathrm{Range}(R) \subseteq \mathrm{Diag}(R)\ (= R \cap id)$
  (this is an important property of epistemic updates)
- an action R is *static* if $\forall x \forall y\, (Rxy \rightarrow y{=}x)$
  i.e., $R \subseteq id$
  (this is an important property of tests guiding a process).

Such special relations may be recognized by their inferential behaviour. For instance, tests are 'monotonic', in that they can be inserted in sequences of premises without disturbing any dynamic conclusions in sense I . And the latter property is also characteristic for them: monotonic relations in this sense must be tests.

Many standard properties of binary relations make sense for the purpose of dynamic classification. For instance, standard propositions P induce both 'tests'

$$?P \quad = \quad \lambda xy \bullet\ y{=}x \wedge Py$$

and 'realizations'

$$!P \quad = \quad \lambda xy \bullet\ Py$$

(cf. Chapter 17) which are stable. But the former has other well-known properties too, such as Symmetry and 'Euclidity': $\forall x \forall y \forall z\ (\ (Rxy \wedge Rxz) \rightarrow Ryz\ )$. The latter kind of relation satisfies $\forall x \forall y \forall z\ (\ Rxy \rightarrow Rzy\ )$.

The more general question is rather how such properties fare under the application of algebraic operations. For instance, some but not all of the basic operations of relational algebra will respect 'stability' or 'stasis'.

16.4 Application and Specialization

The relational perspective offered here is very general, and not at all specific to cognitive or linguistic processing. Nevertheless, operating at this level of generality has some advantages in application. For instance, from the present point of view, one can 'modularize' any dynamic system of logic into a choice of basic actions on suitable states (its 'action repertoire') and a choice of procedural operations (its 'control repertoire'). And this offers many opportunities for conscious variation. For instance, one can add new procedural operations to an existing action repertoire, say in dynamic predicate logic or dynamic modal logic. But also, with the same control structure, one can just as well vary the underlying states. An example is the general 'Tarskian Dynamics' of van Benthem 1990A, which takes the traditional semantic schema

$$(\ \mathbb{D}, I, a) \models \phi$$

of a structure $\mathbb{D}$ with an interpretation function I and an assignment a verifying a formula ϕ , and then considers procedures varying the interpretation function as well as the assignment, and eventually even the underlying structure itself.

Nevertheless, there is also an interesting logical structure to more specific cognitive activities. In particular, what happens in a space of *information states* ordered by the inclusion order $\subseteq$ of Chapter 15? Then, a much richer array of relational operations arises, involving the latter as well, such as taking 'upward parts': $\lambda xy \bullet Rxy \wedge x \subseteq y$.

There are two strategies of formalization here (cf. van Benthem 1990A). One is to provide ordinary relational algebra with an explicit constant relation symbol $\subseteq$ (in addition to the already available identity relation id). The other leaves the inclusion ordering implicit, like accessibility in Modal Logic, and generalizes relational algebra to 'information models', where logical operators no longer satisfy invariance with respect to all permutations of states, but only with respect to $\subseteq$-*automorphisms* (cf. Chapter 12).

Many interesting dynamic properties of propositions emerge only in this setting. An important example is 'updating': i.e., being a subrelation of the inclusion order:

$\forall x \forall y\ (\ (Rxy \rightarrow x \subseteq y).$

Such relations correspond to procedures that can only result in increase of information. Another interesting condition, stressed in Zeinstra 1990, is preservation of inclusion structure by 'considerate' procedures R :

$\forall x \forall y\ (\ (Rxy \wedge x \subseteq x') \rightarrow \exists y'\ (y \subseteq y' \wedge Rx'y')\ )$.

More generally, the question then arises how such properties are gained, modified or lost under the action of various logical operators. For instance, the well-known 'regular operations' $\{ \cup , \bullet , * \}$ (cf. Chapter 17) preserve both the above properties. In particular, this suggests the pursuit of syntactic or semantic criteria for recognizing when a given complex proposition is going to have specific desirable properties. For instance,

Which syntactic forms in the relational algebra over $\subseteq$ are updating ?
Which syntactic operations create, or endanger, this property?

Example. Dynamic Modal Logic Revisited.

One typical system implementing such ideas is the earlier-mentioned family of dynamic modal logics. These describe many features of information processing. Moreover, they have the advantage that they can still be studied by means of standard techniques. For instance, their transition relations still admit of an explicit standard transcription τ :

$\tau(\Diamond)$	$x \subseteq y$
$\tau(q?)$	$Qx \wedge x=y$
$\tau(\text{upd}(q))$	$x \subseteq y \wedge Qy$
$\tau(\mu\text{-upd}(q))$	$x \subseteq y \wedge Qy \wedge \neg\exists z\, (x \subseteq z \wedge z \subset y \wedge Qz)$
$\tau(\phi \wedge \psi)$	$\exists z\ (\tau(\phi)(x, z) \wedge \tau(\psi)\ (z, y))$ for some *new* variable z
$\tau(\neg\phi)$	$y=x \wedge \neg\exists z\ \tau(\phi)(x, z)$

And similar clauses exist for downdates and other extras.

This transcription already explains many general features of dynamic versions of modal logic. For instance, it can only be made to work with *three* variables over states (instead of the standard two), but effective axiomatizability is not endangered by this. More specific questions, e.g., concerning *decidability*, are still open.

16.5 An Alternative: Modal Arrow Logic

Relational Algebra is not the only vehicle for bringing out general procedural structures in dynamic logic. There are obvious alternatives, encoding more intermediate structure into succesful transitions than just their source and target. For instance, in computer science, one often uses 'traces' recording all atomic actions performed. Trace models have also been proposed in the dynamics of reasoning, for instance, when explaining procedural aspects of proofs (indicated by special particles like "if" and "fi" : cf. Vermeulen 1989). Therefore, the theory of this Chapter cannot be the last word yet.

This observation fits in with another concern. If one is really serious about the dynamic perspective, then there might be good reasons for taking transitions *themselves* to be basic objects in their own right: and Relational Algebra would become a theory of these transitions in a more intensional sense, not necessarily reducible to ordered pairs of states.

A more down-to-earth reason for variation is that Relational Algebra itself has its problems. Notably, the given set of basic principles is not a complete axiomatization of the class of all validities on set-representable relation algebras (the latter class is known to be larger, and indeed non-finitely axiomatizable). Moreover, even as it stands, the calculus is *undecidable*. And so, a paradox threatens. We are switching to dynamic systems of inference, because of their greater faithfulness to human cognitive practice than 'standard logic'. But, what we are getting is higher complexity for this more natural 'procedural logic'.

Thus, it may be useful eventually to adopt a somewhat less orthodox form of Relational Algebra. In particular, one might think of transition relations more abstractly, as being sets of 'arrows'. And if we do that, an earlier perspective returns (cf. Chapter 15).

The point is that the calculus of Boolean operations, composition $\bullet$ and converse $^\cup$ may be viewed as a *modal logic*. (Note that $\bullet$ and $^\cup$ both satisfy the basic modal principle of Distribution.) So, we can introduce corresponding relations

Cxyz : 'arrow z is the composition of x and y '
Fxy : 'arrow y is the converse of x '
as well as one special property
Ix : ' x is an identity arrow' .

The various axioms of Relational Algebra then express conditions on what may be called 'arrow frames'

(W, I, F, C) .

Example. Frame Correspondences for Axioms of Relational Algebra.
Here is the list of relevant relational conditions for our algebraic axioms:

- $(R \cap S)^\cup \geq R^\cup \cap S^\cup$: $\forall xyz\,(Fxy \wedge Fxz \rightarrow y = z)$
- $(-R)^\cup \geq -R^\cup$: $\forall x \exists y\; Fxy$

  i.e., F is a function;
- $R^{\cup\cup} \leq R$: $\forall xy\,(Fxy \rightarrow Fyx)$

  i.e., F is idempotent.

Then, C is associative:

- $((R \bullet S) \bullet T) = (R \bullet (S \bullet T))$: $\forall xyzuv\,(Cxyz \wedge Czuv \rightarrow \exists w\,(Cyuw \wedge Cxwv))$

  in conjunction with

  $\forall xyzuv\,(Cyzu \wedge Cxuv \rightarrow \exists w\,(Cxyw \wedge Cwzv))$

Moreover, it interacts with F as follows:

- $(R \bullet S)^\cup = S^\cup \bullet R^\cup$: $\forall xyz\,(Cxyz \rightarrow C\,F(y)\,F(x)\,F(z))$.

And, the rather forbidding final axiom expresses a principle in the same vein:

- $R^\cup \bullet -(R \bullet S) \leq -S$: $\forall xyz\,(Cxyz \rightarrow C\,F(x)\,z\,F(y))$.

◆

At least, these correspondences give us a more concrete view of the meaning of the earlier set of basic axioms for Relational Algebra.

Again, this modal logic of relations can be studied in greater technical detail (cf. van Benthem 1989B). For instance, the preceding example is a special case of a general definability theorem for modal formulas expressing first-order conditions on relational frames, which was formulated in Chapter 15. Moreover, the basic axiom system given above turns out to be *decidable* (I. Németi, personal communication). Further logical properties of the calculus may be found in Vakarelov 1990.

The exact connection between the modal perspective on information structures and that on transition relations remains to be explored. But see also Chapter 17 for a unified presentation.

Remark. A Mathematical Alternative.
The modal perspective is not the only one aiming at a more general view of relations as sets of arrows. Another natural alternative would be to use that of *Category Theory*, viewing points as objects, and arrows as morphisms.

16.6 A Comparison with Language Models

Finally, we turn to an obvious comparison with an earlier theme, namely, the language models studied in Chapter 14. After all, the emergence of various formal analogies between several kinds of models for logics without structural rules introduced so far invites comparison.

First, in one direction, there is a natural embedding.

From L-Models to R-Models

Given a family of languages on a universe of expressions, one can map each language to a binary relation as follows

$$\rho(L_a) = \{ (x, xy) \mid y \in L_a \} .$$

Moreover, let us restrict the universe of admissible pairs to those of the form (x, y) where x is an initial segment of y. Then, we can make the following observation:

Fact. ρ is a *homomorphism* with respect to $-$, $\cap$, $\bullet$, $\langle\rangle$ and $/$.

Illustration:

- $\rho(L_{a\cap b}) \;=\; \rho(L_a \cap L_b) = \{(x, xy) \mid y \in L_a \;\&\; y \in L_b\} \;=\; \rho(L_a) \cap \rho(L_b)$,
- for the case of $\rho(L_{-a})$, the special restriction on admissible pairs is needed,
- $\rho(L_{a \bullet b}) \;=\; \{(x, xyz) \mid y \in L_a ,\ z \in L_b\} \;=\; \rho(L_a) \bullet \rho(L_b)$,
- $\rho(L_{a/b}) \;=\; \{(x, xy) \mid y \in L_{b/a}\} \;=\; \rho(L_b) / \rho(L_a)$, by a simple calculation,
- $\rho(\{\langle\rangle\}) \;=\; \{(x, x) \mid \text{all expressions } x\} \;=\; id$.

●

Since ρ is also injective, this yields an isomorphic embedding, and we have found at least one connection.

Proposition. For the $\{ -, \cap, \bullet, \langle\rangle, / \}$ fragment,
the universal first-order theory of R-models is contained in that of L-models.

One immediate question is if this result can be extended so as to include the converse implication $\backslash$, for which the above representation does not work.

Moreover, the analogy does not necessarily extend to additional operators.

Example. Converse and Inversion.
If we map the earlier inversion on languages to relational converse, then, e.g., the identity $(x \bullet y)^{\cup} = y^{\cup} \bullet x^{\cup}$ will be valid in both cases, but, e.g., $x^{\cup} \bullet -(x \bullet y) \leq -y$ will not: a linguistic counter-example is $x = \{a\},\ y = \{b, ab\}$. ●

Next, we consider the opposite direction.

From R-Models to L-Models

Given any relation, we can choose to re-interpret it as a set of 'symbols' (x, y) , which can be 'concatenated' in the usual way. But, no homomorphic preservation of structure takes place here. Typically, the problem is that the composition $R \bullet S$ will not correspond to the concatenation product of R and S : as certain arrows may not match. Therefore, a representation will only succeed for very special 'uniform' relations. Or on the other side, one might have to consider syntax models allowing for possible *restrictions* on concatenation (certain symbols would not admit of juxtaposition). The latter idea might have some independent interest all the same.

In any case, there are principles which are valid on all L-models, but not on all R-models, reflecting the above difference:

$L_a \bullet L_b = \varnothing$ implies that $L_a = \varnothing$ *or* $L_b = \varnothing$,

but

$R_a \bullet R_b = \varnothing$ does not imply $R_a = \varnothing$ *or* $R_b = \varnothing$.

Now this is still a 'higher' example, above the level of algebraic identities. (We can make our model comparisons, of course, at different levels of their logical 'description languages'.) But, given the earlier emphasis on Gentzen sequents, i.e., Horn clauses, i.e., algebraic identities, here is a more telling illustration.

Example. R-Models are Richer.
The following principle is valid in all L-models, but not in all R-models:

$((-(x \bullet x) \cap x) \bullet (-(x \bullet x) \cap x)) \cap id = \bot$.

On L-models.

Suppose that some string a is in the intersection. I.e.,

$a = \langle\rangle \in (-(x \bullet x) \cap x) \bullet (-(x \bullet x) \cap x)$: so that $\langle\rangle \in -(x \bullet x) \cap x$:

$\langle\rangle \in x$: whence $\langle\rangle = \langle\rangle\langle\rangle \in x \bullet x$, which is a contradiction.

On R-models.

Here is a relational counter-example:

$R = \{\langle 1, 2\rangle, \langle 2, 1\rangle\}$, $id = \{\langle 1, 1\rangle, \langle 2, 2\rangle\}$.

In fact, it is possible to rework this example into one not containing id : so that it is already the combination of • with Boolean operations which creates the difficulty. ◆

The upshot of this discussion seems to be that R-models form the more general class of structures for our analysis of information and its processing.

Remark. Numerical Models.
It would be of interest, nevertheless, to extend the above comparisons so as to include the case of the earlier numerical N-models, and informational I-models. A case in point are the earlier analogies between the previous uses of Modal Logic: once as a theory of information models, and once as a generalized form of Relation Algebra. ◆

16.7 Appendix 1 **Relations and Functions**

Some current dynamic frameworks are couched in terms of *functions* rather than relations, reflecting the intuitive idea of 'transformations' acting on states. In principle, there is no conflict with our approach so far. In one direction, functions are nothing but *deterministic total* relations, and hence the functional perspective is subsumed in the present one. But also conversely, every binary relation R on S induces a function R* from pow(S) to pow(S) , by setting

$$R^*(X) = R[X] \qquad (= \{ y \in S \mid \exists x \in X\ Rxy \}).$$

Moreover, nothing is lost in this larger setting, as these 'lifted' functions can still be uniquely retrieved via a well-known mathematical property (cf. Chapters 10, 11):

Fact. A function $F : pow(S) \rightarrow pow(S)$ is of the form R^* for some binary relation R on S if and only if F is *continuous*, in the sense of commuting with arbitrary unions of its arguments.

The reason is that continuous maps F can be computed 'locally', from their values at singleton arguments only. The existence of such 'liftings' and 'lowerings' between various levels of set-theoretic representation can also be studied more systematically (see van Benthem 1986A, chapter 3). As another example, here is a similar reduction of maps on sets of states to 'propositions' in the standard style, being subsets of S :

Fact. A unary operation F on pow(S) is both continuous and *introspective* (i.e., $F(X) \subseteq X$ for all $X \subseteq S$) if and only if it represents some unary property P via the rule

$$F(X) = X \cap P .$$

16.8 Appendix 2 **Determining Varieties of Dynamic Consequence**

What the above dynamic notions of inference all had in common was the use of relational *composition*, representing sequential processing of sequences of premises. Thus, we can describe their difference with the classical notion also as one of 'text semantics': they have a different interpretation for the *commas* separating these premises. For, the classical notion interprets these commas in a more parallel fashion via *Boolean intersection*.

Even so, basic dynamic consequence (type I) and standard consequence still share many Boolean properties. For instance, classical and dynamic consequence are both *anti-additive* in their left-hand arguments:

$$X, R_1 \cup R_2, Y \models Z \quad \text{iff} \quad X, R_1, Y \models Z \ \textit{and} \ X, R_2, Y \models Z$$

and both are *multiplicative* in their right-hand argument:

$$X \models R_1 \cap R_2 \quad \text{iff} \quad X \models R_1 \quad \textit{and} \quad X \models R_2 .$$

Remark. Two Senses of Monotonicity.
There is a possible confusion here which should be pointed out. Note that the above properties imply both left and right Monotonicity in the obvious Boolean sense, as applied to single premises or conclusions. This is to be distinguished from the 'non-monotonicity' exhibited by dynamic consequence I with respect to adding premises in the left-hand sequence. The latter property is only equivalent to the former in the presence of the classical structural rules.

Now, let us consider the variety of notions of inference more systematically than has been done so far.

To begin with, all possible candidates satisfying the above two constraints of anti-additivity and multiplicity can be analyzed still further, using the earlier mathematical property of *permutation invariance*. Arguably, proposed notions of inference should be permutation-invariant 'meta-relations' between the binary relations corresponding to their component propositions. Then, only a finite number of options will be seen to remain for the comma: most prominently, Boolean conjunction, disjunction and relational composition:

Proposition. In any set algebra of binary relations, the only permutation-invariant anti-additive and multiplicative inference relations are those definable by a conjunction of schemata of the form

$$\forall x_1 y_1 \in \phi_1 \ \dots \ \forall x_k y_k \in \phi_k : \ \alpha \rightarrow \beta ,$$

where α is some Boolean condition on identities in the variables x_i, y_j and β is some atom of the form $\psi x_i y_j$.

Still, the above two notions of classical and dynamic consequence are characteristic of anything that may be encountered in this general area. For, many of the above schemata are reducible to these two, using suitable algebraic compounds of the ϕ_i and ψ. Moreover, their shape suggests one further plausible restriction. All of the above schemata are *universal clauses* that are preserved under passing to submodels having fewer states. But intersective or compositive inference is even defined in the more restricted natural *Horn clause* format, being preserved under direct products of state models as well. For instance, with two premises, they read as follows:

$\forall xy \in \phi_1 \ \forall zu \in \phi_2 : (x=z \wedge y=u) \to \psi xy$,

$\forall xy \in \phi_1 \ \forall zu \in \phi_2 : y=z \to \psi xu$.

And all candidates in this Horn format are algebraically reducible to just these two.

Another way of stating the special status of classical intersective and dynamic compositive inference arises via purely structural rules only, without any Boolean connectives.

Standard consequence is characteristic for the usual set of structural rules, which lump premises together into mere sets of formulas (cf. Chapter 4), namely:

Reflexivity	Cut
Monotonicity	Contraction .

We omit Permutation here, as it is derivable from these four.

Proposition. The four classical structural rules axiomatize precisely the theory of the set-theoretic relation $a_1 \cap \ldots \cap a_k \subseteq b$.

Proof. The reason is the following representation result. Let R be any abstract relation between finite sequences of objects and single objects satisfying the classical structural rules. Now, define

$a^* \quad = \quad \{ A \mid A$ is a finite sequence of objects such that $A\ R\ a \}$.

Then, it is easy to show that

1 if $a_1, \ldots, a_k\ R\ b$, then $a_1{}^* \cap \ldots \cap a_k{}^* \subseteq b^*$,
using Cut and Contraction, while

2 if $a_1{}^* \cap \ldots \cap a_k{}^* \subseteq b^*$, then $a_1, \ldots, a_k\ R\ b$,
follows by Reflexivity and Monotonicity. ■

Remark. Actually, a similar result could have been proved for the earlier notion of inference III , where premises and conclusions are merely being tested at one single state. Thus, in a sense, the latter is 'classical' too.

On the other hand, dynamic consequence is characteristic for the opposite extreme, leaving the ordering of premises completely intact:

Proposition. Reflexivity and Cut axiomatize precisely the theory of the set-theoretic relation $a_1 \bullet \dots \bullet a_k \subseteq b$.

Proof. This time, a relational representation is needed for any abstract relation R like above satisfying only the two mentioned constraints. Again, consider the universe of all finite sequences of objects B , and set

$$a^{\#} \quad = \quad \{ (B, BA) \mid \text{for any finite sequence } A \text{ such that } A\,R\,a \} .$$

Then, by Cut alone, we have that, if $a_1, .., a_k$ R b , then $a_1{}^{\#} \bullet \dots \bullet a_k{}^{\#} \subseteq b^{\#}$. Moreover, Reflexivity already implies the converse, when applied to the sequence of transitions $(\langle\rangle, a_1), (a_1, a_1a_2), \dots, (a_1 \dots a_{k-1}, a_1 \dots a_k)$.

One might interpret this outcome to mean that the dynamic notion of inference is the more fundamental of the two.

The preceding results determined the structural behaviour of two natural notions of inference. But also conversely, given certain natural clusters of structural rules, one may determine some corresponding semantic notion of inference exemplifying just these.

To see this, consider the central system of earlier Parts of this Book, namely the undirected Lambek Calculus LP , whose structural rules were Reflexivity, Cut and Permutation. Recall the numerical models of Chapter 14, with their addition operation over families of vectors. Now, think of *bags* (multi-sets) of information pieces, that can be added via multiset union + . Propositions P will then denote families [[P]] of such bags (their 'verifiers'). Now, define an inference relation as follows:

$A_1 , \dots , A_k$ R B iff

$[[A_1]] + \dots + [[A_k]] \subseteq [[B]]$ under each interpretation [[ ]] ,

where the addition + is over arbitrary selections of k successive bags from the premise denotations.

Proposition. Reflexivity , Cut and Permutation axiomatize precisely the theory of this set-theoretic relation $a_1 + ... + a_k \subseteq b$.

Proof. Here, one maps an object a to the family of multisets

$\{ A^\$ \mid A R a \}$,

where $A^\$$ is the multiset of all objects occurring in the sequence A .

If $a_1 , ..., a_n \; R \; b$, then $a_1^\$ + ... a_n^\$ \subseteq b^\$$, by Cut.

The converse follows by Reflexivity and Permutation.

●

Remark. An Excusable Error.

It might seem reasonable a priori to identify propositions directly with bags of verifiers, rather than sets of these. But, the resulting notion of inference, stating that the multi-set union of the bags $[[A_i]]$ should be contained in the bag $[[B]]$, does not just have the above three principles for its characteristic structural rules. For instance, this notion of inference would also support the following curious rule:

from X, Y, Z R b *to* X, Z R b .

●

Remark. Further Natural Kinds of Inference.

Other natural clusters of structural rules occur in other areas of reasoning. For instance, in Conditional Logic and in Artificial Intelligence, there has been an interest in 'preferential reasoning', a non-classical kind of inference according to the following semantic schema:

The conclusion holds (not in all, but)

in all *most preferred* models of the premises.

Here, preferences may arise from probabilistic or qualitative heuristic considerations. This is a natural notion, and one which is genuinely different from those considered in this Chapter (cf. van Benthem 1989E for further background).

All structural properties of preferential reasoning may be determined too. This time, the main issue is not treating premises like sets, but rather one of Cut: 'transmission of validity' is not generally plausible in this case. Likewise, Monotonicity has no general backing here. But interestingly, certain modifications of these principles still hold, and hence one obtains a more refined view of posible structural rules than that which has served in this Book (cf. Makinson 1988): their adoption is not an all-or-nothing matter, and interesting variants exist which did not surface in classical Proof Theory.

Proposition. The structural properties of preferential consequence are axiomatized by

- Permutation Contraction Expansion
- 'Strong Reflexivity'

  i.e., $X \models B$ whenever $B \in X$,

as well as two modifications of Cut and Monotonicity, being the dual pair:

- 'Cautious Cut'

  from $A_1, ..., A_n \models B$ and $A_1, ..., A_n, B \models C$
  to infer $A_1, ..., A_n \models C$

  'Cautious Monotonicity'

  from $A_1, ..., A_n \models B$ and $A_1, ..., A_n \models C$
  to infer $A_1, ..., A_n, B \models C$.

Proof Sketch. The key idea of the required representation is this. Objects are now mapped to the family of all 'states' to which they belong. Here, a *state* is any set X of objects for which we have 'harmony':

$b \in X$ iff $X \models b$ for all objects b.

The relevant preference order over such states is then just ordinary set-theoretic inclusion ('the smaller the better'), and it is easy to show that there is one minimal state containing $A_1, ..., A_n$, namely $\{ B \mid A_1, ..., A_n \models B \}$. ◆

17 Dynamic Logic

Information processing has various aspects. Chapter 15 gave us a theory of information structures and Chapter 16 added an account of general procedures over these. It remains to integrate the two into one single perspective. In line with the general approach of this Book, we want to achieve this in a type-theoretic setting. As a side benefit, such a perspective will allow for peaceful coexistence between more classical views of propositions and logical inference and the newer dynamic ones.

17.1 Intensional Type Theory and Dynamic Logic

Consider a standard type theory as in Chapter 12, with primitive types

t	(truth values)
s	(indices, states, possible worlds)
e	(entities)

allowing for the formation of functional and product types.

First, take the { s, t } fragment only ('propositional dynamic logic' in higher types). Classical propositions have type

(s, t) 'Propositions as Statements'

being 'static' sets of states, whereas 'dynamic' relational propositions have the type

(s, (s, t)) 'Propositions as Programs' .

The two systems will have different flavours. The classical statement-oriented view leads to propositional logic, where Boolean operators, and on top of those, modal operators form the main connectors (cf. Chapters 11, 12 and 15). In the realm of propositions as programs, however, important operations for compounding will be notions of *control* : such as

$\bullet$	sequential *composition*
$\cup$	(indeterministic) *choice* .

And the logical paradigm for the latter is not so much Boolean Algebra as the Relation Algebra of the preceding Chapter. Still, the Boolean operations make sense in the new type (s, (s, t)) too: and hence we expect (and indeed find) a rather richer structure of logical constants in current systems of dynamic semantics than was usual in standard systems of logic.

Remark. Strategies of Dynamization.

In Chapter 12, various strategies were considered for 'intensionalizing' an extensional, that is, an {e, t} - based type theory. In particular, one move was to reinterpret old propositions in the truth value type to properties of states:

t goes to (s, t) .

This time, there is another strategy too, viz. to make them into dynamic actions on states:

t goes to (s, (s, t)).

Now, instead of choosing between these two perspectives, we can - and should - have both. Indeed, the above situation is familiar from the field of *Dynamic Logic* , viewed here as a semantic theory of cognitive computation, rather than calculation with mechanical devices. (See Harel 1984 for a survey.) For, one important feature of dynamic logics is precisely the *interaction* between two subsystems, statements and programs, as effected by various cross-categorial operators:

- *test* takes statements ϕ to test programs

  $$? \phi = \{ (x, x) \mid \phi(x) \}$$

- *modality* <> takes programs π to projection operators $<\pi>$ on statements ϕ

  $$<\pi> \phi = \{ x \mid \exists y: (x, y) \in [[\pi]] \ \& \ \phi(y) \}$$

- some versions of Dynamic Logic also have a *fix-point* operator, being the diagonal map taking cognitive programs to their 'truth set'

  $$\Delta(R) = \{ x \mid (x, x) \in R \}$$

  i.e., those states where R has no dynamic effect.

It is this interplay which is involved, e.g., in making so-called 'correctness statements' about the execution of programs π :

$\phi \rightarrow [\pi] \psi$ ('from ϕ-states, program π always leads to ψ-states')

or statements expressing 'enabling' :

$[\pi_1] <\pi_2> \psi$ ('action π_1 enables action π_2 to achieve ψ ').

And a similar interplay occurs with information processing and action in general.

Thus, we acknowledge two distinct domains, and study two additional types of transformation between the two kinds of proposition:

((s, (s, t)), (s, t)) : static *projections* for dynamic propositions ,

((s, t), (s, (s, t))) : dynamic *modes* of using static propositions .

Accordingly, the world of dynamic logic for information processing looks like this:

		static projections <———		
static logical operators	(s, t)		(s, (s, t))	dynamic logical operators
		dynamic modes ———>		

The two transformation arrows in the above types ((s, (s, t)), (s, t)) and ((s, t), (s, (s, t))) represent various *mechanisms* for switching from one perspective to another, which are of independent interest. For instance, classical propositions P provide 'descriptive contents' for different kinds of action on states, such as

testing P	?P	$\lambda xy \bullet\ y{=}x \wedge Py$
realizing P	!P	$\lambda xy \bullet\ Py$.

Such different modes are ubiquitous in cognition and action. For instance, when playing a game, there is a continuing alternation of actions performed ("serving a ball") and statements to be tested ("the ball is in"). But also, even within ordinary argument, there can be clearly felt shifts in mode. Intuitively, the transition "so" from premises to conclusion often indicates a change from sequential processing of the premises to a testing mode for the conclusion in the state reached so far.

Likewise, in the opposite direction, there are various natural projections from programs to standard statements recording the behaviour of the former, such as

domain R	$\lambda x \bullet\ \exists y\, Rxy$
diagonal R	$\lambda x \bullet\ Rxx$.

A richer picture of such mechanisms arises with various dynamic modes that may be acquired by a standard proposition: it can serve as a content for a variety of cognitive activities such as *testing*, *updating*, *downdating* (and the latter two both in 'liberal' or 'minimal' variants). To define the latter, however, one has to endow the base domain D_S with at least the partial order structure of Chapters 12, 15 and also 16. Then we can have informational 'modal modes' such as

$$\begin{array}{lll} \text{upd } \phi & = & \{(x, y) \mid x \subseteq y \ \&\ \phi(y)\} \\ \text{downd } \phi & = & \{(x, y) \mid y \subseteq x \ \&\ \neg\phi(y)\} \\ \mu\text{-upd } \phi & = & \{(x, y) \mid x \subseteq y \ \&\ \phi(y) \ \&\ \neg\exists z\text{: } x \subseteq z \subsetneq y \ \&\ \phi(z)\} \\ \mu\text{-downd } \phi & = & \{(x, y) \mid y \subseteq x \ \&\ \neg\phi(y) \ \&\ \neg\exists z\text{: } y \subsetneq z \subseteq x \ \&\ \neg\phi(z)\} . \end{array}$$

And Modal Logic of information flow will now be the dynamic logic of structures and operators such as these.

Of course, in this enterprise, the questions of crucial interest to us need not be the standard technical concerns inherited from our founding fathers, as we shall see.

17.2 Logical Issues at Two Levels

There is further fine-structure to be investigated in the above type-theoretic framework, especially, concerning the interplay of classical and dynamic viewpoints. The following discussion provides a first illustration, drawing on notions and concerns from earlier Chapters of this Book (further details are in van Benthem 1989C).

- What is an appropriate choice of *logical constants* at various levels?

In general, semantic notions should now be formulated in a way which will apply to all relevant types at once, so as not to miss useful analogies. And for this purpose, one may return to the general semantic perspective of Part IV.

For a start, there is one general set-theoretic notion which made sense as a constraint on logicality across arbitrary types, viz. *invariance for permutations* of the base domain D_S. What this means is that, given any permutation π of D_S, an operation F on (s, t) propositions should commute with it:

$$F(\pi[P]) = \pi[F(P)] ,$$

and likewise with more than one argument, or with operations on (s, (s, t)). And what we know from Chapters 10, 16 is that

- the only permutation-invariant operators on (s, t) type propositions are the set-theoretic *Boolean* ones,
- on the type (s, (s, t)), such operators include all the usual notions of Relation Algebra (in particular, all Booleans, composition, diagonal and converse).

On top of this, one can then study the effects of further denotational constraints. The type-theoretic generality then brings to light analogies in behaviour across different semantic domains. For instance, we can determine all permutation-invariant operators of the above 'transformer' types which also respect inferential structure, in that they are Boolean *homomorphisms*.

Proposition. The only permutation-invariant Boolean homomorphism in type $((s, (s, t)), ((s, t))$ is the earlier diagonal operator

$$\lambda R\bullet \lambda x\bullet Rxx .$$

In type $((s, t), (s, (s, t)))$ the only two examples are the functions

$$\lambda P_s\bullet \lambda x_s\bullet \lambda y_s\bullet P(x_s)$$

$$\lambda P_s\bullet \lambda x_s\bullet \lambda y_s\bullet P(y_s) .$$

Proof. The first assertion was already proved in Chapter 11, in an analysis of Reflexivization of transitive verbs, using 'deflation' of Boolean homomorphisms in any type $((a, t), (b, t))$ to ordinary functions of type (b, a).

What this method implies for the second assertion is that we need only search for permutation-invariant items in type $(s\bullet s, s)$. And these are just

$\lambda x_{s\bullet s}\bullet \pi_{left}(x)$, $\lambda x_{s\bullet s}\bullet \pi_{right}(x)$:

which explains the above outcome. ◆

The other transformations mentioned above, such as *domain* or *test* are not entirely without preservation behaviour either. Although not homomorphisms, both are at least *continuous* maps, in the earlier sense of commuting with arbitrary unions of their arguments. And, as was already shown in Chapter 12, of such logical continuous mappings, there are only finitely many too, that can be enumerated explicitly. The continuity property makes these operations related to many ordinary logical operators, such as union, intersection, composition and converse on dynamic propositions.

In the modal setting of information models with an inclusion pattern $\subseteq$, this style of analysis will shift somewhat, as we have seen in Chapter 12. Significant intensional operators need not be invariant for all permutations of information states: after all, the latter may destroy relevant information about their inclusion ordering. But, what they should be invariant for are those permutations of D_s which preserve the ordering structure: i.e., the *inclusion automorphisms*.

Moreover, it makes sense to generalize various elements in the earlier modal analysis of Chapter 15 to the present type-theoretical framework. One important example is the notion of *bisimulation invariance* . We would like to express that certain higher modal operations in our dynamic picture are bisimulation invariant. And this is precisely what may be done using the analysis of bisimulation in higher types developed in Chapter 12, which applied to dynamic propositions as well as to static ones, but also to propositional operators or transformations.

Of course, not all important modal constructions pass this test: after all, it was too restrictive even for ordinary Modal Logic eventually, witness Chapter 15. For instance, a propositional mode like updating is not bisimulation-invariant: because zigzag relations will not preserve inclusion in a suitably strict fashion. Another counter-example is the behaviour of a relational operator like Composition. But then, as before, we can introduce stronger notions of invariance to deal with such cases. These will not be pursued here, however.

- Our next concern is the *interplay* of dynamic and static information.

The shift toward the newer dynamic format does not mean an essential break with earlier static formalisms at the meta-level. For, the behaviour of programs can always be described 'statically' in terms of transition predicates having the proper arity. Indeed, the analysis via Relation Algebra supports precisely the opposite conclusion:

Dynamic Logics can always be reduced to Classical Ones.

One telling piece of evidence for this is the existence of routine standard translations from Dynamic Logic into standard logical formalisms, just as for Modal Logic: for details, see Part VI.8.2. What remains to be seen, of course, is just when such reductions are useful. Here is one example where this happens.

In cognitive practice, dynamic procedural effects may be 'local', leaving no traces in short-term memory, while classical propositions may be closer to eventual stored content that can be recalled afterwards. Thus, operating jointly with two (or even more) levels of 'short-term logic' and 'long-term logic' may be just what is needed.

One way of working in tandem uses a well-known computational technique, namely the tracing of programs as *predicate transformers*. In computer science, behaviour of programs is often explained in terms of their transforming 'preconditions' into 'postconditions' . From our dualist point of view, this amounts to measuring standard 'unary' informational content along the way. In particular, starting from some set of states described by a predicate P , execution of a program π moves us to a new set of states (the 'image of P under π '), defined by the *strongest postcondition* of P under π :

$$SP\ (P, \pi)\ .$$

This is just the functional representation of relations described in Appendix 1 to Chapter 16. Conversely, one may also compute *weakest preconditions* for a resulting predicate under a program: WP (π, P) . The two are related by conversion:

$$SP\ (P, \pi) \quad = \quad WP\ (\pi^{\smile}, P)\ .$$

Now, weakest preconditions describe nothing but the inverse images of relations with respect to certain sets. But, this is precisely what is defined by the central notion of propositional dynamic logic, being $<\pi>\phi$. And then, the key axioms of the latter system may be seen as recursive clauses for computing weakest preconditions. Indeed, this will work only for a limited fragment of regular operations on programs, such as composition or choice. There is no similar decomposition for such Boolean combinations as $<\pi_1 \cap \pi_2>$ or $<-\pi>$, which explains their somewhat more complex status.

Now, the same idea may be applied to earlier systems of dynamic interpretation and inference. For instance, for the dynamic version of predicate logic presented in Chapter 16, we obtain a simple effective mechanism for computing changing 'classical' informational contents during the dynamic process.

Proposition. The following decompositions hold for weakest preconditions:

$$WP\ (At, P) \quad = \quad P \wedge At$$

$$WP\ (x{:=}\,\text{-}, P) \quad = \quad \exists x\ P$$

$$WP\ (\phi_1 \bullet \phi_2, P) \quad = \quad WP\ (\phi_1, WP\ (\phi_2, P))$$

$$WP\ (\neg\phi, P) \quad = \quad P \wedge \neg WP\ (\phi, T) \qquad \text{(with 'T' for 'true')}\ .$$

Proof. This may be computed directly or derived via transcription into propositional dynamic logic. ■

For strongest postconditions then, the above reduction WP ($\pi^{\cup}$, P) works, since dynamic predicate logic is closed under conversion:

- both atomic actions are symmetric

  $(At)^{\cup} \quad = \quad At \qquad\qquad (x{:=}\,\text{-})^{\cup} \quad = \quad x{:=}\text{-}$
- and conversion treats compounds as follows

  $(\phi \bullet \psi)^{\cup} \quad = \quad \psi^{\cup} \bullet \phi^{\cup} \qquad\qquad (\neg\phi)^{\cup} \quad = \quad \neg\phi\ .$

More direct recursion clauses for strongest postconditions would look like this:

$$\begin{aligned} SP\ (P, At) &= P \wedge At \\ SP\ (P, x{:=}\text{-}) &= \exists x\ P \\ SP\ (P, \phi \bullet \psi) &= SP\ (SP\ (P, \phi), \psi) \\ SP\ (P, \neg\phi) &= P \wedge \neg WP\ (\phi, T) \quad (!)\ . \end{aligned}$$

Here is a sample computation on a sequence of dynamic formulas, with static trace points as indicated by bold face numerals:

$\exists x(Ax \bullet Bx)\ .\ \exists x\ Cx\ .\ \neg\exists x\ Dx$

$\exists x\ \mathbf{1}\ (Ax\ \mathbf{2} \bullet Bx\ \mathbf{3})\ .\ \exists x\ \mathbf{4}\ Cx\ \mathbf{5}\ .\ (\neg\exists x\ Dx)\ \mathbf{6}$

SP	0	T	1	$\exists x T\ (= T)$
	2	Ax	3	$Ax \wedge Bx$
	4	$\exists x\ (Ax \wedge Bx)$ *	5	$\exists x(Ax \wedge Bx) \wedge Cx$,
	6	$\exists x(Ax \wedge Bx) \wedge Cx \wedge \neg\exists x\ Dx$.		

At the trace point *, the initial $\exists x$ becomes an ordinary quantifier after all.

Similar calculations are possible for other dynamic systems.

For instance, here is a recipe computing weakest preconditions for the earlier dynamic modal logic, using a suitable temporal formalism with operators

$\mathbb{F}$ ('future'), $\mathbb{P}$ ('past'), $\mathbb{U}$ ('until'), $\mathbb{S}$ ('since') :

WP (q?, A)	=	$q \wedge A$
WP (upd(q), A)	=	$\mathbb{F}\,(q \wedge A)$
WP (μ-upd(q), A)	=	$\mathbb{U}\,(q \wedge A, \neg q)$
WP (downd(q), A)	=	$\mathbb{P}\,(\neg q \wedge A)$
WP (μ-downd(q), A)	=	$\mathbb{S}\,(\neg q \wedge A, q)$
WP ($\phi \wedge \psi$, A)	=	WP (ϕ, WP (ψ, A))
WP ($\neg\phi$, A)	=	$A \wedge \neg$WP (ϕ, T) .

Thus, the issue of 'static versus dynamic' is rather subtle. Dynamic formalisms are not irrevokably beyond the scope of classical systems, but on the other hand, they provide an intrinsically import new perspective on questions which would not easily come to the fore otherwise.

17.3 General First-Order Theory of Information

In the background of our Dynamic Logic, there lies a more standard first-order description language of information structures. As was stated already in Chapter 15, this formalism provides for one-place propositions (via formulas with one free variable), but also for two-place relations (two free variables), etcetera. Now, one can also broaden the scope of the earlier investigation of an informational hierarchy of modal fragments and their semantic characteristics to this wider formalism. To fix one's thoughts, think here of the earlier 3-variable analysis for Relational Algebra: which also seems to be the proper level of complexity for formulating many significant dynamic operations.

First, we extend the notion of bisimulation in a perhaps unexpected new direction, namely, to a notion relating not just single states, but also pairs of states, triples of states, etcetera.

In first-order model theory, there is a well-known concept of *partial isomorphism* between models $\mathbb{M}$ and $\mathbb{N}$, being the existence of a non-empty family $\mathbb{PI}$ of finite partial isomorphisms between their domains, with 'Back and Forth' extension properties:

> If the partial isomorphism (X, Y) (viewed as a pair of matching sequences) is in $\mathbb{PI}$, and a is any object in $\mathbb{M}$, then there exists some object b in $\mathbb{N}$ such that (Xa, Yb) is also in $\mathbb{PI}$. And analogously in the opposite direction.

One important observation is that corresponding sequences X, Y verify the same first-order formulas in the two models.

For k-variable fragments, this notion may be restricted in an obvious manner to partial isomorphisms of length *at most* k, to obtain a notion of k-*partial isomorphism*. And then, a straightforward induction, taking some care as to precise assignments, proves

Proposition. Formulas from the k-variable fragment are invariant for matching sequences in a k-partial isomorphism.

In fact, the basic modal language needs no more than length 2, with the 'action' only occurring at length 1. This explains why the notion of 'bisimulation' in Chapters 12, 15 could get by with coupling individual objects only.

Example. 2-Partial Isomorphism Versus 3-Partial Isomorphism.
The following two models $\mathbb{M}$, $\mathbb{N}$ admit of a 2-partial isomorphism relating the ordered pair (1, 3) in $\mathbb{M}$ to the pair (1, 3') in $\mathbb{N}$, but not of any 3-partial isomorphism doing the same:

$$\mathbb{M} \quad 1 \rightarrow_a 2 \rightarrow_b 3 \qquad\qquad \mathbb{N} \quad 1 \rightarrow_a 2 \rightarrow_b 3$$

$$1' \rightarrow_a 2' \rightarrow_b 3' \qquad\qquad 1' \rightarrow_a 2' \rightarrow_b 3'$$

$$1'' \rightarrow_a 2'' \rightarrow_b 3''$$

What this shows is that the characteristic formula for composition of binary relations cannot be defined using two variables only. It may be useful for the reader to check that not even a 2-partial isomorphism can effect the described matching between the two related models having merely objects { 1, 2, 3 } and { 1, 2, 3, 1', 2', 3' }. ■

There is also a converse to the above Proposition (cf. van Benthem 1990B).

Theorem. Any formula $\phi = \phi(x_1, ..., x_k)$ in the full first-order language (possibly employing other bound variables besides those displayed) which is invariant for k-partial isomorphism is logically equivalent to a formula constructed using $x_1, ..., x_k$ only.

Proof. This may be proved essentially like the characterization of the basic modal fragment in Chapter 15, showing how the invariant formula must follow from its own k-variable consequences. That is, for any invariant formula ϕ, it may be shown that

$$\mathbb{k}(\phi) \models \phi ,$$

where $\mathbb{k}(\phi)$ is the set of all logical consequences of ϕ lying in the k-variable fragment. The desired definability for ϕ then follows by Compactness.

The argument runs as follows.
Consider any model $\mathbb{M}$ for $\mathbb{k}(\phi)$. By standard model-theoretic reasoning, there exists some model $\mathbb{N}$ with the same k-variable theory where ϕ holds. The crucial step is then as follows:

> Any two models $\mathbb{M}, \mathbb{N}$ that are elementarily equivalent with respect to k-variable formulas have saturated elementary extensions $\mathbb{M}^+, \mathbb{N}^+$ that are k-partially isomorphic via the family of all their pairs of sequences up to length k verifying the same type in the k-variable language.

Hence, the following statements are true, successively:

$\mathbb{N} \models \phi$	(assumption)	$\mathbb{N}^+ \models \phi$	(elementary extension)
$\mathbb{M}^+ \models \phi$	(invariance)	$\mathbb{M} \models \phi$	(elementary descent) .

Van Benthem 1990B supplies some further details. ●

Together, these results provide a complete model-theoretic characterization for the k-variable fragments of a full first-order language over information models.

Remark. Temporal Logic over Linear Orders.
By this analysis, the well-known Functional Completeness of the 3-variable fragment of a monadic first-order language over *linear orders* may be understood as follows. Between such structures, 3-partial isomorphism implies full partial isomorphism (by a simple argument resting on linearity): and hence, any first-order formula $\phi(x, y, z)$ on these models is already definable by one from the 3-variable fragment. ●

More specifically, the various 'extension patterns' needed to induce the back-and-forth properties up to length k yield an obvious choice for a functionally complete set of operators in a corresponding variable-free modal notation (which must exist by the analysis of Chapters 15, 16).

Example. Three-Dimensional Modal Logic.
The following would at least be a useful set for the special case of k=3:

- Boolean operations both on unary and on binary predicates
- Relational algebraic operations of composition, converse and diagonal
- Unary forward and backward modalities F, P
- A modality of "Betweenness" from unary predicates ϕ to binary ones:
  $\lambda x \bullet \lambda y \bullet\ \exists z \bullet (x \subseteq z \subseteq y \wedge \phi(z))$,
- Some book-keeping operators introducing or removing argument places.

Remark. Life Inside Fragments.
It was already observed in Chapter 15 that strict confinement to k-variable fragments may result in loss of useful properties from full predicate logic. Here is a simple question, concerning semantic preservation, whose answer appears to be unknown.

Let some k-variable sentence be preserved under the formation of *submodels*.

Will it then also have a syntactically *universal* equivalent in the same fragment?

By the Los-Tarski Theorem, there must be some universal first-order equivalent, but the latter may have a long prenex form involving more than k variables. In fact, our fragments need not be closed under the prenex operations, whence a modified version of 'universal' form is needed in any case (being a construction using quantifier-free formulas, $\wedge, \vee$ and $\forall$ only).

All this has been a general analysis of k-variable fragments, along the lines of Section 3.1. But what about the additional effects of various semantic notions of bisimulation, inducing restrictions to still further subfragments, as introduced in Chapter 15? These may now be understood as follows.

Basic modal bisimulation is a case of 2-partial isomorphism where the back-and-forth condition only applies to objects lying in a specified position with respect to those already matched. More specifically, one only considers $\subseteq$-successors. Similarly, one may carve out interesting parts of the 3-variable fragment by means of such pattern restrictions. Here is one example, involving a correspondence between *linear* searches that could be performed in two information patterns $\mathbb{M}_1$, $\mathbb{M}_2$:

A *trisimulation* between two models $\mathbb{M}_1$, $\mathbb{M}_2$ is a relation C between states, but also between ordered pairs and between ordered triples of states which are linearly ordered by $\subseteq$, satisfying the following conditions:

1. if $w_1 \mathrm{C} w_2$, then w_1, w_2 verify the same proposition letters
2. if C relates two items of length smaller than 3, then these satisfy the back-and-forth property under extension to longer linear sequences (in particular, with length 2, this allows both 'extension at the ends' and 'interpolation')
3. if two pairs or triples are related by R, then so are their restrictions to subsequences of lower length.

As to the corresponding invariance, it is not hard to see that

Proposition. All transcriptions of formulas from the modal language with "Until" ($\mathbb{U}$) and "Since" ($\mathbb{S}$) are invariant for trisimulation.

An explanation for this observation may be found in the following result, that may be proved along the earlier lines.

Theorem. The trisimulation invariant first-order formulas $\phi = \phi(x)$ are precisely those which are definable from unary atoms using $\neg, \wedge, \vee$ as well as restricted quantifiers

$$\exists y\,(x \subseteq y \wedge \alpha(x, y)),\quad \exists y\,(y \subseteq x \wedge \alpha(x, y)),\quad \exists z\,(x \subseteq z \subseteq y \wedge \alpha(x, y, z)),$$
$$\exists z\,(x \subseteq y \subseteq z \wedge \alpha(x, y, z)),\quad \exists z\,(z \subseteq x \subseteq y \wedge \alpha(x, y, z)),$$

modulo logical equivalence.

These unary formulas can all be written using 3 variables in all, as is easily seen by inspection of their syntax. In fact, they form a fragment including all transcriptions of formulas from the language of Since and Until.

Likewise, it would be possible to analyze those fragments of the three-variable language corresponding to the translations of various systems of dynamic modal logic as to their characteristic invariances.

Example. Pure Testing.

The earlier pure test formalism is invariant for matching states and pairs of states via a relation C in such a way that

1. matching states verify the same proposition letters
2. matching states allow of back-and-forth extension with respect to arbitrary objects
3. matching pairs allow of 'decomposition' with respect to third objects: if (a_1a_2, b_1b_2) is in C , then for each object a , there exists some object b such that (a_1a, b_1b) and (aa_2, bb_2) are both in C .

Clause 2 takes care of strong failure, clause 3 of sequential conjunction. ◆

In this setting, semantic concerns arise like those of Chapter 15. For instance,

Which syntactic forms of definition will guarantee which desired semantic behaviour?

Or, to repeat a question from Chapter 16,

Which modal schemata define binary relations among information states that are *progressive* , in the sense of being included in $\subseteq$?

This is one dynamic counterpart to the determination, in the unary standard case, of all upward persistent statements. Note, e.g., that the progressive relations are closed under Boolean $\wedge$, $\vee$, relational $\bullet$, as well as the backward modality $\mathbb{P}$.

This concludes our exploration of the basic model theory of a dynamic modal logic of information. We pass on to a

17.4 General Discussion

A number of general points behind the story of this Chapter (and this Part) may be worth setting out separately, now that we have come to the end.

There is perhaps one obvious omission to be addressed right away. Although the word "information" has occurred throughout, it must have struck the reader that we have had nothing to say on what information *is*. In this respect, our theories may be like those in physics: which do not explain what "energy" is (a notion which seems similar to "information" in several ways), but only give some basic laws about its *behaviour* and *transmission*.

The eventual recommendation made here has been to use a broad type-theoretic framework for studying various more classical and more dynamic notions of proposition in their interaction. This is not quite the ideology advocated by many current authors in the area, who argue for a whole-sale *switch* from a 'static' to a 'dynamic' perspective on propositions. Our personal opinion is that the classical notion of proposition remains natural and useful: so much so, that, even if one had started from a dynamic viewpoint all along, it would have to be *invented*, as a receptacle of invariant 'content' across a number of different modes. This is not the place, however, to survey the philosophical arguments for and against the more radical move.

The above presentation still leaves many questions about possible reductions from one perspective to another. For instance, it would seem that classical systems ought to serve as a 'limiting case', which should still be valid after procedural details of some cognitive process have been forgotten. There are various ways of implementing the desired correspondence: e.g., by considering extreme cases with $\subseteq$ equal to identity, or, in the pure relational algebra framework by considering only pairs (x, x). What we shall want then are reductions of dynamic logics, in those special cases, to classical logic. But perhaps also, more sophisticated views are possible. How do we take a piece of 'dynamic' prose, remove control instructions and the like, and obtain a piece of 'classical' text, suitable for inference 'in the light of eternity'? Postconditions are one implementation of this two-level world picture:

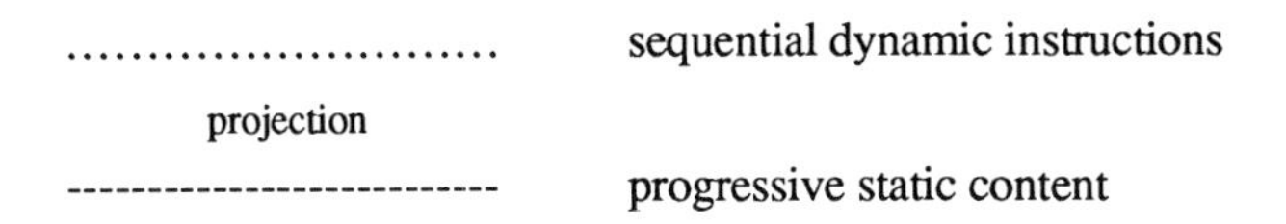

It remains to investigate more in detail how logical connections at the upper level project to the poorer propositional system at the lower level. Here is where the earlier mathematical observations come in, concerning the effects of projections when mapping a relation algebra to a Boolean algebra.

There is also a more technical side to the matter of 'reduction'. By now, Logic has reached such a state of 'inter-translatability' that almost all known variant logics can be embedded into each other, via suitable translations. In particular, once an adequate semantic has been given for a new system, this usually induces an embedding into standard logic: as we know, e.g., for the case of Modal Logic. Likewise, all systems of dynamic interpretation or inference proposed so far admit of direct embedding into an ordinary 'static' predicate logic having explicit transition predicates (cf. van Benthem 1988B). Thus, our moral is this. The issue is not whether the new systems of information structure or processing are essentially beyond the expressive resources of traditional logical systems: for, they are not. The issue is rather which interesting phenomena and questions will be put into the right focus by them.

The next broad issue concerns the specific use of the perspective proposed here, vis-à-vis concrete proposals for information-oriented or dynamic semantics. The general strategy advocated here is to locate some suitable base calculus and then consider which 'extras' are peculiar to the proposal. For instance, this is the spirit in which modal S4 would be a base logic of information models, and intuitionistic logic the special theory devoting itself to upward persistent propositions. Or, with the examples in Chapters 16, 17, the underlying base logic is our relational algebra; whereas, say, ordinary updates then impose special properties, such as 'stability':

$$xRy \;\Rightarrow\; yRy\,.$$

Does this kind of application presuppose the existence of one distinguished base logic, of which all others are extensions? This would be attractive - and some form of relational algebra or linear logic might be a reasonable candidate. Nevertheless, the enterprise does not rest on this outcome. What matters is an increased sensitivity to the 'landscape' of dynamic logics, just as with the Categorial Hierarchy in Parts II, III, where the family of logics with their interconnections seems more important than any specific patriarch.

Finally, perhaps the most important issue in the new framework is the possibility of *new* kinds of questions arising precisely because of its differences from standard logic. At least, given the option of regarding propositions as *programs*, it will be of interest to consider systematically which major questions about programming languages now make sense inside logic too.

Example. Questions About Programs.
Here are some well-known concerns from Computer Science.
Correctness

When do we have

$$[[\pi]]\,(\,[[A]]\,)\;\subseteq\;[[B]]$$

for (s, t) propositions A, B and a dynamic $(s, (s, t))$ proposition π ?

Program Synthesis

Which dynamic proposition will take us from an information state satisfying A to one satisfying B ?

(This question needs refinement, lest there be trivial answers.)

Determinism

Which propositions as programs are 'deterministic',

in the sense of defining single-valued functions from states to states?

Querying

What does it mean to ask for information in the present setting?

(Presumably, individual types referring to e will be crucial here.)

Some relevant suggestions may be found in the Appendix to this Chapter.

At least if one believes that 'dynamics' is of the essence in cognition (rather than a mere interfacing problem between the halls of eternal truth and the noisy streets of reality), the true test for the present enterprise is the development of a significant new research program not merely copying the questions of old.

17.5 Appendix **Cognitive Programming**

If we are to take the above intentions seriously, then we shall have to pay attention to actual algorithms and *procedures* in our logical semantics. And indeed, there are some ways of introducing such concerns into Logic, mainly using tools from *Automata Theory.*

One natural question is whether the earlier logical constants on information models admit of a procedural explication, in terms of instructions for searching through the information pattern. Such explications have been given for logical quantifiers in van Benthem 1986A, using 'semantic automata' surveying the universe of relevant individuals in some arbitrary order. But in the present case, search should probably proceed along the built-in relation of inclusion among information states. Semantic automata which are suitable for this kind of task have been studied, e.g., in the model theory of Temporal Logic (cf. Thomas 1989) - but also, for more general linguistic purposes, in van Benthem 1988C. Here is an example of the latter kind of approach.

Let us assume that information models are *finite trees*, with truth or falsity of atomic propositions marked at their nodes. Our automata will work progressively upward, processing a node only when all its children have been processed. The core machine is a set of instructions which first reads the atomic information on the current node, then determines which routine to run on the set of state markers left on its children at the end of some previous cycle, and following the outcome of that, prints another such state marker on the current node. In the simplest case, this routine will be a finite state transducer.

For instance, to check whether some extension of the distinguished node (i.e., the top node of the tree) has property q , the machine will print some suitable state marker q on q-nodes, which gets passed on to their parents, and so on, until the top. (The central routine here just checks an existential quantifier on children.) This is merely one illustration of a general phenomenon (cf. van Benthem 1989C):

Theorem. All forward-looking basic *modal* properties of trees can be computed in this way by finite state procedures.

But not just unary modalities like ◇ can be computed in this format, also essentially binary ones like $\mathbb{U}pq$. (Here, the idea is to mark nodes having p , and then to pass up some special state marker on their parents having q , etcetera.) Thus, the whole Modal Hierarchy on information models of Chapter 15 can also be analyzed in terms of machine instructions for this kind of search.

Still, this perspective is not yet fully satisfactory. For one thing, the 'bottom up' direction of search does not reflect intuitive inspection of truth conditions, which rather seem to work 'top-down'. Still, this is not essential - and we can rework the above into a top-down set of recursive instructions for checking our desired semantic properties. The problem is rather that information models need be neither finite nor tree-like: and we may have to inspect, e.g., non-well-founded graphs. Or even assuming well-foundedness, checking for $\mathbb{U}pq$ may involve non-linear inspection of patterns like the following:

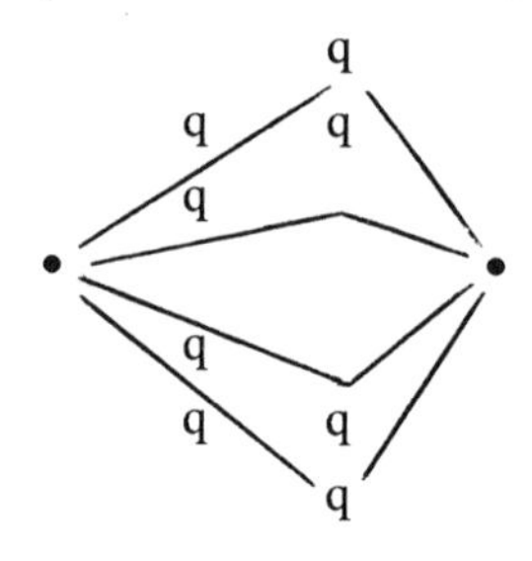

Here, traversing just one single unbroken q-path up to a p-state is not enough.

So, we have to conclude with a *Question*:
What would be an appropriate kind of graph automaton on information models?

But, there are also other ways in which such procedural considerations can enter. Notably, there is the issue of programming cognitive transitions. Given some relation among information states, can we write an explicit program in our dynamic logic effecting just these transitions? Or conversely, what is the class of relations expressible by various classes of modal programs?

Here, we can still consider a great variety of programs: arising by choices of control instructions (finitary ones, or also infinitary) and of basic actions.

To be more concrete, one could take all the regular operations:

$$; \quad \cup \quad *$$

and allow three kinds of atomic action:

a test ? on some proposition letter ,
a move S to some successor of the current node ,
a move P to some predecessor of the current node .

This allows us to program, e.g., the following transitions on *trees*:

$$S^* ; q? \quad : \quad \lambda x \bullet \lambda y \bullet \; x \subseteq y \wedge \neg\!\!\!\!\!\; q y$$
$$(\neg q? ; S)^* ; q? \quad : \quad \lambda x \bullet \lambda y \bullet \; x \subseteq y \wedge q y \wedge \forall z \, (x \subseteq z \subsetneq y \rightarrow \neg q z) \; .$$

But in general, one also obtains non-first-order relations, say, demanding an odd number of intermediate q positions for an "Until pq" statement.

Remark. Acyclic Finite Automata.
First-orderness is guaranteed when we abandon the infinitary iteration * : using, e.g., only the earlier operations of Relation Algebra. But in addition, some simple, 'acyclic' programming structures involving * will be first-order too. Recall the notion of 'first-order languages' and their corresponding acyclic finite-state recognizers in Chapter 8.

Remark. Complexity of Operations.

There are some interesting issues of *complexity* here: not so much for the yield in classes of functions or relations being computed, but rather for programming constructs themselves. In which sense can one say, e.g., that the regular operations are 'less complex' than the others supplied by the full Relational Algebra? ♠

Without going into details here, it may be observed that the present analysis does suggest an interesting kind of (non-deterministic) finite automaton for the above programs which has some Turing machine-like behaviour. Its format will be this:

from *state* plus *test on propositional atoms*
to *move* (S , P or "stay") plus *new state* .

Here are the diagrams for the above two examples:

$1 \xrightarrow{q?} 2$, with a loop S at 1 :

1	q?	⇒	–	2 (*accepting*)
1	TRUE	⇒	S	1

$1 \xrightarrow{q?} 2$, with a loop S, $\neg q?$ at 1 :

1	q?	⇒	–	2 (*accepting*)
1	¬q?	⇒	S	1

But one can easily vary. For instance, the following machine will search for q in the whole information pattern generated from the current node:

1	q?	⇒	–	2 (*accepting*)
1	TRUE	⇒	S	1
1	TRUE	⇒	P	1 .

Finally, a more realistic theory would have automata not just testing facts at fixed information states, but also e.g. have them *construct* new states, by adding individuals or facts. Basic actions would then include such instructions as

"create x such that ..." ,

"see to it that Px becomes true" .

VII A TOUR OF BASIC LOGIC

This book is a logical investigation in the following sense. *Logic* is the study of reasoning, both in the sense of description of reasoning practice and the sense of design of reasoning systems. And as such, it is a discipline in its own right, without being a handmaiden to any special employer. Since sustained reasoning is the life blood of science, logical notions and tools may be useful across many other fields. In particular, this has turned out to be the case in Mathematics, Philosophy, Linguistics and Computer Science. In all these cases, 'utility' does not just mean the introduction of logical jargon and notations, but also, a logical contribution to theory formation in the relevant field. For that reason, it does not suffice to present some convenient formalisms or definitions for consumption: logical *theorems* and methods of proving them are also important, sometimes because of what they say directly, sometimes at least for the kind of research question or attitude that they embody.

This being the intention of the present treatment, it presupposes a certain familiarity with a broad range of logical subjects. Some relevant material has been collected in this final Part, for easy reference. Inevitably, some idiosyncratic choices have been made here. Larger exhaustive collections may be found in such compendia as the "Handbook of Mathematical Logic" (Barwise, ed., 1977), the "Handbook of Philosophical Logic" (Gabbay & Guenthner, eds., 1982-1989), the "Handbook of Logic in Computer Science" (Abramsky, Gabbay & Maibaum, eds., to appear) and the "Handbook of Logic in Artificial Intelligence and Logic Programming" (Gabbay, Hogger & Robinson, eds., to appear). Moreover, there are some thorough introductions with wide coverage, such as Bell & Machover 1977 for logic in mathematics, GAMUT 1990 for logic in linguistics.

1 First-Order Predicate Logic

This is the by now canonical system of introductory logic courses, serving as a point of departure for the whole field. It was only invented in the twenties, as a 'fortunate fragment' of the wider type theory of Frege and Russell - and indeed, some features of its contemporary presentation, such as the 'standard' model-theoretic 'Tarskian' semantics, only crystallized out during the sixties. See Hodges 1983 for an excellent presentation, and Goldfarb 1979 for a fascinating history of its genesis.

1.1 Language

1.1.1 Alphabet.

x, y, z, ...	individual variables
c, d, ...	individual constants
f, g, h, ...	function symbols with arities indicated
A, B, ..., P, Q, ...	predicate symbols with arities indicated
=	identity predicate
$\neg, \wedge, \vee, \rightarrow, \leftrightarrow$	propositional connectives
$\exists, \forall$	quantifiers
), (	brackets

1.1.2 Syntax.

Terms

- individual variables and constants are terms
- if f is a function symbol of arity k and $t_1, ..., t_k$ are terms, then $ft_1...t_k$ is a term.

Atomic formulas

are of the form $Pt_1...t_k$ with P of arity k and $t_1, ..., t_k$ terms.

Formulas

- atomic formulas are formulas
- if ϕ, ψ are formulas, then so are $\neg\phi$, $(\phi\wedge\psi)$, $(\phi\vee\psi)$, $(\phi\rightarrow\psi)$, $(\phi\leftrightarrow\psi)$
- if ϕ is a formula and x an individual variable, then $\exists x\phi$, $\forall x\phi$ are formulas.

From now on, we shall usually work with less formalistic examples, such as 'fgxy' or 'Rcz' , where arities are irrelevant, or understood from context.

1.1.3 Grammar.

Formulas are *uniquely readable*, having one construction tree displaying the action of the above formation rules. Then, the *scope* of an occurrence of a logical constant is the occurrence of the unique subformula in front of which it was introduced during that construction. A quantifier $\exists x$, $\forall x$ *binds* all 'free' occurrences of the variable x in its scope, where *free* occurrences of x are those which lie (yet) outside the scope of any such quantifier. Afterwards, these occurrences of x have become *bound*. Replacing some quantifier variable in an occurrence of a quantifier and all free variables bound by it by some new variable gives an *alphabetic variant*, retaining the same meaning, intuitively.

Substitution of a term t for a variable x in a term or formula ϕ means replacing all free occurrences of x in ϕ by one of t. If no variable in t becomes bound at such a position, then t is called *free for* x *in* ϕ. 'Closed' formulas without free occurrences of variables are called *sentences.*

Occurrences of parts of formulas can be 'positive' or 'negative'. In particular, there may be *positive* and *negative occurrences* P of predicate symbols in formulas ϕ:

- P occurs positively in the formula P
- if P is positive (negative) in ϕ, then it is also positive (negative) in $(\phi\wedge\psi)$, $(\psi\wedge\phi)$, $(\phi\vee\psi)$, $(\psi\vee\phi)$, $(\psi\rightarrow\phi)$ and negative (positive) in $\neg\phi$, $(\phi\rightarrow\psi)$
- if P is positive (negative) in ϕ, then it is positive (negative) in $\exists x\phi$, $\forall x\phi$.

Finally, here is a useful measure of complexity. The *quantifier depth* of a formula is the maximum length of a nest of quantifiers occurring in each other's scope within it. Recursively,

$qd(\phi)$	=	0	for atomic formulas ϕ
$qd(\neg\phi)$	=	$qd(\phi)$	
$qd(\phi\#\psi)$	=	maximum ($qd(\phi)$, $qd(\psi)$)	for all connectives #
$qd(Qx\phi)$	=	$qd(\phi)+1$	for both quantifiers Q.

1.2 Semantics

1.2.1 Structures.

Interpretation takes place in semantic *structures* of the form

$$\mathbb{D} = (D, \mathbb{O}, \mathbb{P}),$$

where D is thought of as a domain of individual objects, with $\mathbb{O}$ a family of operations over D (including, perhaps, 'distinguished objects' of arity 0) and $\mathbb{P}$ a family of predicates. These structures may be thought of as mathematical situations, consisting of an algebraic part with operations (think of a 'group') and a more 'geometric part' with relations (think of a 'graph' or 'space'). But also, they may be viewed as, being or modelling, parts of external reality. Thus, the usual text book wisdom of presenting them as *set-theoretic* entities is a convenience, but not an intrinsic part of buying the 'predicate-logical package'.

1.2.2 Interpretation.

In order to connect the formal language with these structures, two linkages are needed. An *interpretation function* I maps individual constants c to distinguished objects I(c), k-ary function symbols f to k-place operations I(f) in $\mathbb{O}$ and k-ary predicate symbols Q to k-place predicates I(Q) in $\mathbb{P}$. Moreover, an *assignment* a maps individual variables x to objects a(x) in D . The division of labour here is that between a more permanent linkage, corresponding to longer-term linguistic conventions that one learns, and a more local 'dynamic' one, that may shift as required by context.

All these ingredients are necessary for interpretation. For instance, a silent observer of a scene may know $\mathbb{D}$ and I, a without feeling the need to produce any piece ϕ of syntactic code recording the situation. Or, someone who buys a Turkish newspaper with a picture about a World Cup soccer match may know the syntactic code ϕ and the described situation $\mathbb{D}$ without having the interpretation function I telling him what is about what.

1.2.3 Truth Definition.

Now, the basic scheme of semantic interpretation, conveniently attributed to Tarski for the purpose of historical simplification, is as follows

$\mathbb{D}, I, a \models \phi$ 'ϕ is *true* in $\mathbb{D}$ under I and a'.

Note that 'truth' is a relation here, not a property, as is still iterated in much of the philosophical literature. The explanation is by recursion on the syntactic construction of formulas, as an instance of Frege's Principle of '*Compositionality* of meanings':

- auxiliary notion: *values* of terms

  value (x, $\mathbb{D}$, I, a) = a(x)

  value (c, $\mathbb{D}$, I, a) = I(c)

  value (ft_1t_2, $\mathbb{D}$, I, a) = I(f) (value(t_1, $\mathbb{D}$, I, a), value (t_2, $\mathbb{D}$, I, a))

- $\mathbb{D}, I, a \models Rt_1t_2$ iff I(R) holds of the objects value(t_1, $\mathbb{D}$, I, a), value(t_2, $\mathbb{D}$, I, a) in that order

  $\mathbb{D}, I, a \models t_1=t_2$ iff value(t_1, $\mathbb{D}$, I, a) *is the same as* value(t_2, $\mathbb{D}$, I, a)

  $\mathbb{D}, I, a \models \neg\phi$ iff *not* $\mathbb{D}, I, a \models \phi$

  $\mathbb{D}, I, a \models \phi\wedge\psi$ iff $\mathbb{D}, I, a \models \phi$ *and* $\mathbb{D}, I, a \models \psi$

  and likewise for the other propositional connectives

  $\mathbb{D}, I, a \models \exists x\phi$ iff *there exists some* object d in D such that $\mathbb{D}, I, a[x / d] \models \phi$

  where 'a[x / d]' is the assignment which is like a except for the possible difference that it assigns the object d to the variable x (this relation is also written as '$a =_x b$'),
  and likewise for the universal quantifier.

In practice, one becomes more careless with notation. For instance, the function I may often be taken for granted. Moreover, for 'closed formulas' without free occurrences of variables, assignments may be disregarded because, in the above schema, only values of a for free variables in ϕ really matter. One common notation is $\mathbb{M}, a \models \phi$, where a *model* $\mathbb{M}$ is a pair of a structure and an interpretation function.

This schema suggests various directions of 'variation'. Given a model $\mathbb{M}$, one can investigate all formulas true in it. But conversely, for pieces of text ϕ, one can describe the models verifying them. And even, given a text ϕ and a structure $\mathbb{D}$, one can look at all interpretations making that text true in that situation. All of these perspectives are found in logical research.

1.3 Valid Consequence

One of the key notions in reasoning is that of 'valid' argument. In the simplest format, this involves a set of premises Π and a conclusion γ. There are various perspectives on defining validity, of which the semantic one is quite basic ('transmission of truth'):

- $\Pi \models \gamma$ (' γ is a *valid consequence* from Π ') iff
  for all models $\mathbb{M}$ and assignments a ,
  if $\mathbb{M}, a \models \pi$ for all π in Π, then $\mathbb{M}, a \models \gamma$.

Because of the quantification over the totality of all models, this is a quite abstract notion. But, there exist more concrete methods of testing it, such as 'semantic tableaus'.
On this explanation of validity, statements in the following pairs are mutual consequences, and hence, logical synonyms:

$$\phi\vee\psi \ / \ \neg(\neg\phi\wedge\neg\psi) \qquad \phi\rightarrow\psi \ / \ \neg\phi\vee\psi \qquad \forall x\phi \ / \ \neg\exists x\neg\phi$$

Therefore, attention may be restricted to just the logical constants $\neg$, $\wedge$ and $\exists$. But note that this is only permitted on the semantics given here: these equivalences would not be valid, for instance, in intuitionistic logic.
Other ways of defining validity are more concrete from the start. In particular, there is the *deductive* approach, proceeding from systems of inferential steps, which takes its point of departure in human inferential practice:

- $\Pi \vdash \gamma$ iff there exists a *derivation* for γ from assumptions in Π
  using only permissible rules of some logical calculus.

And finally, in this century, various authors have revived the classical idea of logical activity as a debating game. Thus, Lorenzen 1961 presents an interesting third notion of validity based upon the following idea:

- $\Pi \Vdash \gamma$ iff a defender of γ has a guaranteed *winning strategy* against an opponent granting Π in a logical game of argumentation.

Hintikka 1973 even has a game-theoretical approach to semantic interpretation itself, between a 'verifier' and a 'falsifier', where one draws objects from the domain which can be tested for certain atomic facts.

Usually, semantic notions of validity have served as a touch-stone of adequacy for such further proof-theoretic or game-theoretic proposals. What the latter provide, however, are more vivid ideas about structuring of arguments and procedures for reasoning.

1.4 Alternative Formulations

1.4.1 Variable-Free Notation.

The above presentation is not the only possible one, so that not too much should be read into some features of its presentation. For instance, there are also systems of predicate logic doing without variables altgether, such as that of Quine 1971. Its syntax defines predicates of arbitrary arities, whose construction rules include the following:

- if P is a k-ary predicate,
  then $\neg P$ is a k-ary predicate too 'Negation'
- if P is a k-ary predicate and Q an m-ary one,
  then $P \wedge Q$ is a (k+m)-ary predicate 'Conjunction'
- if P is a (k+1)-ary predicate,
  then $\exists P$ is a k-ary predicate 'Projection'

Here, meanings are as follows:

$$\neg P \quad = \quad \lambda x_1. \dots \lambda x_k \bullet \neg P x_1 \dots x_k$$
$$P \wedge Q \quad = \quad \lambda x_1. \dots \lambda x_k \bullet \lambda y_1. \dots \lambda y_m \bullet P x_1 \dots x_k \wedge Q y_1 \dots y_m$$
$$\exists P \quad = \quad \lambda x_1. \dots \lambda x_k \bullet \exists x_{k+1} P x_1 \dots x_k x_{k+1} .$$

Moreover, there are some administrative operations manipulating argument positions:

$$\Pi 1P \quad = \quad \lambda x_1. \dots \lambda x_k \bullet P x_2 \dots x_k x_1 \qquad \text{'Major Inversion'}$$
$$\Pi 2P \quad = \quad \lambda x_1. \dots \lambda x_k \bullet P x_1 \dots x_{k-2} x_k x_{k-1} \qquad \text{'Minor Inversion'}$$
$$IdP \quad = \quad \lambda x_1. \dots \lambda x_{k-1} \bullet P x_1 \dots x_{k-1} x_{k-1} \qquad \text{'Identification'}$$

This formalism is equivalent in expressive power to ordinary predicate logic, and it may be closer to actual human reasoning in some respects.

1.4.2 Algebraization.

Other variable-free notations are found in the *algebraic* tradition in predicate logic. Starting from the analogy with Boolean Algebra for propositional logic (cf. Sikorski 1969), one can try to algebraize the whole predicate-logical language, as a notation for predicates with logical constants as algebraic operations. For the fragment with only binary relations, this is called *Relational Algebra* (see Jónsson 1987); for the whole language, it is the *Cylindric Algebra* of Henkin, Monk & Tarski 1985. Németi 1990 has a nice survey of different options here, and their mathematical properties.

1.4.3 Dynamic Interpretation.

Finally, the above Tarskian semantic mechanism admits of variation too. For instance, it even has a dynamic mode where processing a formula ϕ in a model $\mathbb{M}$ leads to shifting assignments, or even shifting interpretation functions: see Chapter 16 (and Heim 1982, Barwise 1987, Groenendijk & Stokhof 1989). In such a framework, the above notions of 'scope' or 'binding' all acquire different effects.

One such system of dynamic interpretation for predicate logic in a model $\mathbb{M}$ with assignments a, b works as follows:

$[[Pt]]$	$=$	$\{ (a, a) \mid \mathbb{M} \models Pt\,[a] \}$	atomic test
$[[\exists x]]$	$=$	$\{ (a, b) \mid a =_x b \}$	random assignment
$[[\phi \wedge \psi]]$	$=$	$[[\phi]] \bullet [[\psi]]$	relational composition
$[[\neg\phi]]$	$=$	$\{ (a, a) \mid$ *for no* $b, (a, b) \in [[\phi]] \}$	strong failure

On this view, quantifications $\exists x\phi$ will be read as conjunctions '$\exists x \wedge \phi$' : that is, they obtain their dynamic effect through a composition of random assignment for x and succesful transitions for their matrix ϕ. Thus, quantifiers correspond to basic actions on assignments, not to procedural operations.

Nevertheless, standard predicate logic is able to express this dynamic action too. The following is an explicit definition for transition predicates $TR(\phi, x_1, ..., x_k, y_1, ..., y_k)$ associated with newly interpreted formulas ϕ all of whose variables are among the list $x_1, ..., x_k$:

$$TR(Pt) = Pt \wedge y_1=x_1 \wedge ... \wedge y_k=x_k$$

$$TR(\exists x_i) = \bigwedge \{ y_j=x_j \mid j \neq i \}$$

$$TR(\phi\wedge\psi) = \exists z_1 ... \exists z_k \;(TR(\phi, x_1, ..., x_k, z_1, ..., z_k) \wedge TR(\psi, z_1, ..., z_k, y_1, ..., y_k))$$

$$TR(\neg\phi) = y_1=x_1 \wedge ... \wedge y_k=x_k \wedge \neg\exists z_1 ... \exists z_k \, TR(\phi, x_1, ..., x_k, z_1, ..., z_k) .$$

1.5 Fragments

In many applications, only certain sublanguages of the full predicate-logical formalism are involved. Important examples are

- *monadic predicate logic*, having only 1-ary predicates, and no function symbols,
- *universal formulas*, having only a sequence of universal quantifiers in front, followed by a quantifier-free matrix,
- *Horn clauses*, being universal formulas with a matrix of the special form 'conjunction of atoms implies some atom'.

But many other fragments keep arising. For instance, from the perspective of this Book, there would be an interest in

- *single-bond formulas*, where each quantifier binds only one occurrence of a variable in its scope.

Moreover, Chapters 14, 15, 16 have payed a good deal of attention to

- *k-variable formulas*, being all predicate-logical formulas constructed using just the variables $x_1, ..., x_k$, free or bound.

Dreben and Goldfarb 1979 present a vast survey of fragments, motivated by the complexity of their decision problems.

1.6 Theories

Many important sets of mathematical axioms can be formulated in the language of predicate logic. Here are some examples that are used in this book:

Boolean Algebra

$x+(y+z)$	$=$	$(x+y)+z$	$x\bullet(y\bullet z)$	$=$	$(x\bullet y)\bullet z$
$x+y$	$=$	$y+x$	$x\bullet y$	$=$	$y\bullet x$
$x+x$	$=$	x	$x\bullet x$	$=$	x
$x+(y\bullet z)$	$=$	$(x+y)\bullet(x+z)$	$x\bullet(y+z)$	$=$	$(x\bullet y)+(x\bullet z)$
$x+(x\bullet y)$	$=$	x	$x\bullet(x+y)$	$=$	x
$-(x+y)$	$=$	$-x\bullet -y$	$-(x\bullet y)$	$=$	$-x+-y$
$x+0$	$=$	x	$x\bullet 0$	$=$	0
$x+1$	$=$	1	$x\bullet 1$	$=$	x
$x+-x$	$=$	1	$x\bullet -x$	$=$	0
$--x$	$=$	x	$0\neq 1$		

Relational Algebra

i	all Boolean identities, written here with $\cup$ instead of + and $\cap$ instead of •				
ii	$(R \cup S)^{\cup}$	=	$R^{\cup} \cup S^{\cup}$		
	$(-R)^{\cup}$	=	$-R^{\cup}$		
	$R^{\cup\cup}$	=	R		
	$id^{\cup}$	=	id		
iii	$(R \cup S) \bullet T$	=	$R \bullet T \cup S \bullet T$		
	$R \bullet (S \cup T)$	=	$R \bullet S \cup R \bullet T$		
	$R \bullet (S \bullet T)$	=	$(R \bullet S) \bullet T$		
	$R \bullet id$	=	$id \bullet R$	=	R
iv	$(R \bullet S)^{\cup}$	=	$S^{\cup} \bullet R^{\cup}$		
	$(R^{\cup} \bullet -(R \bullet S))$	$\subseteq$	$-S$,		

where '$A \subseteq B$' abbreviates '$A \cup B = B$'.

Peano Arithmetic

$\forall x \neg Sx = 0$

$\forall x \forall y \ (Sx = Sy \rightarrow x = y)$

$\forall x \ x+0 = x$

$\forall x \forall y \ x+Sy = S(x+y)$

$\forall x \ x \cdot 0 = 0$

$\forall x \forall y \ x \cdot Sy = (x \cdot y)+x$

$(\phi(0) \wedge \forall x(\phi(x) \rightarrow \phi(Sx)) \rightarrow \forall x \phi(x)$

Without multiplication, the resulting system is called *Presburger Arithmetic*.

As far as less algebraic properties are concerned, predicate logic is able to express all the usual properties of binary relations, such as

$\forall x \, Rxx$	reflexivity
$\forall x \forall y \, (Rxy \rightarrow Rxx)$	quasi-reflexivity
$\forall x \, \neg Rxx$	irreflexivity
$\forall x \forall y \, ((Rxy \wedge Ryx) \rightarrow x=y)$	anti-symmetry
$\forall x \forall y \forall z \, ((Rxy \wedge Ryz) \rightarrow Rxz)$	transitivity
$\forall x \forall y \forall z \, ((Rxy \wedge Rxz) \rightarrow Ryz)$	euclidity
$\forall x \forall y \, (Rxy \vee Ryx \vee x=y)$	linearity

Moreover, it can even provide a full set of axioms for *Set Theory* (cf. Barwise 1977).

2 Logical Syntax and Mathematical Linguistics

We now turn to an interface of Logic and Mathematical Linguistics: see Hopcroft & Ullman 1979 for an authoritative text book.

The definition of logical formulas really gave a kind of simple grammar for producing all well-formed expresssions. For instance, restricting attention to the propositional case, it may also be written with production rules of the form:

$F \Rightarrow$ p, q, r	proposition letters
$F \Rightarrow \neg F$	negation
$F \Rightarrow (F \wedge F)$	conjunction

'Formulas' are then all expressions that can be produced by rewriting the 'start symbol' F via these rules to an expression no longer containing the symbol F .

2.1 Rewrite Grammars

Formally, we assume some *alphabet* A of 'symbols', divided into 'terminal symbols' [in the above: p, q, r,), (, ¬, ∧] and 'auxiliary symbols' (in the above only: F) whose finite sequences are called *expressions*. *Rewrite rules* are of the form

$E_1 \Rightarrow E_2$	for arbitrary expressions E_1, E_2 .

A *context-free* rewrite rule is one of the form

$X \Rightarrow E$	where X is an auxiliary symbol and E an arbitrary expression.

A *(left-)regular* rewrite rule has one of the even more special forms

$X \Rightarrow aY$	where a is a terminal symbol and Y
$X \Rightarrow a$	an auxiliary symbol.

A *rewrite grammar* is a finite set G of rewrite rules in some alphabet, with one auxiliary symbol S designated as its *start symbol*. The *language generated by* G is the set of all expressions over the alphabet that arise through some finite sequence of rewritings of S via the rules of G (replacing continuous substrings E_1 by E_2) so as to end with a sequence of terminal symbols only. These grammars come in the so-called 'Chomsky Hierarchy', three of whose important levels are as follows:

• type 0 grammars	no restrictions on rewrite rules
• context-free grammars	only context-free rewrite rules
• (left-) regular grammars	only left-regular rewrite rules.

There is a proper inclusion here, in the upward direction.

2.2 Categorial Grammars

Grammar formalisms can also be organised along different lines, witness the *categorial grammars* of this book, that work with

- a set of *categories*, consisting of basic categories and others constructed out of these using possibly iterated binary slashes (A/B) and (A\B),
- an assignment of a finite set of categories to each terminal expression,
- recognition of those expressions whose terminal symbols admit of at least one corresponding sequence of assigned categories which can be reduced to the distinguishes category S via repeated application of the two combination rules

$$A + (A \backslash B) = B \qquad\qquad (B/A) + A = B\,.$$

Gaifman's Theorem: Categorial grammars and context-free grammars recognize the same class of languages.

2.3 Mathematical Properties

Various language families in the Chomsky Hierarchy have useful mathematical characteristics.

Pumping Lemma for Regular Languages

For each regular grammar G , there exists a natural number n such that, if an expression generated by G has length at least n , then it has the form xyz , where each repetition xy^iz is also generated by G .

Pumping Lemma for Context-Free Languages

For each context-free grammar G , there exists a natural number n such that, if an expression generated by G has length at least n , then it has the form xyzuv , where y, u are not both empty, such that each repetition xy^izu^iv is also generated by G .

The next result requires some definitions. Let the alphabet have k symbols. Then, each expression E has an associated 'occurrence vector' $E^\#$ of length k recording the number of occurrences of each symbol in it. For a grammar G , $E^\#(G)$ is the set of occurrence vectors of all expressions generated by it. Now, some arithmetical definitions. A set of k-vectors is *linear* if it can be written as the set of all vector sums of one fixed vector (its 'base') plus all multiples of some finite set of k-vectors (its 'periods'). Finite unions of linear sets are called *semi-linear*.

Parikh's Theorem

For each context-free grammar G , $E^{\#}(G)$ is semi-linear.

This criterion is quite natural from a logical point of view. Sets of k-vectors of natural numbers may be regarded as extensions of k-place predicates over the latter. Now, the k-ary semi-linear predicates are precisely those that can be defined by formulas with k free variables in the language of Presburger Arithmetic (see Ginsburg 1966).

2.4 Automata

Another mathematical perspective on languages involves recognition via some kind of automatic device.

2.4.1 Finite State Automata.

The simplest one is a *finite state automaton* over some finite alphabet, reading successive symbols, while assuming various states. Formally, this requires specification of

- a finite set of states
- a 'next move' function from states and symbols to states
- a division of states into 'accepting' and 'rejecting' ones.

An expression will be *recognized* by a finite automaton if the latter ends in an accepting state after traversal of all its successive symbols.

Theorem The class of languages recognized by finite state automata is precisely that generated by the regular grammars.

There are also other useful descriptions of these languages. In particular, finite automata may be encoded in *regular expressions*, being all notations formed from atomic symbols a standing for singleton languages {a} using the 'regular operations'

$\cup$ *choice* ; *sequence* * *iteration* .

These refer to the following operations on languages:

$L_1 \cup L_2$	Boolean union
$L_1 ; L_2$	all concatenations of strings from L_1 and L_2, in that order
L^*	all finite concatenations of strings from L .

These expressions serve precisely to encode transition graphs for finite state automata.

Kleene's Theorem The regular languages are precisely those described by regular expressions.

2.4.2 Push-down Store Automata.

The next kind of more powerful machine has a (restricted) memory. A *push-down store automaton* can keep a finite stack during its reading of a string. Formally, it may use a separate stack alphabet, and

- the next move function now indicates, for each state, symbol read, and top symbol on the stack, what will be the next state, as well as how the top symbol on the stack gets rewritten to a new expression in the stack alphabet.

To give a typical example, in an alphabet {a, b}, a finite state automaton cannot determine, in the long run, whether the majority of symbols in an expression read are a . But a push-down store automaton can do so, by storing symbols a, b , ticking off b's against a's .

Theorem The context-free languages are precisely those recognized by push-down store automata, provided that the latter are allowed an *indeterministic* next-move function.

The latter freedom means that the machine may have options at certain stages, and a string will count as being 'recognized' if there exists at least one felicitous set of choices for the machine leading to a recognizing state at the end of its traversal.

2.4.3 Turing Machines.

The most powerful type of symbol-processing machine still performing effective computation is that which emerged from Turing's famous analysis of computability. This time, the string to be read is put on an infinite tape, and a *Turing machine* has the following repertoire: given a state and a symbol read, it can

- change to a new state
- print a new symbol instead of the one just read
- decide to stay at the same position, or shift its reading head one step toward the left or right.

Thus, it can actively transform a potentially infinite memory. For instance, unlike a push-down store automaton, a Turing machine can check whether an expression contains equal numbers of three symbols a, b, c (cf. Chapter 8).

Theorem The languages recognized by Turing machines are precisely those generated by type 0 grammars.

2.5 Logical Syntax

The preceding notions make sense at various places within Logic. Here are some examples, taken from van Benthem 1987$.

- The formulas of predicate logic form a context-free, non-regular language.

The lack of regularity has to do with the long-distance coordination that is required for the bracketing of well-formed formulas. Without this disambiguation device, complexity will be lower (compare the discussion of possible 'regular regions' of natural language in Chapter 4, where 'recursion' can often be simulated by suitable 'iteration').

A similar result holds for most logical formalisms, such as the class of terms of the typed Lambda Calculus in this book. Nevertheless, slight changes may lead to higher grammatical complexity:

- The single-bind Lambek fragment Λ is not context-free.

Checking for single occurrence of variables requires devices that go beyond maintaining a simple stack in memory.

Next, one can also consider grammatical complexity of 'languages' consisting of the theorems of some logical calculus. Here we have, for instance, that, on a finite alfabet of proposition letters:

- The set of classical theorems of propositional logic is context-free.

  The set of theorems of the minimal propositional modal logic is not context-free.

Valid modal principles involve coordination between stacked modalities that cannot be handled by a push-down store.

But considerations of complexity may also be applied to texts, or in Logic, to proofs:

- The set of correct Hilbert-style axiomatic deductions in propositional calculus is a type 0 language that is not context-free.

Finally, automata have been used outside of syntax too, for purposes of logical *semantics* (see Thomas 1989, van Benthem 1986A, 1988D on such 'semantic automata'). As a simple example, one can turn the above devices into *transducers*, traversing a string while outputting expressions in some signal alphabet, that may serve to describe semantic denotations. For instance, a finite state automaton or push-down store automaton might read propositional formulas, together with some suitably encoded valuation for the relevant proposition letters, and then output truth values 0, 1 when appropriate. Here again, van Benthem 1987$ arrives at an outcome similar to the above:

- Correct semantic evaluation for classical propositional logic is possible by means of a push-down store transducer,
- while finite state transducers are bound to make mistakes of two kinds: they will sometimes fail to compute values where they should, but they will also sometimes compute incorrect values.

2.6 Algebraic Theory

There are quite a few further results in mathematical linguistics on language families. Many of these have to do with closure properties under certain natural operations (cf. Chapter 14): Boolean ones, regular ones, etcetera. For instance,

- the regular languages are closed under all Boolean operations,
- the context-free languages are closed under unions, but not under intersections or complements.

Other natural operations are those described in Chapter 14 corresponding to the slashes of Categorial Grammar

$$L_1 \backslash L_2 = \{ x \mid \text{for all } y \in L_1 , yx \in L_2 \}$$
$$L_2 / L_1 = \{ x \mid \text{for all } y \in L_1 , xy \in L_2 \}$$

Regular languages are closed under the latter too.
Thus, languages may be described by complex algebraic terms involving such operations. One can ask for a complete description of 'validity' here, in the following sense:

$t_1, .., t_k \models t$ means that, for any assignment of sets of expressions to variables over languages, the sequential concatenation of the languages denoted by $t_1, .., t_k$ is contained in that denoted by t.

Buszkowski's Theorem The Lambek Calculus L is sound and complete for validity among terms constructed using just the two categorial slashes.

Weaker versions of this result exist for terms involving more of the above kinds of operation, but then with respect to a generalized notion of validity (cf. Dosen 1985 and the discussion in Chapter 14).

3 Model Theory

There is no introductory working text for logical Model Theory covering all widely useful material. But, two good more comprehensive references are Bell & Slomson 1969 and especially, Chang & Keisler 1973.

3.1 Relations and Operations on Models

Independently from any formal language, the universe of structures has its own basic mathematical relations. The most basic of these is 'isomorphism':

Two models $\mathbb{M} = (D, \mathbb{O}, \mathbb{P}, I)$, $\mathbb{M}' = (D', \mathbb{O}', \mathbb{P}', I')$ are *isomorphic* if there exists a bijection F between D and D' which is a homomorphism with respect to corresponding operations g in $\mathbb{O}, \mathbb{O}'$:
i.e., $F(I(g)(d)) = I'(g)(F(d))$ for all d in D,
and which also respects all corresponding predicates in $\mathbb{P}, \mathbb{P}'$:
i.e., $I(Q)(d)$ iff $I'(Q)(F(d))$.

But other relations are important too, such as that of being a 'substructure':

$\mathbb{M}$ is a *submodel* of $\mathbb{M}'$ if D is a subset of D' and all interpretations $I'(g)$, $I'(Q)$ are the set-theoretic restrictions of the corresponding $I(g), I(Q)$ to D'. (In particular, then, D' must be closed under the operations $I(g)$.)

There are also certain important constructions of new models from old ones, such as the formation of 'direct products':

Given a family of models $\{ \mathbb{M}_i \mid i \in I \}$, the *direct product* $\Pi\{ \mathbb{M}_i \mid i \in I \}$ has for its domain the Cartesian product of all domains D_i with the following pointwise definition for its operations and predicates:

$$I(g)(\langle f(i)\rangle_{i\in I}) \quad = \quad \langle I_i(g)(f(i))\rangle_{i\in I}$$

$$I(Q)(\langle f(i)\rangle_{i\in I}) \quad \text{iff} \quad I_i(Q)(f(i)) \text{ for all } i \in I.$$

Slightly more sophisticated is the formation of 'ultraproducts':

Given a family of models $\{ \mathbb{M}_i \mid i \in I \}$ and an ultrafilter U over some index set I, the *ultraproduct* $\Pi_U\{ \mathbb{M}_i \mid i \in I \}$ has for its domain the Cartesian product of all domains D_i divided out by the following equivalence relation:

$$f \sim g \quad \text{iff} \quad \{ i \in I \mid f(i) = g(i) \} \in U.$$

Its predicates and operations are well-defined by the following stipulation:

$$I(g)(\langle f(i)\rangle_{i\in I}) \quad \text{is the } \sim\text{-equivalence class of } \langle I_i(g)(f(i))\rangle_{i\in I}$$

$$I(Q)(\langle f(i)\rangle_{i\in I}) \quad \text{iff} \quad \{ i \in I \mid I_i(Q)(f(i)) \} \in U.$$

3.2 Invariances

One basic measure of expressive power of a formalism, or lack of it, is by finding its characteristic invariances across models, in terms of structural relations. If a language does not 'see' certain differences between structures, this may be taken as a sign of weakness, but it can also be a source of strength: since this phenomenon allows transfer of statements from one situation to another.

For this purpose, one may introduce so-called *Ehrenfeucht Games*. For convenience, we disregard function symbols and operations here. The 'n-round Ehrenfeucht game' between two models $\mathbb{M}, \mathbb{N}$ is played by two players I and II in n successive rounds, each consisting of

selection of a model and an object in its domain by player I ,
selection of an object in the other model by player II .

After n rounds, the partial mapping between the two domains created by these pairs is inspected: if it is a partial isomorphism respecting all predicates, then player II has won, otherwise the win is for player I . One writes II $(\mathbb{M}, \mathbb{N}, n)$ if player II has a *winning strategy* in this game.

Theorem

II $(\mathbb{M}, \mathbb{N}, n)$ if and only if $\mathbb{M}$ and $\mathbb{N}$ verify the same sentences up to quantifier depth n.

There is an equivalent characterization in terms of n-sequences of sets of partial isomorphisms, having a 'back-and-forth' property. Here, we only formulate the latter in the following structural notion:

Two models $\mathbb{M}, \mathbb{N}$ are *partially isomorphic* if there exists a non-empty family PI of finite partial isomorphisms between them such that

- for any partial isomorphism F in PI and any d in the domain of $\mathbb{M}$, there exists an object e in the domain of $\mathbb{N}$ such that F extended with the ordered pair (d, e) is again in PI ,
- and analogously in the opposite direction.

Isomorphism implies partial isomorphism, and the latter again implies II $(\mathbb{M}, \mathbb{N}, n)$ for each number of rounds n : so that $\mathbb{M}$ and $\mathbb{N}$ verify the same predicate-logical sentences (i.e., they are *elementarily equivalent*).

Another structural characterization of first-order predicate logic combines two notions defined so far (cf. Doets & van Benthem 1983), as in the following version of

Keisler's Theorem

A class of models is definable in the form $\{ \mathbb{M} \mid \mathbb{M} \models \phi \}$ for some first-order sentence ϕ iff it is closed under the formation of ultraproducts and partial isomorphism.

Chapter 17 contained a semantic characterization of 'k-variable fragments' of first-order logic using an appropriate modification of these semantic notions.

3.3 Basic Properties

The best-known model-theoretic features of first-order predicate logic are expressed in the following two results.

Compactness Theorem

If each finite subset of a set of formulas Σ has a model plus assignment verifying it, then so does the whole Σ simultaneously.

Löwenheim-Skolem Theorem

If a set of formulas Σ is verified in some model plus assignment, then it is already verified in some *countable* model.

These theorems have many applications in constructing new models for theories, witness so-called 'non-standard models' for arithmetic.

The two properties are characteristic of the whole formalism, in that we even have

Lindström's Theorem

First-order predicate logic is the strongest logic having both the Compactness and Löwenheim-Skolem properties.

Of course, a precise statement and proof of this result requires precisation of the abstract notion of 'logic' used here.

One useful application of Compactness is the existence of special 'saturated' models:

A model $\mathbb{M}$ is ω-*saturated* if, for each countable set Σ of formulas in the free variables $x_1, ..., x_k$ involving only finitely many objects from its domain as parameters, the following holds:

if each finite subset of Σ has some k-tuple of objects satisfying it in $\mathbb{M}$, then there exists some k-tuple of objects satisfying the whole set Σ simultaneously in $\mathbb{M}$.

Theorem

Each model has an ω-saturated *elementary extension*.

The latter means that the original model is a submodel of the extended one which even gives the same truth value to each first-order formula ϕ with free variables $x_1,..., x_m$ at all m-tuples of objects from the old domain. Saturated models were used at various places in Chapters 15 and 17.

3.4 Preservation

Survival of special relations between, or operations upon models, even when not valid for predicate logic in general, may still be a good test for some of its fragments.
Here are some important 'preservation theorems' to this effect:

- A first-order sentence ϕ is *preserved under submodels* if, for all models $\mathbb{M}$, $\mathbb{M} \models \phi$ implies $\mathbb{N} \models \phi$ for all submodels $\mathbb{N}$ of $\mathbb{M}$.

Los' Theorem

A first-order sentence is preserved under submodels iff that formula is logically equivalent to a *universal* one.

Note that at least the effective enumerability of this class of sentences could be predicted a priori. ϕ is preserved under submodels iff the following implication is valid:

$$\phi \models \phi^A ,$$

where ϕ^A is the *syntactic restriction* of ϕ *to* A obtained by relativizing all quantifiers occurring in it to individuals satisfying the predicate A.

- A first-order sentence ϕ with a predicate symbol Q is *monotone in* Q if, for each model $\mathbb{M}$, $\mathbb{M} \models \phi$ implies $\mathbb{N} \models \phi$ for all models $\mathbb{N}$ differing from $\mathbb{M}$ only in having a larger extension for the predicate Q assigned by their interpretation function.

Lyndon's Theorem

A first-order sentence is monotone in Q iff that formula is logically equivalent to one having only positive occurrences of Q.

Monotonicity is important as a general principle of 'surface reasoning' in natural language (see van Benthem 1986A, Sanchez 1990). Moreover, it is an essential mathematical property, for instance, in the theory of inductive definitions, where monotone formulas ϕ define monotone operators ('the next approximation as computed by ϕ') on predicates Q (Moschovakis 1974), and derived from this, approximation for recursive queries (Immerman 1986).

A stronger requirement, which has occurred throughout this book, is the following.

- A first-order sentence ϕ with a predicate symbol Q is *continuous* in Q if, for each model $\mathbb{M}$ with $I(Q) = \bigcup\{ Q_i \mid i \in I \}$, $\mathbb{M} \models \phi$ iff there exists some $i \in I$ such that $\mathbb{M}_i \models \phi$, where $\mathbb{M}_i$ is the model which is exactly like $\mathbb{M}$ except for having $I(Q) = Q_i$.

Theorem

A first-order sentence is continuous in Q iff it is logically equivalent to an 'existential form' $\exists x_1 \ldots \exists x_k \ (Qx_1 \ldots x_k \wedge \phi)$, where ϕ does not contain Q.

- An important weaker notion is ω-*continuity*, which is continuity only for unions of *countable* non-decreasing *chains* of sets Q_i.

The latter notion already suffices for the existence of fixed points for the associated monotone operators on predicates after ω steps.

Theorem

A first-order sentence is ω-continuous in Q iff it is logically equivalent to one constructed from arbitrary Q-free formulas and atoms with predicate Q using only $\wedge, \vee, \exists$.

Working within fragments may raise questions of its own, though. For instance, inside k-variable fragments, there is no simple reduction to universal prenex form, even for conjunctions or disjunctions of universal sentences. And indeed, an appropriate version of Los' Theorem, relativized again to this fragment is still open (cf. van Benthem 1990B).

3.5 Definability

Actual reasoning involves an interplay between performing appropriate arguments and introducing suitable concepts:

A predicate Q is *explicitly definable* in a theory T if there exists a formula ϕ in the remaining vocabulary of T such that $T \models \forall x(Qx \leftrightarrow \phi)$.

A predicate Q is *implicitly definable* in a theory T if there exist no two models for T having the same interpretation for all predicate and function symbols in the language of T except for Q, while having different interpretations for Q.

Padoa's Principle

Explicit definability implies implicit definability.

Beth's Theorem

Implicit definability implies explicit definability.

The latter result is a consequence of the

Interpolation Theorem

If $\phi \models \psi$, then there exists some formula α containing only predicate and function symbols that occur both in ϕ and in ψ (as well as possibly $\top, \bot$) such that $\phi \models \alpha$ and $\alpha \models \psi$.

3.6 Further Directions

Many of the above results may be studied as properties of 'abstract logics'. This has been done in so-called Abstract Model Theory (Lindström 1969, Barwise & Feferman 1985), where various extensions of, and alternatives to, first-order predicate logic are studied as to their possible combinations of meta-properties.
Inspired by Computer Science, there has been a recent development of Finite Model Theory (Gurevich 1985, Makowsky, to appear), where attention is restricted to finite models only. Then, some of the above notions and techniques become irrelevant as they stand (Compactness, Löwenheim-Skolem), while other results require arguments of a more subtle combinatorial nature. In particular, the above preservation theorems for universal and positive sentences fail when restricted to finite models.

4 Proof Theory

Proofs form another basic approach to the logical analysis of reasoning. There are even philosophies of meaning that accord them a central position in human cognition, in preference to semantics (cf. Chapter 13). The latter is true, in particular, for intuitionistic and more generally, constructive logic (Dummett 1977, Martin-Löf 1984).
One interesting feature of proof systems is that they come in different natural formats, and that these formats need not all lead to the same preferred 'base logic'. For instance, some of them favour classical logic, others intuitionistic logic, and yet others linear logic or other occurrence-based calculi. This is in line with 'Bolzano's Program' for Logic (cf. van Benthem 1985*), developed already in the early nineteenth century, conceived as the study of *varieties* of human reasoning, with different standards of validity, rather than focusing upon the 'logical constants', as became the dominant theme afterwards.
In fact, in line with an earlier discussion found in Chapter 6, the main interest is not so much the canon of bare derivable transitions anyway, whether 'classical' or 'intuitionistic'. This is probably an unfortunate focus in modern Logic, which makes a shift between options seem like defection of apocalyptic proportions. (In fact, there are even historical accidents determining 'standard options' here, as well as a good deal of professional inertia in favour of the existing order.) The key issue should rather be 'inferential mechanisms': their lay-out and general properties.

4.1 Formats

4.1.1 Axiomatics.

In Hilbert-style axiomatic deduction, derived from the set-up of Euclidean Geometry, there is a set of axiom schemata, together with rules of inference, and derivations are finite sequences of formulas where each formula is either

- an assumption,
- an axiom,
- a consequence of earlier formulas by a rule.

For predicate logic, in a presentation with $\rightarrow$, $\neg$ and $\forall$ only, these are as follows.

Axioms:

all formulas of the forms

$(\phi \rightarrow (\psi \rightarrow \phi))$

$(\phi \rightarrow (\psi \rightarrow \xi)) \rightarrow ((\phi \rightarrow \psi) \rightarrow (\phi \rightarrow \xi))$

$(\neg\phi \rightarrow \neg\psi) \rightarrow (\psi \rightarrow \phi)$

$\forall x(\phi \rightarrow \psi) \rightarrow (\forall x\phi \rightarrow \forall x\psi)$

$\forall x\phi \rightarrow [t/x]\phi$ provided that t is free for x in ϕ

$\phi \rightarrow \forall x\phi$ provided that x does not occur freely in ϕ,

each possibly prefixed by a number of universal quantifiers

Rules:

Modus Ponens: from ϕ and $(\phi \rightarrow \psi)$ to infer ψ

Notation: $\phi_1, \ldots, \phi_n \vdash \psi$

if there exists an axiomatic derivation whose last formula is ψ

where the only assumptions used are among $\phi_1, \ldots, \phi_n$.

Note the decision insidiously incorporated into this standard phrasing, to allow 'dependence' on assumptions that are not actually used. Requiring genuine dependence would lead to a non-monotonic notion, where allowing arbitrary additions on the left would destroy 'derivability'.

4.1.2 Natural Deduction Trees.

In a natural deduction format, derivations are *trees*, with conclusions at the root, depending on a number of formulas at the leaves. The basic rules either introduce or remove logical constants. This time, only the rules only for conjunction, implication and the universal quantifier will be displayed, these being the most natural representatives of their kind:

Conjunction:

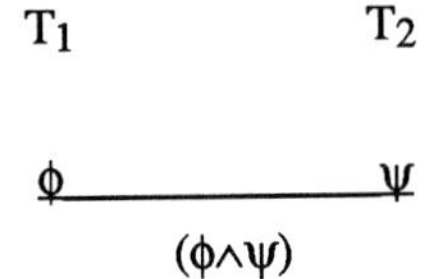

$$\dfrac{\begin{matrix} T_1 & & T_2 \\ \phi & & \psi \end{matrix}}{(\phi \wedge \psi)}$$

if tree T_1 derives ϕ and T_2 derives ψ, then their combination derives $(\phi \wedge \psi)$, with the conclusion dependent on the union of all assumptions for the two conjuncts

Decomposition:

$$\begin{matrix} T \\ \dfrac{(\phi \wedge \psi)}{\phi} \end{matrix} \qquad \begin{matrix} T \\ \dfrac{(\phi \wedge \psi)}{\psi} \end{matrix}$$

in both cases, assumptions remain the same

Conditionalization:

$$\begin{matrix} T \\ \dfrac{\psi}{(\phi \rightarrow \psi)} \end{matrix}$$

if T derives ψ with a set of assumptions possibly including ϕ, then the new tree derives $(\phi \rightarrow \psi)$ with all old assumptions except ϕ

Modus Ponens:

$$\dfrac{\begin{matrix} T_1 & & T_2 \\ \phi & & (\phi \rightarrow \psi) \end{matrix}}{\psi}$$

the conclusion inherits all assumptions from both premises

Generalization:

$$\begin{matrix} T \\ \dfrac{\phi}{\forall x \phi} \end{matrix}$$

provided that the variable x do not occur freely in any assumption on which ϕ depends in T, while all assumptions remain in force for the conclusion

Instantiation:

$$\frac{\begin{matrix} T \\ \\ \forall x \phi \end{matrix}}{[t/x]\phi}$$

provided that t is free for x in ϕ; with a conclusion depending on all earlier assumptions

4.1.3 Gentzen Sequents.

Another format displays *sequents* of the form

$X \Rightarrow \psi$

indicating a conclusion ψ depending on a number of assumptions X.

Alternatively, these may be of the form

$X \Rightarrow Y$

with an intuitive reading to the effect that the conjunction of all premises implies the disjunction of all conclusions.

This time, derivations will be trees again. We display the former format, which favours intuitionistic logic (as long as we work with sets of premises, that is: see Chapter 4).

- Axioms:

  $X \Rightarrow \psi$ where ψ is in X

- Structural Rules:

  $$\frac{X \Rightarrow \psi}{\pi[X] \Rightarrow \psi}$$ for any permutation π of X Permutation

  $$\frac{X \Rightarrow \psi}{X, \phi \Rightarrow \psi}$$ Monotonicity

  $$\frac{X, \phi, \phi \Rightarrow \psi}{X, \phi \Rightarrow \psi}$$ Contraction

Without Monotonicity, one could still have the weaker principle

$$\frac{X, \phi \Rightarrow \psi}{X, \phi, \phi \Rightarrow \psi} \qquad \text{Expansion}$$

$$\frac{X \Rightarrow \phi \qquad Y, \phi \Rightarrow \psi}{X, Y \Rightarrow \psi} \qquad \text{Cut}$$

- Logical Rules:

$$\frac{X, \phi \Rightarrow \psi}{X, (\phi \wedge \alpha) \Rightarrow \psi} \qquad \frac{X, \phi \Rightarrow \psi}{X, (\alpha \wedge \phi) \Rightarrow \psi}.$$

$$\frac{X \Rightarrow \phi \qquad Y \Rightarrow \psi}{X, Y \Rightarrow (\phi \wedge \psi)}$$

$$\frac{X \Rightarrow \psi}{X - \{\phi\} \Rightarrow (\phi \rightarrow \psi)}.$$

$$\frac{X \Rightarrow \phi \qquad Y, \alpha \Rightarrow \psi}{X, Y, (\phi \rightarrow \alpha) \Rightarrow \psi}$$

$$\frac{X \Rightarrow \psi}{X \Rightarrow \forall x \psi}. \qquad \text{provided that } x \text{ do not occur free in any formula in } X$$

$$\frac{X, [t/x]\phi \Rightarrow \psi}{X, \forall x \phi \Rightarrow \psi} \qquad \text{provided that } t \text{ is free for } x \text{ in } \phi$$

4.1.4 Other Formats.

There are many further interesting formats for logical deduction. For instance, manipulating Quine's variable-free notation requires a quite different set-up: see Bacon 1985. The same holds for more algebraically inspired calculi of 'natural logic': see Purdy 1990. But already C. S. Peirce experimented with psychologically more attractive 'existential graphs' (see the modern exposition in Sanchez Valencia 1989), based on basic operations of 'addition', 'deletion' and 'copying' in the propositional case, plus suitable operations of 'linking' and 'interrupting' in the predicate-logical case.

4.2 Properties of Theories

Many common properties of theories are usually expressed in the proof-theoretic format.

- A theory T is *consistent* if, for no formula ϕ, both ϕ and $\neg\phi$ are derivable from it.
- A theory T is *syntactically complete* if, for all formulas ϕ, either ϕ or $\neg\phi$ is derivable from it.

Examples of simple properties (in classical systems) are:

T does not derive ϕ iff $T, \neg\phi$ is consistent

T, ϕ is not consistent iff $T \vdash \neg\phi$

If T is consistent, then so is either T, ϕ or $T, \neg\phi$.

Lindenbaum's Lemma

Each consistent theory has a *maximally consistent* extension.

4.3 Equivalences

There are a number of equivalence results stating that various proof-theoretic formats produce the same outcomes. For instance, in the appropriate formulations of the calculi, we have that, both for classical and intuitionistic logic:

Theorem

Axiomatic deductions, natural deduction trees and sequent proofs all generate the same derivable sequents.

Here, the former two formats are relatively close, and the equivalence proof involves essentially just one non-straightforward observation, namely Tarski's

Deduction Theorem

If $X, \phi \vdash \psi$ is derivable by axiomatic deduction, then there also exists some axiomatic derivation of $(\phi \rightarrow \psi)$ from X only.

The connection with the third format is less straightforward, at least, from a combinatorial point of view. Given the spirit of the sequent rules, that build up conclusions, a derivation rule like Modus Ponens, reflected in the structural rule of Cut, is not obviously admissible. Nevertheless, Gentzen provided an effective method of Cut Elimination, showing that

Gentzen's Hauptsatz

The rule of Cut is an admissible rule of inference in the sequent calculus without it, which does not produce new derivable sequents.

This result has many applications. For, the sequent calculus without Cut has very perspicuous derivations, from which much constructive information can be extracted. For instance, we have the

Subformula Property

A derivable sequent can be proved using only sequents involving occurrences of its subformulas (and hence, in particular, involving only logical constants that appear explicitly in that sequent).

Existence Property

If an existential formula $\exists x\phi$ is derivable without premises in intuitionistic logic, then some explicit instance $[t/x]\phi$ must be derivable already.

For classical logic, we have at least

Herbrand's Theorem

If an existential formula is derivable without premises, then some disjunction of its explicit instances must be derivable already.

Remark. It might be of interest to develop an 'Abstract Proof Theory', by analogy with the above Abstract Model Theory, where it could be investigated which combinations of structural properties of calculi imply which others. For instance, which general features of a system of deduction guarantee the validity of the Cut rule?

4.4 Completeness

Perhaps the most influential result in the development of modern Logic has been Gödel's *Completeness Theorem*, which says that the semantic explication and the proof-theoretic explication of valid consequence coincide for classical first-order predicate logic:

Theorem $\phi_1, ..., \phi_n \vdash \psi$ iff $\phi_1, ..., \phi_n \models \psi$.

This result has been a 'role model' for many others since; so much so that the term "completeness theorem" has changed its grammatical category: from a proper name to a common noun.

5 Computation Theory

5.1 Computability

There are a number of logical approaches to a more precise definition of what is meant by 'effective computability'.

- Turing gave an analysis in terms of *Turing machines*, which leads to a class of functions that are naturally programmable in the machine code of his device.
- One can also try to define the effective functions in the usual mathematical inductive style, starting from simple base functions and then listing some admissible effective means of compounding, such as composition, minimization and primitive recursion. This gives the so-called general *recursive* functions.
- Yet other options occur in computer science, where one can use an imperative higher programming language producing structured programs. For instance, in a language with zero and the arithmetical successor and predecessor functions, the following *while-programs* will do:
    - assignments $x := t$
    - $\pi_1 ; \pi_2$ — Sequence
    - IF ε THEN π_1 ELSE π_2 — Conditional Choice
    - WHILE ε DO π — Guarded Iteration.
- Finally, both in logic and in computer science, one can use declarative systems of definitions for effectively computable functions, such as Horn clause definitions of suitable arithmetical predicates in logic programming.

5.2 Arithmetic

Effective computation is usually related to arithmetic, because all other effective methods, on arbitrary symbolic structures, can be simulated within arithmetic by suitable methods of *coding*. The language most frequently used for the latter purpose is that of Peano Arithmetic, with $0, S, +$ and $\bullet$. Under various coding schemes, for instance, Gödel's 'prime number code' or Gödel's 'β-code', all intuitively 'computable' predicates on symbolic structures will then correspond to simply computable arithmetical relations between the corresponding code numbers.

A useful tool for measuring complexity of arithmetical definitions is the *Arithmetical Hierarchy* of first-order formulas in the Peano language, ordered by their quantifier prefixes. The base level is that of 'recursive' statements, which can be constructed from term equalities using Booleans as well as restricted quantifiers $\exists x(x<y\wedge$, $\forall x(x<y\rightarrow$. This level is called Σ^0_0 or Π^0_0 or Δ^0_0 . Then, a prenex form having k alternating blocks of similar quantifiers in front of a Δ^0_0 matrix is Σ^0_k if the first block has $\exists$, and Π^0_k if the first block has $\forall$. Formulas that are logically equivalent to one of these forms also obtain the corresponding appellation. A formula that is both Σ^0_k and Π^0_k is called Δ^0_k . For instance, a decidable set like the propositional tautologies has an arithmetical counterpart at level Δ^0_0 , while the set of predicate-logical validities lies at level Σ^0_1 . (By the Completeness Theorem, valid consequence is equivalent to derivability, and after arithmetization, 'existence of a derivation' turns out to be a Σ^0_1 notion.)

Also, certain results in the above can be made more precise after arithmetical encoding. For instance, Hilbert & Bernays 1939 presents the

Arithmetized Completeness Theorem

> If a consequence from $\phi_1, ..., \phi_n$ to ψ is not valid, then there exists a counter-example already on the structure of the natural numbers, in which the interpretation function assigns Δ^0_2-predicates to all predicate symbols in $\phi_1, ..., \phi_n, \psi$.

5.3 Decidability and Undecidability

5.3.1 Positive Results.

There are many *decidable* logical calculi, for which the question whether a sequent is valid in them can be answered by an effective method: or after encoding, by a computable function which gives value 1 just in case the consequence is valid, and value 0 in all other cases.

Examples are classical and intuitionistic propositional logic, as well as many modal logics. Other examples were various occurrence logics in the Categorial Hierarchy (see Chapter 7). But also, various fragments of predicate logic have decidable decision problems, such as monadic predicate logic, or universal formulas without function symbols. Famous decidability results concern less obvious cases, such as *Rabin's Theorem* giving a decision method for the *second-order* theory of two successor functions on a universe of sequents ordered by the relation of 'initial segment' (cf. Rabin 1977). Other special decidability results concern special first-order theories, such as Boolean Algebra or Elementary Geometry (Tarski 1959).

5.3.2 Negative Results.

More profound, and influential in logical research, are a number of 'negative' results concerning the deductive power and complexity of logical systems.

Gödel's Incompleteness Theorem

> Peano Arithmetic is incomplete, in the sense of failing to derive certain true arithmetical statements.

Moreover, the theorem is uniform, in that the same incompleteness will recur for all effectively axiomatized mathematical theories of a certain minimal arithmetical strength. Now, syntactically complete effectively axiomatized theories are always decidable (this is 'Post's Theorem'), and hence Gödel's Theorem shows also that in fact, Peano Arithmetic must be *undecidable*. Again, this result holds uniformly for all effectively axiomatized arithmetical theories over a small finite set of principles (called 'Robinson's Arithmetic'). From this, one obtains

Church's Theorem

> The set of valid consequences in first-order predicate logic is undecidable.

To this same circle of ideas belongs

Tarski's Theorem

> The set of true arithmetical sentences (in the language of Peano Arithmetic) is not effectively axiomatizable, and indeed, it is not even definable in the Arithmetical Hierarchy.

These results all refer to arithmetic with both addition and *multiplication*. By contrast, purely additive Presburger Arithmetic axiomatizes the complete theory of the natural numbers with 0, S, + only. Hence, by Post's Theorem, this theory is even *decidable* .

5.3.3 Applications.

Many further outcomes can be obtained by either mimicking Gödel's arguments in new situations, or via *effective reduction* of one of the above undecidable problems to new ones. In this book, reduction to Tarski's Theorem has been used, e.g., in the analysis of Elementary Syntax (Chapter 14).

We only give two examples that may be somewhat surprising:

- The 'elementary meta-theory' of propositional logic is undecidable, and not even arithmetically definable.

This is the first-order theory of the set of all propositional formulas with finite sets of proposition letters, ordered by 'valid consequence' and 'occurrence' (Mason 1985). So, undecidability strikes already on relatively familiar non-quantificational territory.

Likewise, even basic semantic notions from this Book may have an undecidable recognition problem on elementary syntactic forms (cf. van Benthem 1988B):

- Whether a first-order formula is monotone in the predicate symbol Q is an undecidable notion, although it is $\Sigma^0{}_1$ by Lyndon's Theorem.

5.4 Theory of Computability

There is a whole theory of computability on its own, which is comparable in its richness to Model Theory or Proof Theory. One branch is *Recursion Theory* which studies the structure of recursive functions, and higher ones, as a mathematical object in its own right. This theory has produced many tools of wider significance, such as the *Recursion Theorems* (Rogers 1967, Soare 1987). Another more recent branch is *Complexity Theory*, which studies the fine-structure of decision procedures, as to their time- or space-complexity.

We do not survey these topics here, as they have not been very prominent in this Book. For instance, the precise time- or space-complexity of the decision problem for many categorial calculi presented here is unknown. In particular, it is an open question whether derivability in the Lambek Calculus L is decidable in polynomial time. This may turn out to be a serious shortcoming: witness the critical analysis of current paradigms for natural language processing given in Barton, Berwick & Ristad 1987. Computational concerns of this kind would form a natural extension of the scope of our present investigation.

6 Extensions I : Higher-Order Logics

There are many extensions of first-order predicate logic, some of which are relevant to the subject of this Book.

6.1 Second-Order Logic

In this system, an 'asymmetry' of first-order logic is removed: quantification is now also allowed over functions or predicates, in the same semantic structures as before. Examples of important second-order statements are

$\forall x \forall y\, (x = y \leftrightarrow \forall P\, (Px \leftrightarrow Py))$	Leibniz' Law
$\forall x \exists y\, Rxy \leftrightarrow \exists f \forall x\, Rxf(x)$	Skolem's Reduction

This additional expressive power is bought at a price. Valid consequence in second-order logic is no longer effectively axiomatizable (it is not even arithmetically definable, as arithmetical truth can be effectively embedded in the system). Moreover, many first-order model-theoretic results fail, or become trivial. For instance, Beth's Definability Theorem now holds without effort: if Q is implicitly definable in T , then the following second-order sentence defines Q explicitly in T :

$$\forall x\, (Qx \leftrightarrow \exists Q'\, (T(Q') \wedge Q'x))\ .$$

For further information on this logic, see the survey Doets & van Benthem 1983.
One curious illustration of the complexity of this system concerns its semantic relation to first-order logic:

> The question whether a given second-order sentence is definable by means of a first-order one is undecidable, and indeed not even arithmetically definable.

6.2 Lambda Calculus

This system has been surveyed already in Chapter 2.
Here is a complete statement of all rules of inference for the Extensional Typed Lambda Calculus (where syntactic notions like 'freedom' are defined by analogy with the syntax of predicate logic, viewing lambdas as quantifiers):

Lambda Conversion:

$\lambda x \bullet \sigma(\tau) = [\tau / x]\, \sigma$ provided that τ is free for x in σ.

Identity Principles:

$\sigma = \sigma$		
$\sigma = \tau$	implies	$\tau = \sigma$
$\sigma = \tau,\ \tau = \upsilon$	implies	$\sigma = \upsilon$
$\sigma = \tau$	implies	$\sigma(\upsilon) = \tau(\upsilon)$
$\sigma = \tau$	implies	$\upsilon(\sigma) = \upsilon(\tau)$
if $\sigma = \tau$ is derivable, then so is		$\lambda x \bullet \sigma = \lambda x \bullet \tau$

Variants:

$\sigma = \tau$ if σ, τ are alphabetic variants

Extensionality:

$\lambda x \bullet \sigma(x) = \sigma$ if x does not occur freely in σ.

We merely recall some basic results:

Church Rosser Theorem

If σ can be reduced to τ via a number of successive lambda conversions, and so can σ to υ, then there exists some term α to which both τ and υ are conversion-reducible.

Normalization Theorem

For each term, there exists a provably identical term in normal form.

Strong Normalization Theorem

The process of lambda conversion is well-founded:
each reduction sequence terminates after finitely many steps.

As a consequence, provable identity in the typed lambda calculus is a *decidable* notion. For the broader computational significance of these results, the reader is referred to the survey Klop 1987 on term rewriting systems.

The *formulas-as-types* correspondence gives an effective one-to-one mapping between typed lambda terms τ_b with free variables $x_1, ..., x_k$ of types $a_1, ..., a_k$, respectively, on the one hand, and natural deduction derivations of the sequent $a_1, ..., a_k \Rightarrow b$ (under the implicational readings of these types) on the other:

- x_a corresponds to the single-node derivation a
- $\sigma_{(a, b)}(\tau_a)$ corresponds to the derivation obtained by joining the σ-derivation for $(a \rightarrow b)$ and the τ-derivation for a in a final application of Modus Ponens, to obtain the new conclusion b.
- $\lambda x_a \bullet \tau_b$ corresponds to the τ-derivation of b, followed by a final Conditionalization step introducing the conclusion $(a \rightarrow b)$ and removing the dependence on the assumption a.

Conversely, a similar procedure encodes each implicational natural deduction tree by a typed lambda term, starting from different variables for each assumption at its leaves. The categorial calculi of this Book need fresh variables for each *occurrence* in the latter case.

Standard models for the typed lambda calculus were explained in Chapter 2. They consist of families $\{ D_a \mid a \in \text{TYPE} \}$, having arbitrary domains D_a for basic types a, and otherwise $D_{(a, b)}$ equal to the space of all functions from D_a to D_b. *General models* exist too, which do not necessarily require full function spaces: a well-known example are so-called 'term models' (cf. Friedman 1975, Hindley & Seldin 1986). Lambek & Scott 1986 consider yet another class of models, and develop a categorial duality between typed lambda calculi and *Cartesian-Closed Categories*.

An alternative formulation for the above would be a typed version of Combinatory Logic (see Curry & Feys 1958), with basic combinators

$$K_{a,b}\, x_a\, y_b \;=\; x_a$$

$$S_{a,b,c}\, x_{(c,(b,a))}\, y_{(c,b)}\, z_c \;=\; x_{(c,(b,a))}(z_c)\,(y_{(c,b)}(z_c))\,.$$

For detailed comparisons between the two approaches, see Hindley & Seldin 1986.

6.3 Type Theory

The Theory of Finite Types T_ω used in this Book employs a higher-order language with

- all identity statements $\sigma_a = \tau_a$ between pure application terms of the same type (which are themselves counted as being of the truth value type t)
- the ordinary propositional connectives in type t
- quantifiers $\exists x_a$, $\forall x_a$ in all types a .

Standard models are still the function hierarchies for the Lambda Calculus. As T_ω is highly undecidable over the latter, no standard axiomatization exists. Gallin 1975 presents a reasonable base set, which may be paraphrased as follows:

- 'comprehension principles' ensuring the existence of denotations for all terms formed using lambda abstraction
- all principles of the typed lambda calculus
- extensionality in its straightforward form:

$$\forall f \forall g\, (\, f=g \leftrightarrow \forall x\, f(x) = g(x)\,)$$

- an explicit definition of the truth value domain D_t via

$$\forall x_t\;\, g(x_t) \;\leftrightarrow\; (g(T) \wedge g(\bot))\,.$$

A useful re-interpretation is due to Henkin 1950. Type theory can also be viewed as a *many-sorted first-order* theory describing families of domains D_a . Thus, it has *general models* consisting of just such families. Basic axioms will then ensure some basic 'coherence', in terms of a partial application predicate App which relates objects of 'sort' (a, b) and objects of sort a to objects of sort b . Extensionality then reads as follows:

$$\forall x \forall y\, (\, x=y \;\leftrightarrow\; \forall z \forall u\, (App(x, z, u) \leftrightarrow App(y, z, u))\,)\,.$$

For instance, the 'pre-models' of Friedman 1975 are such extensional general models. Further existential *comprehension principles* will then ensure the existence of enough objects of various sorts to guarantee the availability of denotations for all lambda-definable terms of the language.

This move is not just a sign of despair. In many applied situations, there is an intuitive feeling that one does not really 'go up' in the way standard models do. For instance, in natural language, "object" and "property" are both common nouns that intuitively denote disjoint domains living side-by-side rather than one on top of the other.

The functional perspective in Lambda Calculus and Type Theory is elegant, but it sometimes also engenders needless complications in semantics. (For instance, defining 'relativization to submodels', which ought to be a harmless exercise, becomes a formidable technical task.) Therefore, Orey 1959 proposed a *relational* reformulation of type theory, working rather with hierarchies of relations (i.e., 'partial multi-valed functions'). See van Benthem & Doets 1983 for an exposition of this relational version, Gallin 1975 for its mathematical attractions, and Muskens 1989 for a defense of its advantages in the semantics of natural language.

More powerful higher type theories, as mentioned at the end of Chapter 13, are found in Martin-Löf 1984, Hindley & Seldin 1986, Lambek & Scott 1986, Troelstra & van Dalen 1988. Cf. especially the systematic perspective developed in Barendregt & Hemerik 1990.

6.4 Fragments

As with first-order predicate logic, various fragments of full higher-order languages are important too. For instance, Montague 1970 pointed out the significance of 'restricted' second-order sentences, which do retain some pleasant model theory.
One useful fragment is the following. Many results in Chapters 10, 11 of this book have been inspired by the theory of *Generalized Quantifiers*. These are objects in the second-order type ((e, t), t) or more generally, type ((e, t), ((e, t), t)) . Examples in the former type ((e, t), t) are the usual quantifiers $\exists$, $\forall$ that may be seen as properties of predicates or sets, namely, 'non-emptiness and 'universality', or probabilistic quantifiers, such as "for almost all objects (except for a set of measure 0)". In natural language, the general case is the binary pattern ((e, t), ((e, t), t)) ('binary predicates of predicates') exemplified in:

all A B *some* A B *two* A B *most* A B *few* A B .

These do not always reduce to the corresponding unary quantifiers: a counter-example is the determiner *most* . Yet more general generalized quantifiers may be found in Lindström 1966 and van Benthem 1987C.

In mathematical logic, generalized quantifier logics often arise by choosing some particular Q which is then added as a constant of its type to first-order predicate logic, with syntax $Qx\bullet \phi(x)$, to obtain a system EL(Q) . Models then consist of ordinary structures, together with a suitable set of subsets Q* of their domain interpreting Q . Then,

$$\mathbb{M}, a \models Qx\bullet \phi \quad \text{iff} \quad \{ d\in D \mid \mathbb{M}, a[x/d] \models \phi \} \in Q^* .$$

Barwise & Feferman 1985 contains many results on such logics. Westerstahl 1989 provides connections between their mathematical and linguistic manifestations. For instance, adding a monotone unary quantifier Q , whose denotation is always closed under the passage to supersets, leads to a logic which is still axiomatizable by the standard predicate-logical deductive calculi plus the monotonicity schema

$$(\forall x(\phi \rightarrow \psi) \wedge Qx\bullet \phi) \rightarrow Qx\bullet \psi .$$

That this is non-trivial may be seeen by the fact that adding other kinds of quantifier, such as permutation-invariant ones, may result in non-effectively axiomatizable systems. (Doets 1990 has a general result explaining the positive outcome for monotonicity.)

The theory of generalized quantification in natural language has a somewhat different flavour, being concerned directly with denotational constraints on objects in the type domain $D_{((e, t), ((e, t), t))}$. One ubiquitous constraint, that seems to hold for all 'determiner expressions' in this category across human languages is *Conservativity*:

$$Q\,AB \quad \text{iff} \quad Q\,A(B\cap A) .$$

See van Benthem 1986 or Westerstahl 1989 for a body of results, of which we mention a few, serving as a background for Chapter 10:

- Monotonicity and Standard Quantifiers.

A binary quantifier Q has four possible kinds of *monotonicity* behaviour

left upward	$Q\,AB \quad A\subseteq A'$	implies	$Q\,A'B$
left downward	$Q\,AB \quad A'\subseteq A$	implies	$Q\,A'B$
right upward	$Q\,AB \quad B\subseteq B'$	implies	$Q\,AB'$
right downward	$Q\,AB \quad B'\subseteq B$	implies	$Q\,AB'$

Note that the four quantifiers in the traditional 'Square of Opposition' have such double monotonicity properties:

$-+$	all	no	$--$
$++$	some	not all	$+-$

Moreover, they are also *variable*, in the sense that

If A is non-empty, then there exist B, B' such that $Q\,AB, \neg Q\,AB'$.

Proposition

The four classical quantifiers are the only ones that are conservative, doubly monotone and variable.

By contrast, a non-classical quantifier like *most* has only (upward) monotonicity in its right-hand argument, while other non-classical ones lack monotonicity altogether.

- Genuine expressions of 'quantity' must be *permutation-invariant*, that is,

  $Q\,AB$ iff $Q\,\pi[A]\pi[B]$

  for each permutation π of the individual domain D_e.

Such logical conservative quantifiers may be represented numerically as the set of all pairs

$|A-B|$, $|A \cap B|$ for all sets A, B such that $Q\,AB$.

Geometrically, this may be displayed in a 'Tree of Numbers'

0,0

1,0 0,1

2,0 1,1 0,2

etcetera.

Using geometric inspection of the Tree, one can show, e.g., that

Proposition

All left monotone (either upward or downward) conservative logical quantifiers are *first-order definable*, in the monadic first-order language having unary predicates 'A', 'B' and identity $=$.

- First-order definable quantifiers may also be described model-theoretically using the earlier technique of Ehrenfeucht Games. They are invariant for changes in models that respect numbers of individuals in the four slots formed by the two predicates A, B up to some suitable finite threshold. Again, this has a simple geometrical reflection in the Tree.
- Quantifiers may also be analysed as to the complexity of their recognition procedures on finite models. For this purpose, finite strings of objects are fed to some device ('semantic automaton'), with each object marked for one of the two relevant cases: 'in $A-B$', 'in $A \cap B$' . Then we have

Proposition

The first-order quantifiers are precisely those recognized by 'permutation-invariant' 'acyclic' finite state automata having at most one-state loops.

Theorem

The quantifiers computable by push-down store automata are precisely those whose corresponding numerical condition on pairs $|A-B|$, $|A \cap B|$ can be expressed in Presburger Arithmetic.

The latter outcome is of interest because of the *decidability* of purely additive arithmetic. Higher quantifiers with undecidable behaviour do not seem to occur in natural language.

6.5 Infinitary Logics

Frank Ramsey once asked why one should exclude formulas from logic whose only fault is that they are too long to write down on a piece of paper. The infinitary logic $L_{\omega_1\omega}$ admits countable conjunctions and disjunctions over the first-order predicate-logical base. This language has greater expressive power, witness:

Scott's Theorem

Two countable models verify the same sentences of $L_{\omega_1\omega}$ iff they are isomorphic.

Also natural is the further extension $L_{\infty\omega}$ which allows conjunctions and disjunctions over arbitrary sets of formulas. With formulas in some fixed finite set of free variables, we then have the following semantic characterization, in terms of an earlier notion:

Proposition

Two models are partially isomorphic iff they verify the same sentences of $L_{\infty\omega}$.

These languages arise naturally in some computational contexts. For instance, transition predicates for infinitary formalisms, allowing the earlier WHILE construct (or Kleene Iteration) will be expressible in $L_{\omega_1\omega}$ using countable disjunction of the form:

"the register content $x_1, .., x_n$ is transformed into $y_1, .., y_n$ either directly, or via one intermediate π-step, or via two intermediate π-steps, or ..."

Moreover, for the modal bisimulation invariance of Chapter 15, we have this

Proposition

There exists a bisimulation between two possible worlds models $\mathbb{M}, \mathbb{M}'$ relating two states s, s' iff $\mathbb{M}$, s and $\mathbb{M}'$, s' verify the same sentences in the modal language extended with arbitrary set conjunctions and disjunctions.

7 Extensions II : Intensional Logics

Intensional logics arise by adding operators of modality, knowledge, temporality, action, etcetera, to standard logical systems. The unifying semantic idea is 'multiple reference': all these operators involve comparison across different semantic structures ('possible worlds', 'epistemic scenarios', 'points in time', 'states of a physical or cognitive agent').

7.1 Modal Logic

Good surveys are Hughes & Cresswell 1968, 1984 as well as Bull & Segerberg 1984.

7.1.1 Syntax and Semantics

The syntax now has two additional *modal operators*

$\Box$ 'necessarily'

$\Diamond$ 'possibly'.

Semantic structures, for the propositional case, are *possible worlds models*

$$\mathbb{M} = (W, R, V)$$

consisting of a non-empty set W of 'worlds' or 'situations', ordered by a binary relation R of 'accessibility', with a 'valuation' V indicating, for each world w , the truth values of all proposition letters at that world. (The latter may change across different worlds.) The alternative relation R provides a degree of freedom that can be modulated in modelling specific phenomena. In particular, it may be constrained, so as to satisfy various well-known relational conditions: reflexivity, transitivity, symmetry, etcetera. Now, the following truth definition explicates *truth* of a formula at a world within a model $\mathbb{M}$: $\mathbb{M}, w \models \phi$, via the following recursion

$\mathbb{M}, w \models p$	iff	$w \in V(p)$
$\mathbb{M}, w \models \neg\phi$	iff	*not* $\mathbb{M}, w \models \phi$
$\mathbb{M}, w \models (\phi \wedge \psi)$	iff	$\mathbb{M}, w \models \phi$ *and* $\mathbb{M}, w \models \psi$
$\mathbb{M}, w \models \Box\phi$	iff	*for all* v *with* Rwv, $\mathbb{M}, v \models \phi$
$\mathbb{M}, w \models \Diamond\phi$	iff	*for some* v *with* Rwv, $\mathbb{M}, v \models \phi$.

Note how the two modalities are interdefinable by double negation, for instance,

$\Box\phi$ as $\neg\Diamond\neg\phi$.

There is also an algebraic semantics for this language, in terms of *modal algebras*, being Boolean algebras $(A, 0, 1, +, \bullet, *)$ with an additional unary operator $*$ interpreting the modality (say) $\Diamond$. Possible worlds models generate set-based modal algebras via their obvious subset structure with the set-theoretic Boolean operations and the modal operation

$$*(X) = \{ w \in W \mid \exists x \in X: Rwx \} .$$

Conversely, provided that the additional operator satisfies suitable *continuity* properties, modal algebras may be represented as subset algebras over possible worlds models (Jónsson & Tarski 1951) via an extension of the well-known Stone Ultrafilter Representation for Boolean Algebras. Goldblatt 1988 has a general modern treatment.

There are further semantic alternatives in modal logic too.
For instance, recently, *function-based* models have entered the folklore, consisting of a set of worlds with a family of functions over these, with a key semantic clause

$\mathbb{M}, w \models \phi$ iff $\mathbb{M}, f(w) \models \phi$ for all functions f in the model.

Then, various options in modelling arise by letting the function set be a semi-group, a category, or a group, etcetera.

7.1.2 Invariance.

Like first-order predicate logic, modal logic has its characteristic semantic invariances.
A relation C between two possible worlds models $\mathbb{M}, \mathbb{M}'$ is called a *bisimulation* if

- C-related worlds verify the same proposition letters
- if $w\,C\,v$ and Rww', then there exists some v' with Rvv' and $v\,C\,v'$
- conversely in the opposite direction.

Proposition

For all modal formulas ϕ, and worlds w, w' with $w\,C\,w'$,

$\mathbb{M}, w \models \phi$ iff $\mathbb{M}', w' \models \phi$.

Conversely, for instance, there is this

Proposition

If two finite models $\mathbb{M}, \mathbb{M}'$ verify the same modal formulas at w, w', then there exists some bisimulation between $\mathbb{M}, \mathbb{M}'$ connecting w and w'.

These results may be used to show that certain principles are not modally definable, as they do distinguish between bisimulating models. Examples are found in Chapter 15, involving 'betweenness'. Invariance for bisimulation subsumes a number of basic results from the modal literature such as the 'generation theorem' and the 'p-morphism theorem'.

A possible worlds model (W, R) without a valuation is called a *frame* $\mathbb{F}$. Truth of a modal formula at a world in a frame is explained as truth at that world *in all models* on that frame. Truth in the frame means truth at all its worlds. Several preservation properties for modal formulas on frames follow from the above observation:

- $\mathbb{F}$ is a *generated subframe* of $\mathbb{F}'$ if $W \subseteq W'$ and $R = R' \cap (W x W)$.
  In this case, $\mathbb{F}' \models \phi$ implies $\mathbb{F} \models \phi$.
- The *disjoint union* of a family of frames $\{ \mathbb{F}_i \mid i \in I \}$ has for its domain the disjoint union of all W_i and for its relation the disjoint union of all R_i.
  If a modal formula is true in all frames $\mathbb{F}_i$, then it is true in their disjoint union.

- A function f is a p-*morphism* from a frame $\mathbb{F}$ to a frame $\mathbb{F}'$ if
  (i) Rwv implies R'f(w)f(v) , and
  (ii) R'f(w)v implies the existence of some u with Rwu and f(u) = v .
  If a modal formula is true in a frame $\mathbb{F}$, then it is true in all surjective p-morphic images of $\mathbb{F}$.

There are also further preservation results, derived from the above algebraic representation, such as 'anti-preservation under ultrafilter extensions'.

7.1.3 Correspondence.

The modal language may be embedded into first-order predicate logic, by the following *standard translation*

$(p)^* \quad = \quad Px$

$(\neg\phi)^* \quad = \quad \neg(\phi)^*$

$(\phi\wedge\psi)^* \quad = \quad (\phi)^*\wedge(\psi)^*$

$(\Diamond\phi)^* \quad = \quad \exists y\ (Rxy \wedge [y/x](\phi)^*)$ for some *new* variable y

In fact, two variables x, y will do altogether, if one uses translations having either x or y as their one free variable, while 'rotating' at the modality, as in:

$(\phi)^*_y \quad = \quad \exists x\ (Ryx \wedge (\phi)^*_x)$.

Now, a modal formula will be true at a world in a possible worlds model if and only if its translation is satisfied by that world in the standard sense on the corresponding ordinary first-order model.

Through this translation, the modal language becomes a fragment of a first-order language having one binary predicate symbol and a number of unary ones. Thus, it inherits many standard model-theoretic and proof-theoretic properties from the former, such as compactness or effective axiomatizability of valid consequence. What makes it special is the following semantic characteristic.

Theorem

A first-order formula is equivalent to (the translation of) a modal one iff it is invariant for bisimulation.

Truth on frames becomes a second-order notion through our translation:

If $p_1, \ldots, p_n$ are the proposition letters occurring in ϕ ,

$\mathbb{F} \models \phi$ iff $\mathbb{F} \models \forall x \forall P_1 \ldots \forall P_n\ (\phi)^*$.

Nevertheless, the relational properties of frames which are 'modally expressible' in this way often have simpler first-order descriptions. For instance,

$\mathbb{F} \models \Box p \to \Box\Box p$ iff R is transitive,

i.e., $\forall x \forall y \forall z\, ((Rxy \wedge Ryz) \to Rxz)$

$\mathbb{F} \models \Diamond\Box p \to \Box\Diamond p$ iff R is confluent,

i.e., $\forall x \forall y \forall z\, ((Rxy \wedge Rxz) \to \exists u\, (Ryu \wedge Rzu))$.

But higher-order properties also occur essentially:

$\mathbb{F} \models \Box(\Box p \to p) \to \Box p$ iff R is transitive and $R^{\cup}$ is well-founded.

Here are some results from 'Correspondence Theory' (cf. van Benthem 1985), which studies such phenomena more generally:

Theorem

A modal formula defines a first-order frame property iff
it is preserved under the formation of ultrapowers of frames.

Theorem (Goldblatt & Thomason 1974)

A first-order frame property is modally definable iff
it is preserved under the formation of generated subframes, disjoint unions, p-morphic images, and anti-preserved under ultrafilter extensions.

More constructive information is provided by the following

Theorem

All modal formulas of the following 'Sahlqvist form' define first-order frame conditions, and there exists an effective algorithm computing the latter:

$$\phi \to \psi,$$

where ϕ is constructed from proposition letters, possibly prefixed by a number of operators $\Box$, using $\wedge$, $\vee$, $\Diamond$
and ψ is an arbitrary 'positive formula' constructed using proposition letters, $\wedge$, $\vee$, $\Diamond$ and $\Box$.

7.1.4 Axiomatics.

Valid consequence for modal formulas may be defined in various ways, more 'local' or more 'global', but most elegant is the following local notion:

$\Sigma \models \phi$ if, for all models $\mathbb{M}$ and worlds w verifying each formula in Σ, also $\mathbb{M}, w \models \phi$.

The syntactic counterpart is again modal deduction, which can be approached via the same techniques as with first-order logic. Most common is the axiomatic approach.

The *minimal modal logic* K has the following principles.

Axioms:

all valid principles from propositional logic

$\Box(\phi \rightarrow \psi) \rightarrow \Box(\phi \rightarrow \psi)$ Modal Distribution

Definition:

$\Diamond\phi \leftrightarrow \neg\Box\neg\phi$

Rules:

from $(\phi \rightarrow \psi)$ and ϕ to infer ψ Modus Ponens

if ϕ is a theorem, then so is $\Box\phi$ Necessitation.

On top of this, there lies a whole lattice of stronger logics, ascending up to two trivial systems at the upper end:

one has an axiom $\phi \leftrightarrow \Box\phi$ trivializing the modality,

the other has an axiom $\Box\bot$ with similar effects.

In between lie well-known systems such as

- S4 with additional axioms

  $\Box\phi \rightarrow \phi$ Reflexivity

  $\Box p \rightarrow \Box\Box p$ Transitivity

- S5 being S4 plus

  $\Diamond\Box p \rightarrow p$ Symmetry.

- 'Löb's Logic' of arithmetical provability, being K plus the axiom

  $\Box(\Box p \rightarrow p) \rightarrow \Box p$,

  which reflects 'Löb's Theorem' in the foundations of arithmetic.

7.1.5 Completeness.

The most basic modal completeness result is this:

Theorem

A modal formula is derivable in the minimal modal logic K iff it is universally valid in all models (without any restriction on the alternative relation).

Special logics will reflect special frame classes, however. Here are some examples.

Theorem

S4 is complete for validity in all reflexive transitive frames.

S5 is complete for validity in all equivalence relations, or even, in all frames where R is the universal relation.

Theorem

Löb's Logic is complete for validity in finite irreflexive trees.

There are also more general results. For instance,

Sahlqvist's Theorem

All Sahlqvist forms, when added to K , define logics that axiomatize all modal validities over the class of frames satisfying their corresponding first-order condition.

Finally, rather than starting from the side of existing modal calculi, one can also start from natural frame classes, and then search for a corresponding modal logic. For instance,

The modal logic of finite reflexive trees is axiomatized by S4 plus the Grzegorczyk Axiom $\Box(\Box(p \to \Box p) \to p) \to p$.

7.2 Intuitionistic Logic

On this topic, an excellent survey exists, namely Troelstra & van Dalen 1988.

7.2.1 Syntax and semantics.

This time, the *language* is the ordinary one of propositional (or predicate) logic. But now, it is interpreted in a new fashion. *Models* are possible worlds models, with 'worlds' now seen as information stages, ordered by a relation of 'possible growth' which is a *partial order*. Moreover, valuations are 'cumulative':

if a proposition letter holds at a stage, then it holds at all later stages.

This 'heredity' is then inherited by all complex formulas, whose interpretation becomes:

$\wedge, \vee$: interpretation as usual

$\mathbb{M}, w \models (\phi \to \psi)$ iff *for all* v *with* Rwv , *if* $\mathbb{M}, v \models \phi$, *then* $\mathbb{M}, v \models \psi$

$\mathbb{M}, w \models \neg\phi$ iff *for no* v *with* Rwv, $\mathbb{M}, v \models \phi$.

We omit clauses for quantifiers.

This semantics was motivated by the existence of a translation, due to Gödel, from intuitionistic logic into the modal logic S4 , whose clauses generate the above truth conditions. This is the reason why Chapter 15 treated intuitionistic logic as a 'hereditary subpart' of the full modal logic of information.

Without imposing heredity, but retaining the intuitionistic truth conditions, a weaker logical system arises, that shows non-monotonic phenomena (cf. van Benthem 1988A). Imposing rather an additional constraint of 'cofinality', to the effect that, for all formulas ϕ , $\Box\Diamond\phi$ implies ϕ , yields classical logic again, even on these possible worlds models.

There is a completeness theorem which equates validity on the above models with provability in Heyting's propositional logic. This result again has many useful applications, such as quick semantic proofs of the Disjunction Property for intuitionistic deduction:

if $\vdash \phi \vee \psi$, then $\vdash \phi$ or $\vdash \psi$.

7.3 Temporal Logic

This field is well-surveyed too: see van Benthem 1983, 1989H, 1990B.
Basic frames are now 'flows of time', that might also be written as $(T, <)$, where an additional valuation will record the history over time for all proposition letters. The existential modality then becomes 'at least once in the future', and it is quite common to addd a converse 'at least once in the past'.
In this area, extended formalisms have been quite common, enriching the modal base language, in order to account for tenses and other temporal constructions found in natural language. Two well-known examples are the binary operators

$\mathbb{M}, t \models \text{SINCE } \phi\psi$ iff $\exists t' < t : \mathbb{M}, t' \models \phi$ *and for all* t'', *if* $t' < t'' < t$, *then* $\mathbb{M}, t'' \models \psi$

$\mathbb{M}, t \models \text{UNTIL } \phi\psi$ iff $\exists t' > t : \mathbb{M}, t' \models \phi$ *and for all* t'', *if* $t < t'' < t'$, *then* $\mathbb{M}, t'' \models \psi$.

The importance of this enrichment shows in a Functional Completeness result:
Kamp's Theorem

On the Dedekind-continuous linear orders, all one-place formulas in the first-order language over $<$ and unary predicates are definable in the propositional tense logic with SINCE and UNTIL.

This result presupposes the earlier translation into first-order logic, where the SINCE, UNTIL formalism turns out to involve essentially three variables. By an analysis due to Gabbay (cf. Chapters 14, 15, 16), the general situation is as follows:
Theorem

An effective correspondence exists between k-variable fragments of the first-order language and finite temporal operator formalisms with first-order truth conditions.

Theorem

Three variables suffice for defining the full first-order language over even all linear orders.
But no such finite reduction is possible over all partial orders.

The further model theory and proof theory of these extended formalisms can be developed on earlier analogies, for instance, using suitably extended notions of bisimulation (cf. Chapters 15, 17).

7.4 Conditional Logic

Conditional Logic has for its central notion a binary intensional implication $\phi \Rightarrow \psi$.
Its possible worlds semantics is in 'similarity models' (W, C, V) with a ternary relation C of comparative similarity. A minimal structural requirement here is that, from the perspective of each world w, $\lambda xy \bullet C_w xy$ be a *strict partial order*.
The key truth condition is then as follows:

$\mathbb{M}, w \models \phi \Rightarrow \psi$ iff ψ holds in all those worlds which are C_w-closest among the worlds verifying ϕ: .i.e., in those ϕ-worlds which are maximal in the ordering $\lambda xy \bullet C_w xy$.

This clause works on finite models, or infinite models where the comparative order is well-founded. A more complex stipulation is required in the general case.

The basic axiomatic completeness result is this
Theorem

The valid principles on finite similarity models are axiomatized precisely by the following *minimal conditional logic.*

Its principles are these:

- All valid principles of propositional logic
- Replacement of Provable Equivalents
- $\phi \Rightarrow \phi$ — Reflexivity
- $\phi \Rightarrow \psi$ *implies* $\phi \Rightarrow (\psi \vee \xi)$ — Right Monotonicity
- $\phi \Rightarrow \psi$, $\phi \Rightarrow \xi$ *implies* $\phi \Rightarrow (\psi \wedge \xi)$ — Conjunction
- $\phi \Rightarrow \psi$, $\xi \Rightarrow \psi$ *implies* $(\phi \vee \xi) \Rightarrow \psi$ — Disjunction
- $\phi \Rightarrow \psi$, $\phi \Rightarrow \xi$ *implies* $(\phi \wedge \psi) \Rightarrow \xi$ — Cautious Monotonicity.

On top of this, further conditional axioms may be imposed. For instance,

- $((\phi \vee \psi) \Rightarrow \xi) \rightarrow (((\phi \vee \psi) \Rightarrow \phi) \vee (\psi \Rightarrow \xi)))$

  corresponds to 'almost-linearity' of local comparative orderings:

  $\forall x \forall y (C_w xy \rightarrow \forall z (C_w xz \vee C_w zy))$

- $\phi \Rightarrow (\phi \Rightarrow \bot)$, a relative of Löb's Axiom under a suitable modal embedding of conditional logic, corresponds to a more complex condition of compatibility for comparative orders at all worlds, which ensures that the whole pattern can be represented as a single uniform preference, restricted to certain 'subdomains' at each world.

The latter principle relates different comparative orders at different worlds.

8 Computational Logics

Many systems of Logic have been developed for computational purposes over the past two decades. Some of these are concerned with the imperative style, others with a more declarative one. Examples will follow of both kinds.

8.1 Operational Semantics and Correctness

Imperative programs may be regarded as instructions for computing over *data structures*. The latter may be identified with the earlier predicate-logical models, where assignments are now regarded as instantaneous *states*, being a record of momentaneous values for all 'registers' x, y, z, ... Here, the interpretation function I is supposed to map function symbols occurring in programs (i.e., unanalyzed subroutines) to suitable operations on the data structure. Then, in a model $\mathbb{M}$, each program π denotes a binary relation $[[\pi]]$ between assignments. This may be demonstrated for the earlier WHILE programs:

$[[\ x{:=}t\ ]] \quad = \quad \{\ (a, a[x / value(\mathbb{M}, a, t)]\ )\ \mid\ \text{all assignments}\ a\ \}$

$[[\ \pi_1;\pi_2\ ]] \quad = \quad [[\pi_1]]\bullet[[\pi_2]]$

$[[\ \text{IF}\ \varepsilon\ \text{THEN}\ \pi_1\ \text{ELSE}\ \pi_2\ ]] \quad =$

$\{\ (a, b) \mid M, a \models \varepsilon\ \text{and}\ (a, b)\in [[\pi_1]]\ \} \cup \{\ (a, b) \mid M, a \not\models \varepsilon\ \text{and}\ (a, b)\in [[\pi_2]]\ \}$

$[[\ \text{WHILE}\ \varepsilon\ \text{DO}\ \pi\ ]] \quad =$

$\{$ (a, b) $\mid$ some finite sequence of $[[\pi]]$ transitions exists starting from a such that ε holds at each next state, with b being the first state where it fails $\}$

Correctness assertions now say that executing a program π leads from states satisfying 'precondition' ϕ to states satisfying 'postcondition' ψ :

$\mathbb{M} \models \{\phi\}\pi\{\psi\}$ if

for all assignments a, b, *if* $\mathbb{M}, a \models \phi$ *and* $a[[\pi]]b$, *then* $\mathbb{M}, b \models \psi$.

Given any precondition ϕ and program π, this semantics defines a unique class of states where one may end up by executing π starting from some state satisfying ϕ. This class corresponds to the *strongest postcondition* of π with respect to ϕ, which explains the well-known alternative terminology of interpreting programs via 'predicate transformers'. Strongest postconditions for given π, ϕ need not always be explicitly definable in the first-order formalism. Those special structures where such definitions are always available are called *expressive*. ($(\mathbb{N}, +, \bullet)$ is expressive, and so are all finite data structures.)
This formalism may be regarded as an extension of first-order predicate logic to a small fragment of the language $L_{\omega_1\omega}$. In particular, by spelling out transition predicates for WHILE programs, each correctness statement turns out to be equivalent to one single countable disjunction of first-order statements. Using this observation, it may be shown that the model-theoretic gain in expressive power is slight:

Proposition

If two models are elementarily equivalent,
then they also verify the same correctness assertions.

One can also study these assertions as to *preservation* behaviour when shifting the computation from one data structure to another. For instance,

If $\mathbb{M}$ is a submodel of $\mathbb{N}$, while ϕ, ψ are quantifier-free and π involves only quantifier-free tests ε, then $\mathbb{M} \models \{\phi\}\pi\{\psi\}$ implies $\mathbb{N} \models \{\phi\}\pi\{\psi\}$.

No converse Los-type 'preservation theorem' has been established yet.

A simple axiomatic calculus for deducing correctness assertions in models $\mathbb{M}$ is due to Hoare. Its proof rules follow the inductive construction of programs:

Axiom:

$$\{[t/x]\phi\}\ x:=t\ \{\phi\}$$

Rules:

$$\frac{\{\phi\}\ \pi_1\ \{\psi\} \qquad \{\psi\}\ \pi_2\ \{\xi\}}{\{\phi\}\ \pi_1;\pi_2\ \{\xi\}}$$

$$\frac{\{\phi\wedge\varepsilon\}\ \pi_1\ \{\psi\} \qquad \{\phi\wedge\neg\varepsilon\}\ \pi_2\ \{\psi\}}{\{\phi\}\ \text{IF}\ \varepsilon\ \text{THEN}\ \pi_1\ \text{ELSE}\ \pi_2\ \{\psi\}}$$

$$\frac{\{\phi\wedge\varepsilon\}\ \pi\ \{\phi\}}{\{\phi\}\ \text{WHILE}\ \varepsilon\ \text{DO}\ \pi\ \{\phi\wedge\neg\varepsilon\}}.$$

And one final auxiliary principle imports knowledge about the specific model at issue:

$$\frac{\mathbb{M} \models \alpha \rightarrow \phi \qquad \{\phi\}\ \pi\ \{\psi\} \qquad \mathbb{M} \models \psi \rightarrow \beta}{\{\alpha\}\ \pi\ \{\beta\}}$$

This calculus is not complete, since it fails to axiomatize the complete theories of correctness assertions for arbitrary structures (even given the 'oracle' in the last rule). Bergstra & Tucker 1982 show that it is not even complete if we fix some first-order data theory T to be used in the last rule for effectively deriving the $\alpha \rightarrow \phi$, $\psi \rightarrow \beta$, and then ask for an axiomatization of all valid correctness assertions in the whole *class* of data structures (the 'data type') satisfying that T . Nevertheless, we do have

Cook's Theorem

On *expressive* models $\mathbb{M}$, the true correctness assertions are precisely those derivable in the Hoare Calculus over $\mathbb{M}$.

8.2 Dynamic Logic

8.2.1 Language and Semantics.

The core system of propositional dynamic logic (see Harel 1984) has 'formulas' and 'programs', defined via a mutual recursion. Basic operators on formulas are the Boolean connectives, while programs are joined by means of the so-called 'regular operations':

;	sequential composition
$\cup$	Boolean choice
*	Kleene iteration.

Finally, there is a 'test' mode ? taking formulas to programs, and a 'projection' modality $< >$ taking programs π and formulas ϕ to formulas $<\pi>\phi$. (A dual universal modality [ ] may be defined in the usual manner.) This formalism is capable of expressing many standard operators on programs, such as the class of 'WHILE programs' :

IF ε THEN π_1 ELSE π_2	$(\varepsilon? ; \pi_1) \cup ((\neg\varepsilon)? ; \pi_2)$
WHILE ε DO π	$(\varepsilon? ; \pi)^* ; (\neg\varepsilon)?$

Moreover, it expresses various types of statement about execution of programs, not only Correctness, but also 'Termination' or 'Enabling':

$\phi \rightarrow [\pi]\psi$	precondition ϕ implies postcondition ψ after every succesful execution of program π
$<\pi>T$	program π terminates
$[\pi_1]<\pi_2>\phi$	program π_1 'enables' program π_2 to produce effect ϕ.

These kinds of statement also make sense in a general logic of actions, computational or not, and Dynamic Logic has in fact also been used as such.
Semantic interpretation involves possible worlds models $(S, \{R_\pi \mid \pi\}, V)$ consisting of states with binary transition relations R_π over them for each program π . Some key clauses in the joint truth definition for formulas and programs are as follows:

$$\mathbb{M}, s \models \langle\pi\rangle\phi \quad \text{iff} \quad \mathbb{M}, s' \models \phi \text{ for some state } s' \text{ with } R_\pi ss'$$

$$[[\phi?]] = \{ (s, s) \mid \mathbb{M}, s \models \phi \} .$$

The other clauses are the expected ones from propositional logic or relational algebra.

8.2.2 Completeness and Extensions.

Theorem

The set of universally valid formulas over state models is axiomatized completely, on top of the minimal modal propositional logic, by the following axioms:

- $\langle\pi_1 ; \pi_2\rangle \phi \leftrightarrow \langle\pi_1\rangle \langle\pi_2\rangle \phi$
- $\langle\pi_1 \cup \pi_2\rangle \phi \leftrightarrow \langle\pi_1\rangle \phi \vee \langle\pi_2\rangle \phi$
- $\langle\phi?\rangle \psi \leftrightarrow \phi \wedge \psi$
- $\langle\pi^*\rangle \phi \leftrightarrow \phi \vee \langle\pi\rangle \langle\pi^*\rangle \phi$
- $(\phi \wedge [\pi^*] (\phi \rightarrow [\pi] \phi)) \rightarrow [\pi^*] \phi$

Most of these equivalences are direct 'reductions' determining the behaviour of program constructions. The final 'Induction Axiom' for iteration, however, reflects the more complex infinitary behaviour of this notion.
This logical calculus generalizes various earlier systems of reasoning, such as 'regular algebra', 'propositional modal logic' and 'Hoare calculus'. One virtue of the restricted set of regular program operations, as compared to the full relational algebra, is that one can do better than mere axiomatizability. By further analysis and modification of the completeness proof, one can always make do with *finite* counter-examples, and hence:

Proposition

The minimal propositional dynamic logic is *decidable*.

Other programming constructs can be studied too in this framework. For instance, one can add a *converse* operation on programs, running them backwards. This gives rise to what may be called a two-sided 'temporal' variant of the system. Moreover, in the literature, the non-regular Boolean operations $\cap$ and $-$ have been brought in after all, be it at the price of loss of decidability, as well as various less standard operators of 'parallel execution'.

8.2.3 Model Theory.

The modal notion of bisimulation can be directly generalized to the case of many atomic alternative relations. Then we have, in the appropriate first-order language:

Theorem. The first-order formulas $\phi(x)$ that are invariant for bisimulation are precisely those that occur as translations of modal formulas (in a poly-modal logic with possibility operators for each atomic relation).

This allows program constructions like *composition* and *union*: as these admit of obvious reductions. But statements involving Boolean *intersection* or *complement* of relations need not be invariant for bisimulation: they require stronger analogies between models.

Even so, this example shows that the usual notions of Modal Logic may be generalized to a dynamic setting, where they provide interesting constraints on algebraic program operations. There are many other illustrations of this (cf. Harel 1984, van Benthem 1990A). These semantic constraints are of a structural nature, and do not depend on any specific (first-order) formalism. In particular, infinitary operations can share them too, witness the case of bisimulations C between two dynamic models again:

Proposition

- C-corresponding states x, y verify the same formulas ϕ of propositional dynamic logic,
- all regular programs π have the back-and-forth property:

  if xCy and $x[[\pi]]_{M1}x'$,

  then there exists some y' with $y[[\pi]]_{M2}y'$ and $x'Cy'$.

Finally, the literature has a number of results showing how adding various programming constructs to the framework increases expressive power, or how various natural restrictions (e.g., to deterministic programs only) diminish it.

What is still scarce are more theoretical kinds of model-theoretic result. For instance,

> Is there an effective syntactic characterization of all those programs in the full language which are *deterministic* in the sense of always having a partial function for their denotation?
>
> In particular, do the WHILE programs serve as some kind of 'normal form' for the latter class?

To conclude, much of the model-theoretic behaviour of the dynamic formalism may again be understood via its systematic *correspondence* with standard formalisms.

Here is a simultaneous translation from formulas into unary properties of states and programs into binary relations between states, using an infinitary first-order language:

- $(p)^\$ = Px$
  $(\neg\phi)^\$ = \neg (\phi)^\$$
  $(\phi \# \psi)^\$ = (\phi)^\$ \; \# \; (\psi)^\$$ for all binary connectives #
  $(\langle\pi\rangle\phi)^\$ = \exists y \, ((\pi)_£ (x, y) \wedge (\phi)^\$ (y))$ for some *new* variable y
- $(a)_£ = R_a xy$
  $(\pi_1 ; \pi_2)_£ = \exists z \, ((\pi_1)_£ (x, z) \wedge (\pi_2)_£ (z, y))$
  $(\pi_1 \cup \pi_2)_£ = (\pi_1)_£ \vee (\pi_2)_£$
  $(\pi^*)_£ = \bigvee_n \exists z_1 \ldots \exists z_n \, ((\pi)_£ (x, z_1) \wedge \ldots \wedge (\pi)_£ (z_n, y))$
  $(\phi?)_£ = (\phi)^\$ \wedge y=x$.

Of course, there are alternative translations too: for instance, into a second-order language defining the required iteration for the Kleene * .

8.3 Logic Programming

8.3.1 Horn Clauses.

Logic programming involves describing data structures in the Horn clause fragment of first-order predicate logic, and then querying them via at most existentially quantified conjunctions of atomic statements, so that answers may be computed through a perspicuous deductive process, namely *Resolution.* (See Lloyd 1987, Apt 1987 for a full exposition of the framework.)

Horn clauses have the following semantic characteristic:

Theorem

> A predicate-logical formula is equivalent to a universal Horn sentence iff it is preserved under the formation of isomorphic images, direct products and submodels.

Another possible characterization is through the ω-continuity of recursive Horn clause definitions for predicates, which makes querying stabilize at stage ω in the obvious iteration process.

8.3.2 Resolution.

The resolution proof technique arose in the following setting. The full validity problem for first-order predicate logic may be reduced to the following:

> Find out whether the False $\perp$ follows from a certain finite set of universal clauses of the form 'prefix of universal quantifiers, disjunction of atoms and negations of atoms'.

This reduction uses Prenex Normal Forms, Skolem Forms and Conjunctive Normal Forms for the eventual quantifier-free matrix. For the decision problem in this simple form, deduction may be restricted to a very simple format, thanks to Herbrand's Theorem:

Take suitable instantiations of the universal clauses,
then proceed by purely propositional reasoning.

Now, these two steps may be 'mixed' again, leaving essentially only the following *Resolution Rule* of inference:

Consider any two universal clauses of the above disjunctive form,
leaving out quantifiers: say, A and B.
If there is an atom $Qt_1...t_k$ occurring positively in the one and an atom
$Qs_1....s_k$ occurring negatively in the other, such that all t_i, s_i can be
unified simultaneously by some substitution σ for their variables,
then conclude to the universal closure of the matrix $\sigma(A') \vee \sigma(B')$,
where A' arises from A by leaving out the relevant occurrence
of the Q-atom, and B' from B likewise.

The main historical result concerning this mechanism is
Robinson's Theorem

Refutation proofs for predicate-logical clauses can be given entirely by means
of resolution steps.

Here, a *unifying substitution* for a set of terms is one which makes all terms in the set syntactically identical. Moreover, this procedure is decidable:
Proposition

There is an effective terminating Unification Algorithm for locating unifying
substitutions, if they exist. It will even produce *most general unifiers*,
of which all other unifying substitutions are further specifications.

For the special case of Horn clauses, simpler forms of Resolution suffice. With the categorial calculi considered in Chapter 13, the rule even becomes a kind of generalized Modus Ponens:

$$\frac{\forall x_1...\forall x_k\colon Qt_1...t_n\,,\quad \forall z_1...\forall z_m\colon \neg Qs_1...s_n \vee P}{\forall x_1...\forall x_k\forall z_1...\forall z_m\colon \sigma(P)}\,,$$

for any unifying substitution σ as described above.

8.4 Logic in Artificial Intelligence

Current research in Artificial Intelligence has generated several mechanisms of inference that deviate from standard logical systems. Examples are heuristic reasoning with default rules, or probabilistic inference. Many of these new systems lose classical structural rules, in particular, *Monotonicity*. The reasons for this phenomenon are diverse. For instance, conclusions may have to be retracted because they were drawn, not over the universe of *all* models for a given set of premises, but only over their *most preferred* (or, most likely) models, as happens in the well-known technique of 'circumscription' (where 'preference' is usually for smaller domains of individuals and smaller denotations for certain predicates involved in the inference). This preferential minimization operation appeared in Chapters 15, 16 and 17. But also, non-monotonicity may be a side-effect of using certain syntactic mechanisms for drawing quick conclusions. And finally, the categorial or dynamic systems of this Book themselves provide another natural source of non-monotonic phenomena.

Nevertheless, on the positive side, there remain many common structural properties behind many diverse non-monotonic logics (as is brought out in Makinson 1988), which show many analogies with the above minimal conditional logic (as is proved in van Benthem 1989E). Therefore, we think that our categorial dynamics can be profitably studied together with the above systems, in order to throw more light upon the taxonomy of general reasoning.

Certainly, not all consequences of this viewpoint have penetrated into this Book yet. Notably, if one takes the above diversity seriously, then the semantic analysis of natural language in Chapters 4 , 10, 11 is really a rather restricted one, with outcomes tied to one particular mode of inference. For instance, what is the diversity of *readings* for a given expression if the yardstick for 'logical equivalence' is no longer standard logical consequence, but one of the above variants? (For instance, in more audacious heuristic reasoning, some quantifier scopings of type $\forall\exists$ might come to collapse with the stronger $\exists\forall$ type.)

9 General Trends

It is probably a matter of inevitable controversy what are the main trends in Logic to-day. At least, this book exemplifies four identifiable major themes that also seem prominent in the current development of the field:

- Fine-Structure
  A tendency to look at fragments of existing systems, trying to do much with little: low expressive power and limited deductive means.
- Varieties of Reasoning
  Acknowledging different viable options in defining valid consequence or suitable logical constants, without trying to reduce the diversity of human cognition to one standard format.
- Intensional Differences
  Respect for details of semantic or deductive frameworks which may escape attention under the usual regime of equivalence proofs. This may be important as a source of clues toward the 'structuring' of arguments that has remained such an elusive part of human logical competence so far.
- Computational Concerns
  A greater interest in constructive proofs inside Logic, preferably even ones that admit of feasible computation.

BIBLIOGRAPHY

Abramsky, S., D. Gabbay & T. Maibaum, eds.

to appear *Handbook of Logic in Computer Science*, Oxford University Press, Oxford.

Abrusci, V. M.

1988A 'Sequent Calculus for Intuitionistic Linear Propositional Logic', Report 1, Dept. of Philosophy of Science, University of Bari.

1988B 'A Comparison between Lambek Syntactic Calculus and Intuitionistic Linear Propositional Logic', Report 2, Dept. of Philosophy of Science, University of Bari.

Ades, A. & M. Steedman

1982 'On the Order of Words', *Linguistics and Philosophy* 4, 517-558.

Andréka, H.

1990 'The Equational Theories of Representable Positive Cylindric and Relation Algebras are Decidable', Institute of Mathematics, Hungarian Academy of Sciences, Budapest.

Andréka, H., J. D. Monk & I. Németi, eds.

1988 *Algebraic Logic*, Colloq. Math. Soc. J. Bolyai, vol. 54, North-Holland, Amsterdam.

Apt, K.

1987 'Introduction to Logic Programming', Report CS-R8741, Centre for Mathematics and Computer Science, Amsterdam.

Avron, A.

1988 'The Semantics and Proof Theory of Linear Logic', *Theoretical Computer Science* 57, 161-184.

Bach, E.

1984 'Some Generalizations of Categorial Grammars', in F. Landman & F. Veltman, eds., 1986, 1-23.

Bacon, J.

1985 'The Completeness of a Predicate-Functor Logic', *Journal of Symbolic Logic* 50, 903-926.

Bakker, J. de, W. de Roever & G. Rozenberg, eds.

1989 *Linear Time, Branching Time and Partial Order in the Semantics of Concurrency*, Springer, Berlin, 1-49, (Lecture Notes in Computer Science 354).

Barendregt, H.

1981 *The Lambda Calculus. Its Syntax and Semantics*, North-Holland, Amsterdam.

Barendregt, H. & K. Hemerik

1990 'Types in Lambda Calculi and Programming Languages', Technical Report 90-04, Department of Informatics, University of Nijmegen. (To appear in *Proceedings ESOP Conference*, Copenhagen, May 1990.)

Bar-Hillel, Y., H. Gaifman & E. Shamir

1960 'On Categorial and Phrase Structure Grammars', *Bulletin Research Council of Israel* F 9, 1-16.

Barton, G., R. Berwick & E. Ristad

1987 *Computational Complexity and Natural Language*, The MIT Press, Cambridge (Mass.).

Bartsch, R., J. van Benthem & P. van Emde Boas, eds.

1990 *Semantics and Contextual Expression*, Foris, Dordrecht.

Barwise, J., ed.

1977 *Handbook of Mathematical Logic*, North-Holland, Amsterdam.

Barwise, J.

1987 'Noun Phrases, Generalized Quantifiers and Anaphora', in P. Gärdenfors, ed., 1987, 1-29.

Barwise, J. & S. Feferman, eds.

1985 *Model-Theoretic Logics*, Springer Verlag, Berlin.

Bäuerle, R., C. Schwarze & A. von Stechow, eds.

1983 *Meaning, Use and Interpretation of Language*, de Gruyter, Berlin.

Bell, J. & A. Slomson

1969 *Models and Ultraproducts*, North-Holland, Amsterdam.

Bell, J. & M. Machover

1977 *A Course in Mathematical Logic*, North-Holland, Amsterdam.

Benthem, J. van

1983 'The Semantics of Variety in Categorial Grammar', Report 83-26, department of Mathematics, Simon Fraser University, Burnaby B.C. [Reprinted in W. Buszkowski et al., eds., 1988, 37-55.]

1984 'The Logic of Semantics', in F. Landman & F. Veltman, eds., 1984, 55-80.

1985 *Modal Logic and Classical Logic*, Bibliopolis, Napoli / The Humanities Press, Atlantic Heights.

1985* 'The Variety of Consequence, According to Bolzano', *Studia Logica* 44, 389-403.

1986A *Essays in Logical Semantics*, Reidel, Dordrecht.

1986B 'Partiality and Non-Monotonicity in Classical Logic', *Logique et Analyse* 29, 225-247.

1987* 'Categorial Equations', in E. Klein & J. van Benthem, eds., 1987, 1-17.

1987A 'Categorial Grammar and Lambda Calculus', in D. Skordev, ed., 1987, 39-60.

1987$ 'Logical Syntax', *Theoretical Linguistics* 14, 119-142.

1987B 'Meaning: Interpretation and Inference', *Synthese* 73, 451-470.

1987C 'Polyadic Quantifiers', Report 87-04, Institute for Language, Logic and Information, University of Amsterdam. (Also in *Linguistics and Philosophy* 12 (1989), 437-464.)

1987D 'Strategies of Intensionalization', in I. Bodnar et al., eds., 1988, 41-59.

1988A *A Manual of Intensional Logic*, Chicago University Press, Chicago, (2d revised edition).

1988B 'Logical Constants across Varying Types', Report LP-88-05, Institute for Language, Logic and Information, University of Amsterdam. (Appeared in *Notre Dame Journal of Formal Logic* 30:3 (1989), 315-342.))

1988C 'The Lambek Calculus', in R. Oehrle et al., eds., 1988, 35-68.

1988D 'Towards a Computational Semantics', in P. Gärdenfors, ed., 1988, 31-71.

1989A 'Categorial Grammar and Type Theory', *Journal of Philosophical Logic* 19, 115-168.

1989* 'Categorial Grammar Meets Unification', in J. Wedekind, ed., to appear.

1989B 'Language in Action', Report LP 89-04, Institute for Language, Logic and Information, University of Amsterdam. (To appear in the *Journal of Philosophical Logic*.)

1989C 'Modal Logic and Relational Algebra', Institute of Language, Logic and Information, University of Amsterdam. (To appear in *Proceedings Malc'ev Conference on Algebra*, Institute of Mathematics, Siberian Academy of Sciences, Novosibirsk.)

1989D 'Modal Logic as a Theory of Information', Report LP 89-05, Institute for Language, Logic and Information, University of Amsterdam. (To appear in *Proceedings Prior Memorial Conference*, Institute of Philosophy, Christchurch, New Zealand.)

1989E 'Semantic Parallels in Natural Language and Computation', in H-D Ebbinghaus et al., eds., 1989, 331-375.

1989F 'Semantic Type Change and Syntactic Recognition', in G. Chierchia et al., eds., 1989, 231-249.

1989G 'The Fine-Structure of Categorial Semantics', Report LP 89-01, Institute for Language, Logic and Information, University of Amsterdam. (To appear in M. Rosner, ed., *Computational Linguistics and Formal Semantics. Lugano 1988*, Cambridge University Press, Cambridge.)

1989H 'Time, Logic and Computation', in J. de Bakker, W. de Roever & G. Rozenberg, eds., 1989, 1-49.

1990A 'General Dynamics', *Theoretical Linguistics*, to appear. (Ph. Luelsdorff, ed., special issue on 'Complexity in Natural Language'.)

1990B 'Temporal Logic', in D. Gabbay et al., eds., to appear, *Handbook of Logic in Artificial Intelligence and Logic Programming*, Oxford University Press, Oxford.

Bergstra, J. & J. Tucker

1982 'Expressiveness and the Completeness of Hoare's Logic', *Journal of Computer and Systems Sciences* 25, 267-284.

Bodnar, I., A. Mate & L. Polos, eds.

1988 *Intensional Logic, History of Philosophy, and Methodology. To Imre Ruzsa on the Occasion of his 65th Birthday*, Department of Symbolic Logic, Budapest.

Boolos, G.

1984 'Don't Eliminate Cut', *Journal of Philosophical Logic* 13, 373-378.

Bull, R. A. & K. Segerberg

1984 'Basic Modal Logic', in D. Gabbay & F. Guenthner, eds., 1984, 1-88.

Burgess, J.

1982 'Axioms for Tense Logic I. "Since" and "Until" ', *Notre Dame Journal of Formal Logic* 23, 367-374.

Buszkowski, W.

1982 *Lambek's Categorial Grammars*, Dissertation, Mathematical Institute, Adam Mickiewicz University, Poznan.

1986 'Completeness Results for Lambek Syntactic Calculus', *Zeitschrift fuer mathematische Logik und Grundlagen der Mathematik* 32, 13-28.

1987 'Discovery Procedures for Categorial Grammars', in E. Klein & J. van Benthem, eds., 1987, 35-64.

1988 'Generative Power of Categorial Grammars', in R. Oehrle et al., eds., 1988, 69-94.

Buszkowski, W., W. Marciszewski & J. van Benthem, eds.

1988 *Categorial Grammar*, John Benjamin, Amsterdam.

Buszkowski, W. & G. Penn

1989 'Categorial Grammars Determined from Linguistic Data by Unification', Department of Computer Science, The University of Chicago.

Calder, J., E. Klein & H. Zeevat

1987 'Unification Categorial Grammar', *Edinburgh Papers in Cognitive Science* I, Centre for Cognitive Science, University of Edinburgh, 195-222.

Chang, C. & H. Keisler

1973 *Model Theory*, North-Holland, Amsterdam. (Second revised edition 1990.)

Chierchia, G., B. Partee & R. Turner, eds.

1989 *Properties, Types and Meaning*, vol. I (Foundational Issues), vol. II (Semantic Issues), Reidel, Dordrecht.

Cohen, J.

1967 'The Equivalence of Two Concepts of Categorial Grammar', *Information and Control* 10, 475-484.

Cresswell, M.

1973 *Logics and Languages*, Methuen, London.

Crossley, J., ed.

1975 *Algebra and Logic*, Springer, Berlin, (Lecture Notes in Mathematics 450).

Curry, H.

1961 'Some Logical Aspects of Grammatical Structure'. In 'Structure of Language and its Mathematical Aspects', *Proceedings of the Symposia in Applied Mathematics* 12, American Mathematical Society, Providence, 56-68.

Curry, H. & R. Feys

1958 *Combinatory Logic*, vol. I, North-Holland, Amsterdam.

Dalla Chiara, M-L

1985 'Quantum Logic', in D. Gabbay & F. Guenthner, eds., 1985, 427-469.

Damas, L. M.

1985 *Type Assignment in Programming Languages*, Dissertation, Department of Computer Science, University of Edinburgh.

Davidson, D. & G. Harman, eds.

1972 *Semantics of Natural Language*, Reidel, Dordrecht.

Diego, A.

1966 *Sur les Algèbres de Hilbert*, Gauthiers-Villars, Paris.

Doets, K.

1990 'Axiomatizing Universal Properties of Quantifiers', *Journal of Symbolic Logic*, to appear.

Doets, K. & J. van Benthem

1983 'Higher-Order Logic', in D. Gabbay & F. Guenthner, eds., 1983, 275-329.

Dosen, K.

1985 'A Completeness Theorem for the Lambek Calculus of Syntactic Categories', *Zeitschrift für mathematische Logik und Grundlagen der Mathematik* 31, 235-241.

1988 'Sequent Systems and Groupoid Models. I', *Studia Logica* 47, 353-385.

1989 'Sequent Systems and Groupoid Models. II', *Studia Logica* 48, 41-65.

1990 'A Brief Survey of Frames for the Lambek Calculus', Bericht 5-90, Zentrum für Philosophie und Wissenschaftstheorie, Universität Konstanz.

Dosen, K. & P. Schroeder-Heister, eds.

to appear *Logics with Restricted Structural Rules*, Proceedings Workshop on Lambek Calculus, Linear Logic, Relevant Logic and BCK Logic, Seminar für natürlich-sprachliche Systeme, University of Tübingen.

Dowty, D., R. Wall & S. Peters

1981 *Introduction to Montague Semantics*, Reidel, Dordrecht.

Dreben, B. & W. Goldfarb

1979 *The Decision Problem: Solvable Classes of Quantificational Formulas*, Addison-Wesley, Reading (Mass.).

Dummett, M.

1976 'What is a Theory of Meaning?', in G. Evans & J. McDowell, eds., *Truth and Meaning*, Oxford University Press, Oxford, 67-137.

1977 *Elements of Intuitionism*, Oxford University Press, Oxford.

Dunn, M.

1985 'Relevance Logic and Entailment', in D. Gabbay & F. Guenthner, eds., 1985, 117-224.

1990 'Gaggle Theory', lecture presented at JELIA Meeting, September 1990, Centre for Mathematics and Computer Science, Amsterdam.

Ebbinghaus, H-D et al., eds.

1989 *Logic Colloquium. Granada 1987*, North-Holland, Amsterdam.

Ejerhed, E. & K. Church

1983 'Finite State Parsing', in F. Karlsson, ed., *Papers from the Seventh Scandinavian Conference of Linguistics*, Department of General Linguistics, University of Helsinki.

Emms, M.

1989 'Quantifiers with Variable Types', Abstract, *7th Amsterdam Colloquium*, Institute for Language, Logic and Information, University of Amsterdam.

Fenstad, J-E, ed.

1971 *Proceedings of the Second Scandinavian Logic Symposium*, North-Holland, Amsterdam.

Fenstad, J-E, P-K Halvorsen, T. Langholm & J. van Benthem

1987 *Situations, Language and Logic*, Reidel, Dordrecht.

Fitch, F. B.

1952 *Symbolic Logic: an Introduction*, New York.

Frege, G.

1879 *Begriffsschrift. Eine Formelsprache des reinen Denkens*, Halle.

Friedman, H.

1975 'Equality between Functionals', in R. Parikh, ed., 22-37.

Friedman, J. & R. Venkatesan

1986 'Categorial and Non-Categorial Languages', Department of Computer Science, Boston University.

Fuhrmann, A.

1990 'On the Modal Logic of Theory Change', Zentrum Philosophie und Wissenschaftstheorie, Universität Konstanz.

Gabbay, D.

1981 'Functional Completeness in Tense Logic', in U. Mönnich, ed., 91-117.

Gabbay, D. & F. Guenthner, eds.

1983 *Handbook of Philosophical Logic*, vol. I (Elements of Classical Logic), Reidel, Dordrecht.

1984 *Handbook of Philosophical Logic*, vol. II (Extensions of Classical Logic), Reidel, Dordrecht.

1985 *Handbook of Philosophical Logic*, vol. III (Alternatives to Classical Logic), Reidel, Dordrecht.

1989 *Handbook of Philosophical Logic*, vol. IV (Topics in the Philosophy of Language), Reidel, Dordrecht.

Gabbay, D., C. Hogger & J. Robinson, eds.

1990 *Handbook of Logic in Artificial Intelligence and Logic Programming*, Oxford University Press, Oxford.

Gabbay, D., A. Pnueli, S. Shelah & Y. Stavi

1980 'On the Temporal Analysis of Fairness', *ACM Symposium on Principles of Programming Languages*, 163-173.

Gallin, D.

1975 *Systems of Intensional and Higher-Order Modal Logic*, North-Holland, Amsterdam.

GAMUT, L. T. F.

1990 *Logic, Language and Meaning*, Chicago University Press, Chicago.

Gärdenfors, P.

1988 *Knowledge in Flux: Modelling the Dynamics of Epistemic States*, Bradford Books / MIT Press, Cambridge (Mass.).

Gärdenfors, P., ed.

1987 *Generalized Quantifiers. Linguistic and Logical Approaches*, Reidel, Dordrecht.

Gärdenfors, P. & D. Makinson

1988 'Revision of Knowledge Systems Using Epistemic Entrenchment', in M. Vardi, ed., 1988, 83-95.

Gargov, G., S. Passy & T. Tinchev

1987 'Modal Environment for Boolean Speculations', in D. Skordev, ed., 1987, 253-263.

Gazdar, G. & G. Pullum

1987 'Computationally Relevant Properties of Natural Languages and their Grammars', in W. Savitch et al., eds., 387-437.

Geach, P.

1972 'A Program for Syntax', in D. Davidson & G. Harman, eds., 483-497. (Also in W. Buszkowski et al., eds., 1988, 127-140.)

Ginsburg, S.

1966 *The Mathematical Theory of Context-Free Languages*, McGraw-Hill, New York.

Girard, J-Y

1987 'Linear Logic', *Theoretical Computer Science* 50, 1-102.

Goldblatt, R.

1987 *Logics of Time and Computation*, CSLI Lecture Notes, vol. 7, Center for the Study of Language and Information, Stanford University.

1988 'Varieties of Complex Algebras', Department of Mathematics, Victoria University, Wellington. (Appeared in *Annals of Pure and Applied Logic* 44 (1989), 173-242.)

Goldblatt, R. & S. Thomason

1974 'Axiomatic Classes in Propositional Modal Logic', in J. Crossley, ed., 1974, 163-173.

Goldfarb, W.

1979 'Logic in the Twenties: the Nature of the Quantifier', *Journal of Symbolic Logic* 44, 351-368.

Groenendijk, J. & M. Stokhof

1988A 'Dynamic Predicate Logic', Institute for Language, Logic and Information, University of Amsterdam. (To appear in *Linguistics and Philosophy*.)

1988B 'Type Shifting Rules and the Semantics of Interrogatives', Institute for Language, Logic and Information, University of Amsterdam. (Also in G. Chierchia et al., eds., 1989, vol. II, 21-68.)

Groenendijk, J., D. de Jongh & M. Stokhof, eds.

1986 *Studies in Discourse Representation Theory and the Theory of Generalized Quantifiers*, Foris, Dordrecht.

Gurevich, Y.

1985 Logic and the Challenge of Computer Science, Report CRL-TR-10-85, Computing Research Laboratory, University of Michigan, Ann Arbor.

Harel, D.

1984 'Dynamic Logic', in D. Gabbay & F. Guenthner, eds., 1984, 497-604.

Heim, I.

1982 *The Semantics of Definite and Indefinite Noun Phrases*, Dissertation, Department of Linguistics, University of Massachusetts, Amherst.

Hendriks, H.

1987 'Type Change in Semantics. The Scope of Quantification and Coordination', in E. Klein & J. van Benthem, eds., 1987, 95-119.

1989 'Flexible Montague Grammar', Institute for Language, Logic and Information, University of Amsterdam.

Henkin, L.

1950 'Completeness in the Theory of Types', *Journal of Symbolic Logic* 15, 81-91.

1963 'A Theory of Propositional Types', *Fundamenta Mathematicae* 52, 323-344.

Henkin, L., D. Monk & A. Tarski

1985 *Cylindric Algebras*, Parts I and II, North-Holland, Amsterdam.

Henkin, L., P. Suppes & A. Tarski, eds.

1959 *The Axiomatic Method, with Special Reference to Geometry and Physics*, North-Holland, Amsterdam.

Hepple, M.

1990 'Normal Form Theorem Proving for the Lambek Calculus', Centre for Cognitive Science, University of Edinburgh.

Heylen, D. & T. van der Wouden

1988 'Massive Disambiguation of Large Text Corpora with Flexible Categorial Grammar', *Proceeedings COLING 1988*, Budapest.

Hilbert, D. & P. Bernays

1939 *Grundlagen der Mathematik*, Springer Verlag, Berlin.

Hindley, J. & J. Seldin

1986 *Introduction to Combinators and Lambda Calculus*, Cambridge University Press, Cambridge.

Hintikka, J.

1973 'Quantifiers versus Quantification Theory', *Dialectica* 27, 329-358.

Hintikka, J. & J. Kulas

1983 *The Game of Language*, Reidel, Dordrecht.

Hodges, W.

1983 'Elementary Predicate Logic', in D. Gabbay & F. Guenthner, eds., 1983, 1-131.

Hoeksema, J.

1984 *Categorial Morphology*, Dissertation, Nederlands Instituut, Rijksuniversiteit, Groningen.

1988 'The Semantics of Non-Boolean "And" ', *Journal of Semantics* 6, 19-40.

Hopcroft, J. E. & J. D. Ullman

1979 *Introduction to Automata Theory, Languages and Computation*, Addison-Wesley, Reading, Mass.

Hughes, G. & M. Cresswell

1968 *An Introduction to Modal Logic*, Methuen, London.

1984 *A Companion to Modal Logic*, Methuen, London.

Immerman, N.

1986 'Relational Queries Computable in Polynomial Time', *Information and Control* 68, 86-105.

Immerman, N. & D. Kozen

1987 'Definability with Bounded Number of Bound Variables', *Proceedings IEEE* 1987, 236-244.

Jónsson, B.

1987 'The Theory of Binary Relations', Department of Mathematics, VanderBilt University, Nashville, Tenn. (Also in H. Andréka et al., eds., 1988, 245-292.)

Jónsson, B. & A. Tarski

1951 'Boolean Algebra with Operators I, *American Journal of Mathematics* 73, 891-939.

Keenan, E.

1988 'Semantic Case Theory', Department of Linguistics, University of California at Los Angeles. (Also in R. Bartsch et al., eds., 1990, 33-56.)

Keenan, E. & L. Faltz

1985 *Boolean Semantics of Natural Language*, Reidel, Dordrecht.

Keenan, E. & Y. Stavi

1986 'A Semantic Characterization of Natural Language Determiners', *Linguistics and Philosophy* 9, 253-326.

Keisler, H.

1971 *Model Theory for Infinitary Logic*, North-Holland, Amsterdam.

Klein, E. & J. van Benthem, eds.

1987 *Categories, Polymorphism and Unification*, Centre for Cognitive Science / Institute for Language, Logic and Information, University of Edinburgh / University of Amsterdam.

Klop, J-W

1987 'Term Rewriting Systems: A Tutorial', Note CS-N8701, Centre for Mathematics and Computer Science, Amsterdam.

König, E.

1988 'Parsing as Natural Deduction', *Proceedings* COLING, Vancouver.

Koymans, R.

1989 *Specifying Message Passing and Time-Critical Systems with Temporal Logic*, Dissertation, Department of Computer Science, Technological University, Eindhoven.

Kracht, M.

1988 'How to Say "It" ', Mathematical Institute, Free University, Berlin.

Krifka, M.

1989A 'Boolean and Non-Boolean "And" ', Seminar für Natürlich-Sprachliche Systeme, Universität Tübingen.

1989B 'Nominal Reference and Temporal Constitution: Towards a Semantics of Quantity', in R. Bartsch et al., eds., 1990, 75-115.

Kruskal, J.

1972 'The Theory of Well-Quasi-Ordering: a Frequently Discovered Concept', *Journal of Combinatorial Theory* (A) 13, 297-305.

Lafont, Y.

1988 'The Linear Abstract Machine', *Theoretical Computer Science* 59, 157-180.

Lambek, J.

1958 'The Mathematics of Sentence Structure', *American Mathematical Monthly* 65, 154-170. (Also in W. Buszkowski et al., eds., 1988, 153-172.)

1988 'Categorial and Categorical Grammars', in R. Oehrle et al., eds., 297-317.

Lambek, J. & P. Scott

1986 *Introduction to Higher-Order Categorial Logic*, Cambridge University Press, Cambridge.

Landman, F. & F. Veltman, eds.

1984 *Varieties of Formal Semantics*, Foris, Dordrecht.

Langholm, T.

1987 *Partiality, Truth and Persistence*, Dissertation, Department of Philosophy, Stanford University. (Also appeared as CSLI Lecture Notes, vol. 15, Center for the Study of Language and Information / The Chicago University Press, 1988.)

Levin, H.

1982 *Categorial Grammar and the Logical Form of Quantification*, Bibliopolis, Naples.

Lewis, D.

1972 'General Semantics', in Davidson & Harman, eds., 1972, 169-218.

Lincoln, P., J. Mitchell, A. Scedrov & N. Shankar

1990 'Decision Problems for Propositional Linear Logic', Department of Computer Science, Stanford University. (To appear in *Proceedings 31st Annual IEEE Symposium on Foundations of Computer Science*, St. Louis, Missouri, October 1990.)

Lindström, P.

1966 'First-Order Predicate Logic with Generalized Quantifiers', *Theoria* 32, 186-195.

1969 'On Extensions of Elementary Logic', *Theoria* 35, 1-11.

Lloyd, J.

1987 *Foundations of Logic Programming*, Springer Verlag, Berlin.

Lorenzen, P.

1961 'Ein dialogisches Konstruktivitätskriterium', *Infinitistic Methods*, Pergamom Press, Oxford.

Maddux, R.

1983 'A Sequent Calculus for Relation Algebras', *Annals of Pure and Applied Logic 25*, 73-101.

Makinson, D.

1988 'General Non-Monotonic Logic', to appear in D. Gabbay et al., eds., 1990.

Makovsky, J.

to appear *Model Theory and Data Bases*, Technion, Haifa.

Martin-Löf, P.

1984 *Intuitionistic Type Theory*, Bibliopolis, Naples.

Mason, I.

1985 'The Metatheory of the Classical Propositional Calculus is not Axiomatizable', *Journal of Symbolic Logic* 50, 451-457.

Meyer, R.

1976 'A General Gentzen System for Implicational Calculi', *Relevance Logic Newsletter* 1, 189-201.

Mönnich, U., ed.

1981 *Aspects of Philosophical Logic*, Reidel, Dordrecht.

Montague, R.

1970 'Pragmatics and Intensional Logic', *Synthese* 22, 68-94.

1974 *Formal Philosophy*, Yale University Press, New Haven.

Moortgat, M.

1988 *Categorial Investigations. Logical and Linguistic Aspects of the Lambek Calculus*, Foris, Dordrecht.

1989A 'Cut Elimination and Normalization', Abstract, *7th Amsterdam Colloquium*, Institute for Language, Logic and Information, University of Amsterdam.

1989B 'Quantifier Scope. A Decidable Extension of L ', Onderzoeksinstituut voor Taal en Spraak, Rijksuniversiteit Utrecht.

1990 'Unambiguous Proof representations for the Lambek Calculus', Onderzoeksinstituut voor Taal en Spraak, Rijksuniversiteit Utrecht.

Morrill, G.

1989 'Intensionality, Boundedness and Categorial Domains', Centre for Cognitive Science, University of Edinburgh. (To appear in *Linguistics and Philosophy*.)

Moschovakis, Y.

1974 *Elementary Induction on Abstract Structures*, North-Holland, Amsterdam.

Mundici, D.

1989 'Functions Computed by Monotone Boolean Formulas with No Repeated Variables', *Theoretical Computer Science* 66, 113-114.

Muskens, R.

1989 *Meaning and Partiality*, Dissertation, Philosophical Institute, University of Amsterdam. (To appear with Foris, Dordrecht.)

Nash-Williams, C.

1963 'On Well-Quasi-Ordering Finite Trees', *Proceedings of the Cambridge Philosophical Society* 59, 833-835.

Németi, I.

1990 'Algebraizations of Quantifier Logics. An Introductory Overview', Institute of Mathematics, Hungarian Academy of Sciences, Budapest.

Oehrle, R., E. Bach & D. Wheeler, eds.

1988 *Categorial Grammars and Natural Language Structures*, Reidel, Dordrecht.

Ono, H.

1988 'Structural Rules and a Logical Hierarchy', Faculty of Integrated Arts and Sciences, Hiroshima University. (To appear in P. Petkov, ed., 1990.)

1990 'Phase Structures and Quantales - a Semantical Study of Logics Without Structural Rules', Faculty of Integrated Arts and Sciences, Hiroshima University.

Ono, H. & Y. Komori

1985 'Logics Without the Contraction Rule', *Journal of Symbolic Logic* 50, 169-201.

Orey, S.

1959 'Model Theory for the Higher Order Predicate Calculus', *Transactions of the American Mathematical Society* 92, 72-84.

Orlowska, E.

1987 'Relational Interpretation of Modal Logics', Department of Informatics, Polish Academy of Sciences, Warsaw.

Parikh, R., ed.

1975 *Logic Colloquium. Boston 1972-1973*, Springer Verlag, Berlin.

Partee, B.

1986 'Noun Phrase Interpretation and Type-Shifting Principles', in J. Groenendijk et al., eds., 1986, 115-143.

Partee, B. & M. Rooth

1983 'Generalized Conjunction and Type Ambiguity', in R. Bäuerle, C. Schwarze & A. von Stechow, eds., 1983, 361-383.

Pereira, F.

1990 'Semantic Interpretation as Higher-Order Deduction', lecture presented at JELIA Meeting, September 1990, Centre for Mathematics and Computer Science, Amsterdam.

Petkov, P., ed.

1990 *Mathematical Logic. Proceedings Heyting 88*, Plenum Press, New York.

Plotkin, G.

1980 'Lambda Definability in the Full Type Hierarchy', in J. Seldin & J. Hindley, eds., 1980, 363-373.

Ponse, A.

1987 'Encoding Types in the Lambek Calculus', in E. Klein & J. van Benthem, eds., 1987, 262-276.

Pratt, V.

1990 'Action Logic = Homogeneous Intuitionistic Dynamic Logic', lecture presented at JELIA Meeting, September 1990, Centre for Mathematics and Computer Science, Amsterdam.

Prawitz, D.

1965 *Natural Deduction*, Almqvist & Wiksell, Stockholm.

Prijatelj, A.

1989 'Intensional Lambek Calculus: Theory and Application', Report LP-89-06, Institute for Language, Logic and Information, University of Amsterdam.

Purdy, W.

1990 'A Logic for Natural Language', *Notre Dame Journal of Formal Logic*, to appear.

Schellinx, H.

1990 'Representable Lambda Terms in the Pair Model', Institute for Language, Logic and Information, University of Amsterdam.

Schwichtenberg, H.

1977 'Proof Theory: Some Applications of Cut Elimination', in J. Barwise, ed., 867-895.

Scott, D.

1980 'Relating Theories of the λ-Calculus', in J. Seldin & J. Hindley, eds., 403-450.

Seldin, J. & J. Hindley, eds.

1980 *To H. B. Curry. Essays on Combinatory Logic, Lambda Calculus and Formalism*, Academic Press, New York.

Shehtman, V.

1978 'An Undecidable Superintuitionistic Propositional Calculus', *Doklady Akademia Nauk SSSR*, Moscow, 549-552.

Sikorski, R.

1969 *Boolean Algebras*, Springer Verlag, Berlin.

Singletary, W. E.

1964 'A Complex of Problems Posed by Post', *Bulletin of the American Mathematical Society* 70, 105-109.

Skordev, D., ed.

1987 *Mathematical Logic and its Applications*, Plenum Press, New York.

Soare, R.

1987 *Recursively Enumerable Sets and Degrees*, Springer Verlag, Berlin.

Spohn, W.

1988 'Ordinal Conditional Functions: A Dynamic Theory of Epistemic States', in W. Harper et al., eds., 1988, *Causation in Decision, Belief Change and Statistics* II, Kluwer, Dordrecht, 105-134.

Stalnaker, R.

1972 'Pragmatics', in D. Davidson & G. Harman, eds., 1972, 380-397.

Statman, R.

1980 'On the Existence of Closed Terms in the λ-Calculus. I', in J. Seldin & J. Hindley, eds., 1980, 329-338.

1982 'Completeness, Invariance and λ-Definability', *Journal of Symbolic Logic* 47, 17-26.

Steedman, M.

1988 'Combinators and Grammars', in R. Oehrle et al., eds., 1988, 417-442.

Szabolcsi, A.

1988 'Bound Variables in Syntax. (Are There Any?)', Institute of Linguistics, Hungarian Academy of Sciences, Budapest. (Also in R. Bartsch et al., eds., 1990, 295-318.)

Tarski, A.

1959 'What is Elementary Geometry?', in L. Henkin, P. Suppes & A. Tarski, eds., 1959, 16-29.

Thomas, W.

1989 'Computation-Free Logic and Regular ω-Languages', in J.W. de Bakker, W.P. de Roever & G. Rozenberg, eds., 690-713.

Troelstra, A. & D. van Dalen

1988 *Constructivism in Mathematics: an Introduction*, North-Holland, Amsterdam.

Urquhart, A.

1972 'Semantics for Relevant Logics', *Journal of Symbolic Logic* 37, 159-169.

Uszkoreit, H.

1986 'Categorial Unification Grammars', *Proceedings 11th International Conference on Computational Linguistics*, Bonn, August 1986, 187-194.

Vakarelov, D.

1990 'Arrow Logic', Laboratory of Applied Logic, University of Sofia.

Vardi, M., ed.

1988 *Proceedings 2d Conference on Theoretical Aspects of Reasoning about Knowledge*, Morgan Kaufmann Publishers, Los Altos.

Veltman, F.

1989 'Defaults in Update Semantics', Institute for Language, Logic and Information, University of Amsterdam.

Wansing, H.

1989 'The Adequacy Problem for Sequential Propositional Logic', report LP-89-07, Institute for Language, Logic and Information, University of Amsterdam.

Wedekind, J., ed.

to appear *Proceedings Titisee Workshop on Unification Formalisms*, Institut für Maschinelle Sprachverarbeitung, Universität Stuttgart.

Westerstahl, D.

1989 'Quantifiers in Formal and Natural Languages', in D. Gabbay & F. Guenthner, eds., 1989, 1-131.

Wheeler, D.

1981 *Aspects of a Categorial Theory of Phonology*, Dissertation, Department of Linguistics, University of Massachusetts, Amherst.

Zeevat, H.

1990 *Categorial Unification Grammar*, forthcoming Dissertation, Department of Computational Linguistics, University of Amsterdam.

Zeinstra, L.

1990 'Reasoning as Discourse', master's thesis, Mathematical Institute, University of Utrecht.

Zielonka, W.

1981 'Axiomatizability of Ajdukiewicz-Lambek Calculus by Means of Cancellation Schemes', *Zeitschrift für mathematische Logik und Grundlagen der Mathematik* 27, 215-224.

Zucker, J. I. & R. S. Tragesser

1978 'The Adequacy Problem for Inferential Logic', *Journal of Philosophical Logic* 7, 501-516.

Zwarts, F.

1986 *Categoriale Grammatica en Algebraische Semantiek*, Dissertation, Nederlands Instituut, Rijksuniversiteit, Groningen. (To appear with Reidel, Dordrecht.)

INDEX

This index lists a number of key terms used in this Book, together with their most significant pages of occurrence.